Muscle Relaxants
in Anaesthesia

Muscle Relaxants in Anaesthesia

Edited by

Nigel J N Harper MB, ChB, FRCA
Consultant Anaesthetist
Manchester Royal Infirmary, Manchester, UK

and

Brian J Pollard B Pharm, MD, FRCA
Senior Lecturer
Honorary Consultant Anaesthetist
Manchester Royal Infirmary, Manchester, UK

Edward Arnold
A member of the Hodder Headline Group
LONDON BOSTON SYDNEY AUCKLAND

First published in Great Britain 1995 by
Edward Arnold, a division of Hodder Headline PLC,
338 Euston Road, London NW1 3BH

Distributed in the USA by
Little, Brown and Company
34 Beacon Street, Boston, MA 02108

© 1995 Edward Arnold

All rights reserved. No part of this publication may be reproduced or transmitted in any form or by any means, electronically or mechanically, including photocopying, recording or any information storage or retrieval system, without either prior permission in writing from the publisher or a licence permitting restricted copying. In the United Kingdom such licences are issued by the Copyright Licensing Agency: 90 Tottenham Court Road, London W1P 9HE

Whilst the advice and information in this book is believed to be true and accurate at the date of going to press, neither the author nor the publisher can accept any legal responsibility or liability for any errors or omissions that may be made. In particular (but without limiting the generality of the preceding disclaimer) every effort has been made to check drug dosages; however, it is still possible that errors have been missed. Furthermore, dosage schedules are constantly being revised and new side effects recognized. For these reasons the reader is strongly urged to consult the drug companies' printed instructions before administering any of the drugs recommended in this book.

British Library Cataloguing in Publication Data
A catalogue record for this book is available from the British Library

ISBN 0 340 55155 0 (Pb)

1 2 3 4 5   95 96 97 98 99

Typeset in 10/11pt Palatino by Phoenix Photosetting, Chatham, Kent
Printed in Great Britain by St Edmundsbury Press, Bury St Edmunds

Contents

List of Contributors   vii

Preface   ix

1. **Physiology; nerve, junction and muscle**
   F Donati   1

2. **Pharmacology of neuromuscular blocking drugs**
   BJ Pollard   13

3. **Pharmacokinetics of neuromuscular blocking drugs**
   CJR Parker   26

4. **Suxamethonium**
   NJN Harper   55

5. **Non-depolarizing muscle relaxants**
   BJ Pollard   77

6. **Monitoring neuromuscular blockade**
   NJN Harper   97

7. **Techniques of administering muscle relaxants**
   F Donati   127

8. **Reversal of neuromuscular blockade**
   NJN Harper   135

9. **Relaxants in specific clinical situations**
   NJN Harper   156

10. **Drug interactions**
    BJ Pollard 177

11. **Muscle relaxants in paediatric anaesthesia**
    CC McLoughlin, RK Mirakhur 198

12. **Muscle relaxants in the Intensive Care Unit**
    BJ Pollard 221

13. **Adverse reactions to muscle relaxants: the role of the laboratory in clinical investigations**
    J Watkins 234

Index 261

Contributors

François Donati
Harold Griffith Professor, Department of Anaesthesia, Royal Victoria Hospital, Montreal, Quebec, Canada

Nigel JN Harper
Consultant Anaesthetist, Department of Anaesthesia, Manchester Royal Infirmary, Manchester, UK

Cormac C McLoughlin
Consultant Anaesthetist, Department of Anaesthetics, Belfast City Hospital, Belfast, Northern Ireland

Rajinder K Mirakhur
Senior Lecturer, Department of Anaesthetics, Queens University of Belfast, Belfast, Northern Ireland

Christopher JR Parker
Senior Lecturer, University Department of Anaesthesia, Royal Liverpool University Hospital, Liverpool, UK

Brian J Pollard
Senior Lecturer, University Department of Anaesthesia, Manchester Royal Infirmary, Manchester, UK

John Watkins
Deputy Director, Surparegional Protein Reference Unit, Department of Immunology, Northern General NHS Trust, Sheffield, UK

Preface

Advances in the development of new neuromuscular blocking drugs have been so rapid during the last decade that the anaesthetist is required to keep up to date with an ever-expanding field. Individual preferences have evolved rapidly such that a new generation of anaesthetists favours the shorter-acting muscle relaxant drugs and may have little or no experience of tubocurarine, alcuronium or even pancuronium. At the time of writing, mivacurium has been introduced recently into anaesthetic practice and rocuronium is on the threshold of approval for clinical use. A great deal of information concerning these two drugs has been published in the form of research papers but has not previously been made available in a book aimed at a wide anaesthetic readership.

The introduction of agents with rapid onset and rapid offset offers considerable advantages but there remains a need to understand their pharmacology if neuromuscular blockade is to be manipulated to the greatest advantage. The requirement to administer neuromuscular blocking drugs to increasing numbers of patients at the extremes of age and patients with organ failure in the operating theatre and the Intensive Care Unit imposes great demands on the ability of the clinician to understand the pharmacokinetics and pharmacodynamics of these agents and it is appropriate to consider these aspects in some detail.

More than 20 years have elapsed since the description of Train-of-Four stimulation which enabled the anaesthetist to measure the depth of neuromuscular blockade in the clinical setting. Since that time, Post-Tetanic Count and Double Burst stimulation have expanded our ability to quantify profound blockade during surgery and to decide what dose of reversal agent, if any, is necessary at the end of surgery. Residual neuromuscular blockade in the recovery room is surprisingly common and has not yet been eradicated by the introduction of new muscle relaxant drugs.

When an anaphylactoid or anaphylactic reaction occurs during anaesthesia, neuromuscular blocking agents are probably more likely to be responsible than any other group of drugs. An understanding of the immediate management of the crisis and the subsequent identification of the causative agent is crucial if mortality is to be reduced but there are several areas of continuing controversy.

We have set ourselves the task of setting out the relevant information in a format that is useful to those studying for postgraduate anaesthesia examinations as well as the experienced anaesthetist who may wish to browse its pages as part of continuing medical education. We have called freely on the knowledge and experience of experts in specific fields and we express our gratitude to our co-contributors.

Nigel NJ Harper
Brian J Pollard

# 1

Physiology: Nerve, Junction and Muscle

François Donati

Muscle is essential in maintaining life, and in the performance of any mechanical function. The autonomic nervous system (sympathetic and parasympathetic) controls specialized muscles in a variety of organs, (heart, blood vessels, kidney, bladder, eye, gastrointestinal tract, lungs) which are not normally under voluntary control. This system will not be treated here. The skeletal (or voluntary) muscle system is concerned with movement and maintenance of posture. It is essential to life because the muscles of respiration are within this group.

The motor unit

The nerves controlling skeletal muscle, or motor neurones, are very long cells which extend from their origin in the spinal cord (spinal nerves), or the midbrain (cranial nerves), to the muscle cell, which may lie as much as 1 m further. The metabolic centre of the nerve cell is the cell body, located near the origin, which receives information (action potentials) from small appendages called dendrites. This information is integrated in the cell body, modulated by connections (synapses) from other cells, and transmitted down a long, cylindrical structure called the axon. The diameter of a typical motor axon is 10–20 μm. The speed of transmission of the action potential is increased by the presence of a myelin sheath, which consists of many layers of cell membrane wrapped tightly on top of each other. Myelin acts as an insulator. It is interrupted periodically by gaps, called nodes of Ranvier, which are excitable and participate in the generation and propagation of the action potential[1] (Fig.1.1).

Before reaching the neuromuscular junction, the axon branches into nerve terminals, each of which comes into contact with one muscle cell. The nerve structure closest to the muscle is the synapse, which lies opposite a specialized area of muscle, the end-plate. At this point, information is transmitted chemically, instead of electrically. A nerve action potential triggers the release of a neurotransmitter, acetylcholine, which binds selectively to receptors located at the end-plate.

2   *Physiology: nerve, junction and muscle*

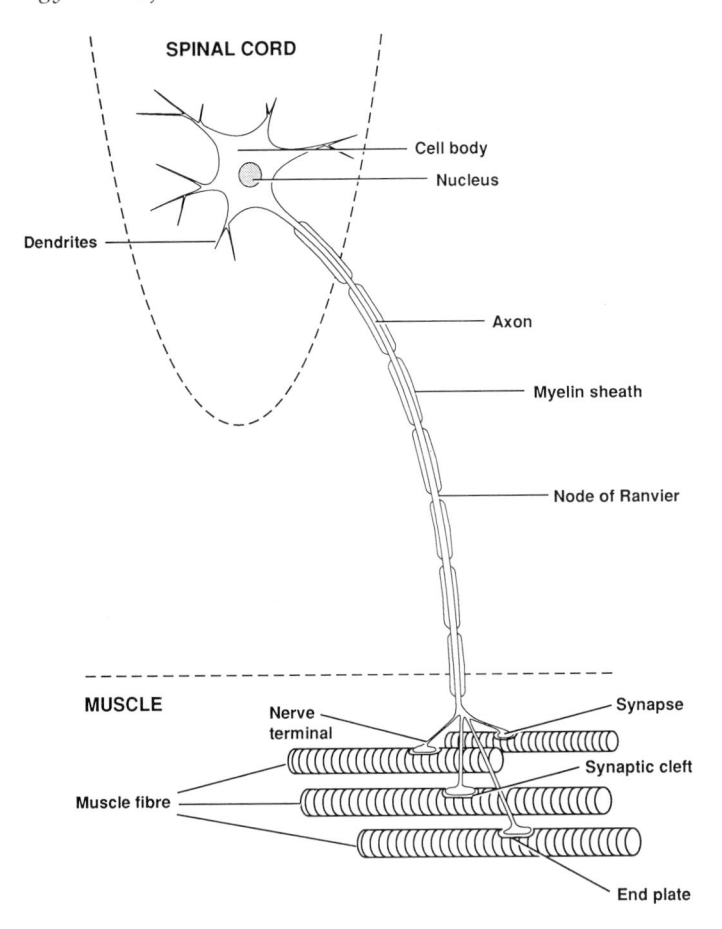

Fig. 1.1   Schematic representation of the motor unit (not drawn to scale).

A nerve axon branches into many terminals and thus innervates many muscle cells. However, except for special cases, a muscle cell has one neuromuscular junction and is innervated by only one nerve. A nerve and the muscle cells it innervates make up a motor unit. The number of muscle cells per motor unit varies from a few to a few thousand depending on the function of the muscle. It is largest in strong, bulky muscles designed for coarse movement, and least in small muscles performing delicate movements.

Muscle cells, or fibres, are also excitable. They are designed for the propagation of action potentials. In this case, however, the action potential does not carry information to other cells, but triggers a contraction process. Propagation of the action potential is much slower in muscle (1–5 m s^{-1}) than in nerve (50–100 m s^{-1}) because muscle is not myelinated. Its membrane also has deep invagations which puts it in close contact with intracellular reservoirs of calcium, the endoplasmic reticulum. Action potentials trigger release of calcium from the endoplasmic reticulum into the muscle cell proper, and this step is crucial in the contraction process.

Excitable membranes

Both nerve and muscle have excitable membranes, that is they respond to electrical stimuli and transmit electrical impulses. This property allows information to travel, within a relatively short time, to distant parts of the body.

Excitable cells are made up of an electrically conductive ionic medium, surrounded by an insulating membrane, which is itself bathed in an ionic medium. The membrane is made up of a phospholipid bilayer, into which proteins are imbedded. The proteins regulate the movement of ions and nutrients.

Resting potential

At rest, the electrical potential (voltage) of the inside of the nerve or muscle cell is negative with respect to the outside. This *resting potential* is usually −70 to −90 mV. A potential difference exists between both sides of the cell membrane because of the different ionic composition of the inside compared with the outside, and the selective permeability of the membrane. The inside of the cell is rich in potassium (K^+) ions but poor in sodium (Na^+) ions. The converse is true of the outside: its Na^+ concentration is high and its K^+ concentration is low (Fig. 1.2). At rest, the cell membrane is permeable to

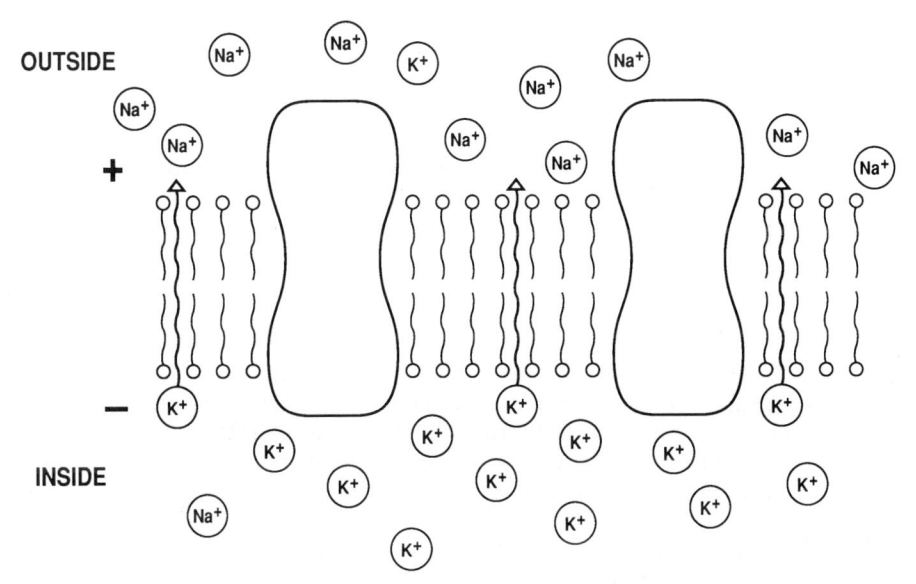

Fig. 1.2 Excitable membrane at rest. It is a phospholipid (represented by a small circle with a long tail) bilayer into which are imbedded proteins (the large structures). Some of these proteins are ionic channels which are closed at rest. The membrane is more permeable to potassium (K^+) than to sodium (Na^+) ions. This outward movement of K^+ makes the outside more electrically positive than the inside. There is debate regarding how ions cross membranes at rest, and it is uncertain whether they seep through the bilayer membrane (as shown), or through protein channels.

K^+ and virtually impermeable to Na^+. Thus, K^+ tends to exit the cell, along its concentration gradient, but movement of K^+ ions creates a slight excess of positive ions on the outside and a deficit of the same ions on the inside. Thus, this outward movement establishes a potential difference, the inside being negative, and this situation puts a brake on further outward movement of positively charged K^+ ions.

Action potential

The situation is altered if the electrical potential is changed by external means, chemical or electrical. If the membrane is depolarized, that is if the inside is made less negative, the permeability to sodium (Na^+) ions increases and becomes greater than that for potassium (K^+). Then, the membrane potential is determined by differences in Na^+ concentration. The Na^+ ions tend to move to the inside of the cell, making its potential positive with respect to the outside. This increased permeability is due to the opening of membrane channels which are specific for Na^+. These channels consist of proteins shaped like a rosette or a doughnut.[2] They are voltage sensitive and ion specific, because they open when voltage across the membrane is depolarized, or is made less negative, and let Na^+ ions pass through more easily than K^+ ions.[2]

This opening of Na^+ channels is, however, short lived. They close after a certain time interval. In addition, the change in potential induced by Na^+ channel opening triggers the opening of other channels which are specific for K^+. These K^+ channels allow an outward movement of K^+ ions, which restores the membrane potential to its resting level. This rapid change in potential is called the *action potential*.[2] Its duration depends on the type of excitable cell involved: typically 1 ms or less in nerve, 5–10 ms in skeletal muscle and 200–300 ms in cardiac muscle.

When an action potential is generated in one area of the cell, it influences neighbouring areas. Because the inside of the cell is a good electrical conductor, action potentials tend to spread. The potential across the membrane area next to the action potential rises above its resting value, thus causing voltage-sensitive Na^+ channels to open, and this in turn leads to an action potential. The process is repeated along the full length of the cell, at a rate which depends on how well the ions are kept within the cell, i.e. how electrically insulated the cell membrane is. Insulation is best in *myelinated* nerves. These nerve cells are wrapped in several layers of membrane, called myelin, and action potentials propagate at speeds reaching 50–100 m s^{-1}. Motor neurones, i.e. nerves supplying skeletal muscles, are of this type. Muscle action potentials propagate more slowly (1–5 m s^{-1}), partly because of the absence of myelin.

Nerve stimulation

Under physiological conditions, a nerve action potential is generated in the cell body, and propagates to the periphery. However, it is also possible to produce an action potential by applying an electrical current at any point along the axon (Fig. 1.3). The intensity and duration of the current must be

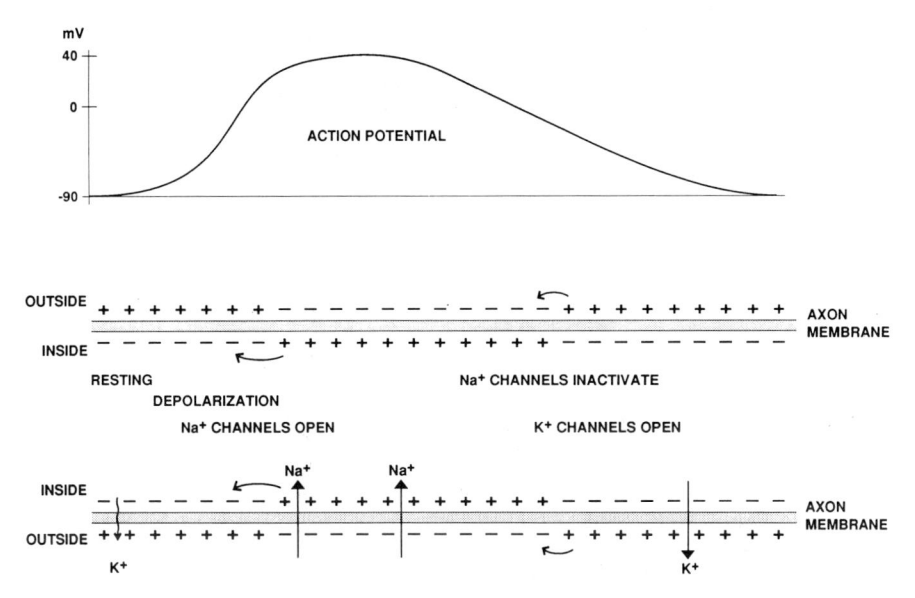

Fig. 1.3 Propagation of the action potential (from right to left). The graph shows potential changes (inside vs outside) versus distance along the axon (top) and the corresponding events which occur at the membrane level (bottom).

sufficient to depolarize the nerve membrane to threshold. Normally, it is unnecessary to deliver stimuli greater than 0.2–0.3 ms in duration. If shorter stimuli are used, a greater intensity may be required to reach threshold. If current is applied for a long period of time, multiple action potentials may be generated, because a nerve axon becomes ready to fire again only a short time interval, called the refractory period, after an action potential has been generated. The refractory period in nerve is 1–2 ms.

The neuromuscular junction

Nerve and muscle come in close contact to each other, but do not fuse. Thus, action potentials do not travel directly from nerve to muscle. The signal transmission from nerve to muscle occurs via the neurotransmitter acetylcholine (Fig. 1.4).

Acetylcholine synthesis and storage

Acetylcholine is synthesized in the synapse from acetyl and choline moieties. This chemical activity requires the presence of enzymes and energy. Only the cell body contains the genetic material needed for the formation of enzymes. Thus, these proteins are transported along the whole length of the axon to the synapse, where they are required. The presence of numerous mitochondria indicates the high degree of metabolic activity within the synapse. Choline is actively taken up by the synapse. The other building block for

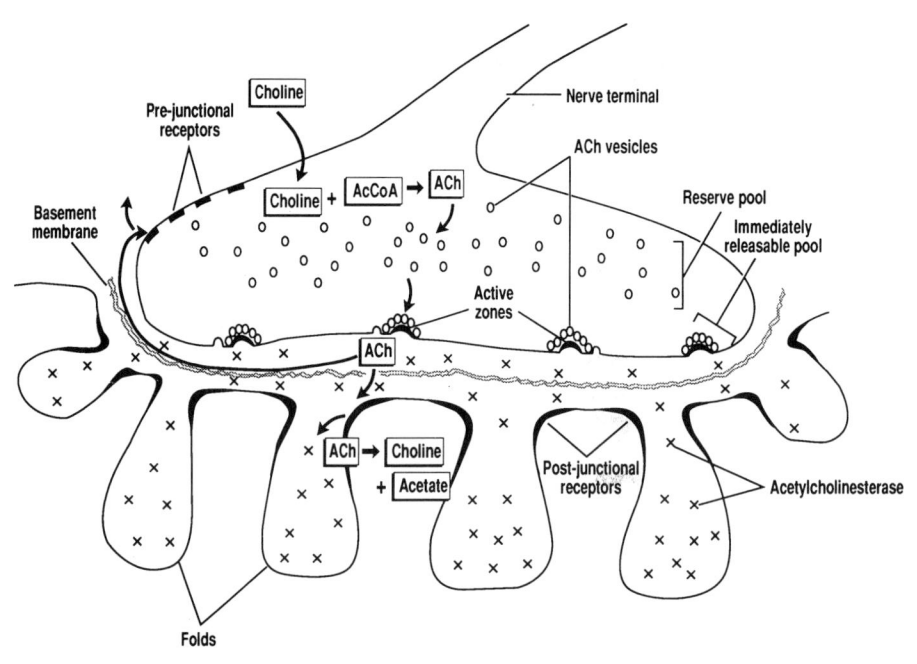

Fig. 1.4 Schematic representation of the neuromuscular junction (not drawn to scale), with emphasis on acetylcholine systhesis, storage, release, interaction with receptors and breakdown.

acetylcholine is acetyl coenzyme A (acetyl CoA), which is part of the glucose metabolic pathway. Synthesis requires the enzyme choline-0-acetylase.[1] An important proportion of the synthesized acetylcholine is incorporated into small vesicles, approximately 60 nm in diameter. A large number of these vesicles are concentrated near the part of the synaptic membrane closest to the muscle end-plate.[1,3] These vesicles are bound by a bilayer membrane similar to that surrounding cells and each vesicle contains approximately 10 000 molecules of acetylcholine.

Acetylcholine release

When an action potential invades the nerve terminal, acetylcholine is released, most probably through the emptying of a large number of vesicles into the space between the synapse and the end-plate, called the synaptic cleft. Then acetylcholine diffuse over a short distance, approximately 50 nm (0.05 μm), to reach the end-plate. The release process involves fusion of the vesicle membrane with the synaptic membrane, and requires calcium, the intracellular concentration of which is increased when the nerve terminal is depolarized.[4] Although there is debate regarding what happens to the fused membrane, most experts agree that it is probably recycled into new acetylcholine vesicles.

At rest, intracellular concentrations of calcium are lower than extracellular concentrations. When an action potential invades the nerve terminal, calcium channels in the membrane open. This allows interaction of calcium with many intracellular proteins, some of which are important in the storage and release process. Among these regulatory proteins, *synapsin I* and *synaptophysin* appear to play a key role. Synaptophysin is a vesicular membrane protein. Calcium is thought to facilitate its binding with a docking protein located on the synaptic membrane opposite the end-plate. This interaction would trigger fusion of the vesicular and synaptic membranes and be the first step in the release process. Synaptophysin is another regulatory protein, which probably anchors the vesicles to intracellular neurofilaments, at some distance from the synaptic membrane. Phosphorylation by calcium probably inactivates synaptophysin, thus allowing the acetylcholine vesicle to move towards the active zone. This process would re-supply the immediately releasable pool.

Under resting conditions, there is random, quantal release of acetylcholine, as indicated by the small (0.5–1 mV) changes in potential which can be observed if an electrode is inserted into the end-plate.[5] These *miniature end-plate potentials* (MEPP) are most likely the result of the spontaneous release of the contents of one acetylcholine vesicle. The size of the MEPP, however, is insufficient to depolarize the end-plate beyond threshold. With depolarization of a nerve terminal produced by an action potential, a large number of acetylcholine vesicles discharge their contents simultaneously, and the *end-plate potential* (EPP) is large enough to reach threshold and trigger muscle concentration.

In addition to quantal release, acetylcholine may also reach the end-plate through a leakage process.[6] There is evidence that under resting conditions, leakage is quantitatively more important than quantal release.[6] Although consistent with almost all experimental evidence, the vesicular hypothesis is not accepted by all experts. Some argue that extravesicular acetylcholine might be released through pores created by depolarization of the synaptic membrane.[7] Both theories are not mutually exclusive.

The end-plate

The end-plate is a small specialized area of muscle which is extremely rich in receptors specific for acetylcholine. It is oval in shape, and typically 20–30 μm across. The end-plate is characterized by a large number of folds in the membrane, and the receptors are located preferentially at the crest of these folds, just opposite the acetylcholine vesicles in the synapse (Fig. 1.4). Receptor density is very high at the end-plate (10 000–20 000 receptors μm^{-2}), and the total number of receptors is 1–10 million per end-plate. Receptors are also found on the muscle membrane in areas other than the end-plate. The density of these *extrajunctional receptors* is much less (approximately 20 μm^{-2}) than that of junctional receptors.

Acetylcholine receptor

The acetylcholine receptor is made up of five protein subunits, arranged as a rosette or doughnut (Fig. 1.5). All subunits lie across the full thickness of the cell membrane, and extend both to the inside and especially to the outside of the cell. Two of these subunits, named alpha, are identical. The others are

8 *Physiology: nerve, junction and muscle*

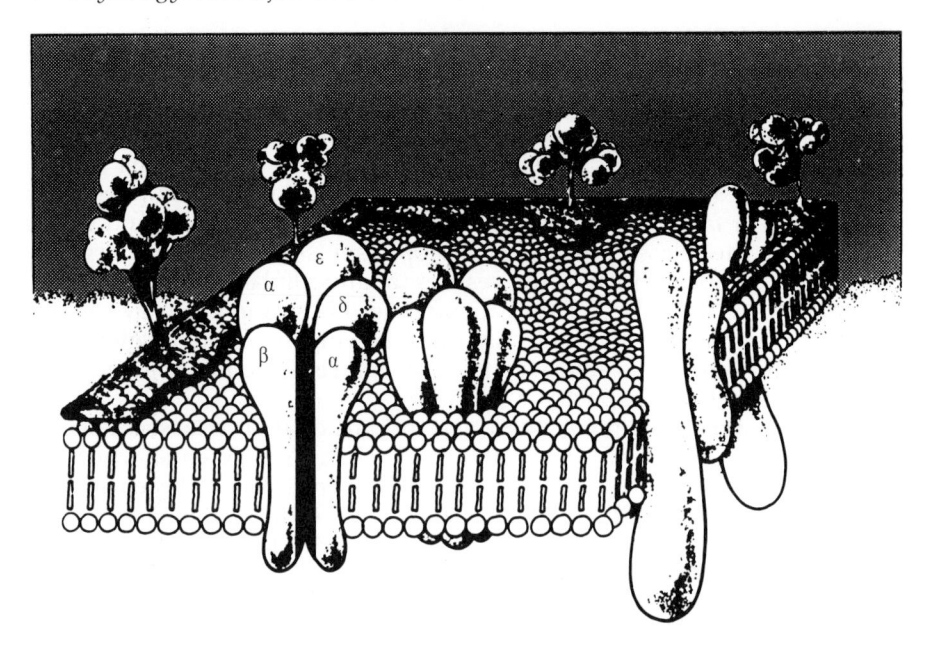

Fig. 1.5 Artist's representation of the receptor at the end-plate. The receptor contains five units. Other proteins are also embedded into the lipid bilayer membrane. The balloon-shaped structures with stalks attached to the membrane are the acetylcholinesterase complexes. (From Standaert FG. Neuromuscular physiology. In Miller RD ed. *Anesthesia* 3rd edn, 1970, New York: Churchill Livingstone, pp. 659–684.)

called beta, delta and gamma or epsilon.[1,8,9] The gamma subunit is part of foetal receptors and extrajunctional receptors. The epsilon subunit is seen in adult junctional receptors. In the resting state the subunits are twisted in such a way that the receptor is closed, i.e. the hole in the doughnut is blocked. When two molecules of acetylcholine bind simultaneously to the alpha subunits, a conformational change is induced in the proteins, that is the subunits twist in such a way as to open the hole. This activation of receptors is blocked if any one or both alpha subunits are blocked by a neuromuscular blocking drug.

The size of this opening (approximately 0.65 nm, 0.00065 μm) is large enough to let all cations pass through indiscriminately.[8] Thus, K^+ ions leak out from inside the cell to the outside, but this movement is minor compared with the flow of Na^+ ions from outside to inside. Because the inside of the cell is negative with respect to the outside, positive Na^+ ions are attracted to the inside and make it more positive. Thus, a depolarization, or change towards a less negative level, of the membrane occurs, and if a certain threshold is reached, an action potential is generated in the muscle. The amount of acetylcholine released following a nerve action potential is far in excess of what is needed to reach threshold at the end-plate. It is estimated that only 6–25% of the acetylcholine released normally is required to reach threshold.[10] Thus, the release process has a wide margin of safety.

The activated acetylcholine receptor stays open for approximately 1 ms, during which time some 10^5 Na^+ ions are allowed to go through.[11] The receptor can also exist in the desensitized state, that is acetylcholine binding to alpha subunits does not produce the conformational change required for channel opening. Drugs can also interact with the receptor by plugging it open.[8] However, this open channel blockade is probably relatively unimportant in the case of non-depolarizing muscle relaxants. These agents produce their effect by binding to one or both of the alpha subunits, thus preventing acetylcholine from reaching the same site in a competitive manner.

Post-synaptic acetylcholine receptors are of the nicotinic type, and are different from the muscarinic acetylcholine receptors found in parasympathetic effector organs. Nicotinic receptors of different subtypes are found in skeletal muscle, autonomic ganglia and the central nervous system.

Fate of acetylcholine

The duration of post-synaptic receptor opening is of the order of 1 ms. If acetylcholine binds again to the receptor, it will open again. However, this does not occur normally, because acetylcholine is broken down rapidly by acetylcholinesterase, located in the synaptic left and within the junctional folds[1] (Figs. 1.4 and 1.5). This enzymatic hydrolysis takes only a few microseconds. As a result each acetylcholine molecule probably has a chance to bind to a receptor only once. Many molecules probably do not even reach the receptor, because they are broken down *en route* by acetylcholinesterase within the synaptic cleft.[1]

Pre-synaptic receptors

Under physiological conditions, the nerve may fire at high frequencies and the neuromuscular junction must be capable of responding to this rapid rate of stimulation. It is estimated that there is enough acetylcholine in the synapse for many thousands of stimulations. However, there must be a mechanism which stimulates acetylcholine production to match demand. In addition, not all acetylcholine stored in the nerve terminal is immediately available for release. Thus, there must be a mechanism for the transfer of the neurotransmitter from the large 'storage pool' to the much smaller 'readily releasable pool'. The regulation of the quantity of acetylcholine immediately available for release is most certainly mediated through pre-synaptic acetylcholine receptors. Current evidence suggests that nicotinic acetylcholine receptors, activated by acetylcholine and blocked by small doses of nondepolarizing relaxant, may be part of a positive feedback loop, the function of which is to maintain acetylcholine release.[1,12,13] Blocking this feedback loop would lead to decreased acetylcholine release. This might be the basis for 'Train-of-Four' and tetanic fade observed with non-depolarizing relaxants. To complicate the picture even more, there appears to be pre-synaptic inhibitory nicotinic and muscarinic receptors.[12,13] Acetylcholine release is modulated even further by pre-synaptic alpha and beta adrenoreceptors.[13]

10 *Physiology: nerve, junction and muscle*

Events in muscle

Action potential

Depolarization of the end-plate leads to activation of sodium channels in the perijunctional area. A muscle action potential is then initiated and propagated from the neuromuscular junction to both ends of the muscle fibre. The electromyogram (EMG) is sum of all electrical activity in the fibres of a given muscle (Chapter 6).

Excitation – contracting coupling

The muscle action potential is the trigger of the contractile process. Muscle fibres contain filaments which are the result of the interdigitation of two proteins, actin and myosin. At rest, interaction between both proteins is

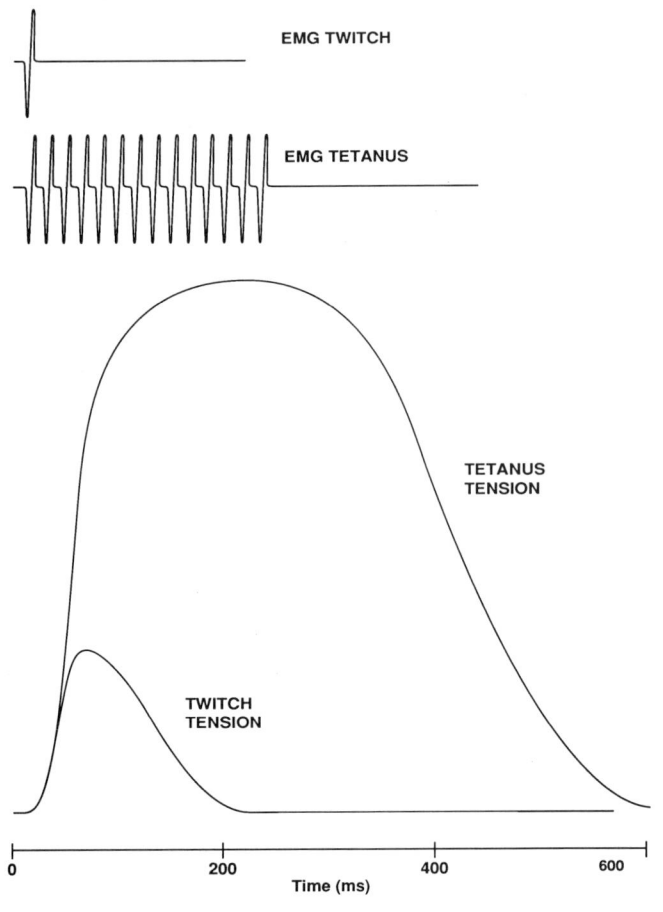

Fig. 1.6 Electrical (EMG) and mechanical (tension) events in muscle after single twitch and tetanic stimulations.

inhibited by a regulatory protein, troponin. This inhibition is removed in the presence of relatively high concentrations of calcium. In a muscle fibre, calcium concentrations are very low because the ion is pumped actively into specialized reservoirs, the sarcoplasmic reticulum, or outside the cell. With depolarization, i.e. an action potential, calcium is released into the muscle fibre, down its concentration gradient, the action of troponin is inhibited, and nothing prevents the interaction between actin and myosin. Thus bridges form between the interdigitating proteins and a contraction, i.e. shortening of the muscle fibre, occurs.

Twitch and tetanic contractions

A single muscle action potential gives rise to a short contraction, or twitch. The duration of the mechanical event, however, is much longer (100–200 ms) than the electrical event (less than 10 ms) (Fig. 1.6). When there is high frequency stimulation of the nerve (and the muscle), there is insufficient time for muscle relaxation between stimuli. Thus, contractions add up and fusion of contractions occur. As a result, the strength of such a tetanic contraction may be many times that of a twitch (Fig. 1.6).

The force developed by a whole muscle is the algebraic sum of the forces developed by each of its individual fibres. This is normally measured by a force transducer. A quantitative assessment of muscle response can also be made by recording its electrical activity (EMG). When there is no disturbance in the relationship between electrical and mechanical activity, in other words when excitation-contraction coupling is normal, both force and EMG can be used interchangeably as an index of muscle performance (Chapter 6).

Summary

Nerve and muscle are excitable cells, that is they can generate and propagate electrical impulses called action potentials. A nerve cell terminates by branching into many nerve terminals, which ends as a specialized structure, called synapse. A synapse contains a neurotransmitter, acetylcholine, which is released extracellularly in response to an action potential. A very narrow space, the synaptic cleft, separates nerve from the end-plate, a specialized area of muscle which contains a high density of acetylcholine sensitive receptors. These receptors are made up of five proteins subunits. Binding of acetylcholine to these receptors produces a conformational change which produces opening of the channel. The inflow of sodium ions induces depolarization of the end-plate area, which produces a muscle action potential, and causes calcium to be released into the muscle fibre. This neutralizes the inhibitory effect of troponin on actin and myosin, allowing both proteins to slide past each other and produce a contraction.

References

1. Bowman WC. *Pharmacology of Neuromuscular Function* 2nd edn, 1990, London: Wright.
2. Catterall WA. Cellular and molecular biology of voltage-gated sodium channels. *Physiological Reviews* 1992, **72:** 515–548.
3. Ellisman MH, Rash JE, Staehelin A, Porter KR. Studies on excitable membrane II. A comparison of specializations at neuromuscular junctions and nonjunctional sarcolemmas of mammalian fast and slow twitch muscle fibers. *Journal of Cell Biology* 1976; **68:** 752–774.
4. Llinas PR. Depolarization release coupling: an overview. *Annals of the New York Academy of Sciences* 1991; **635:** 3–17.
5. Fatt P, Katz B. Spontaneous subthreshold activity at motor nerve endings. *Journal of Physiology* 1952; **117:** 109–128.
6. Katz B, Miledi R. Transmission leakage from motor nerve endings. *Proceedings of the Royal Society* 1977; **B196:** 59–72.
7. Dunant Y, Israel M. The release of acetylcholine. *Scientific American* 1985; **252**(4): 58–66.
8. Sastry BVS. Nicotinic receptor. *Anaesthetic Pharmacology Review* 1993; **1:** 6–19.
9. Guy HR, Hucho F. The ion channel of the nicotinic acetylcholine receptor. *Trends in Neuroscience* 1987; **10:** 318–321.
10. Paton WDM, Wand DR. The margin of safety of neuromuscular transmission. *Journal of Physiology* 1967, **191:** 59–90.
11. Neher E, Sakmann B. Single channel currents recorded from membrane of denervated frog muscle fibres. *Nature* 1976; **260:** 766–802.
12. Bowman WC, Prior C, Marshall IG. Presynatic receptors in the neuromuscular junction. *Annals of the New York Academy of Sciences* 1990; **604:** 69–81.
13. Wessler I. Pre-synaptic neuromuscular block. *Anaesthetic Pharmacology Review* 1993; **1:** 69–76.

2

Pharmacology of Neuromuscular Blocking Drugs

Brian J Pollard

The subject of pharmacology concerns the action of drugs on the body. This chapter will consider general aspects of the pharmacology of the muscle relaxants only. More specific details with respect to each drug on an individual basis are included in Chapters 4 and 5.

The relevance of structure and function

The anatomy and physiology of the neuromuscular junction has been described in detail in Chapter 1. It is appropriate, however, to review those aspects of structure and function which are important in the actions of drugs at the neuromuscular junction.

The motor nerves to striated muscle have a cell body in the central nervous system (CNS), a single long axon and dendrites which make contact with many adjacent cells within the CNS. The terminal branches of the axon each innervate one muscle fibre at a neuromuscular junction (NMJ). The nerve does not make direct contact with the muscle fibre, being separated by a narrow gap, the junctional cleft.

The NMJ consists of three parts: the nerve ending (prejunctional region), the junctional cleft and the motor end-plate (postjunctional region) (Fig. 2.1). There are aspects of all of these which are important in the pharmacology of neuromuscular transmission. The nerve terminal contains the systems which are concerned with the synthesis, storage, mobilization, release and recycling of acetylcholine (ACh). The synaptic cleft is filled with a mucopolysaccharide basement membrane material in which acetylcholinesterase (AChe) is distributed, although it is particularly concentrated in the folds of the postjunctional membrane. The postjunctional region contains the recognition and binding sites for the transmitter ACh and the mechanism to translate this signal into an electrical impulse with the capability to activate the contractile mechanism of the muscle.

14 *Pharmacology of neuromuscular blocking drugs*

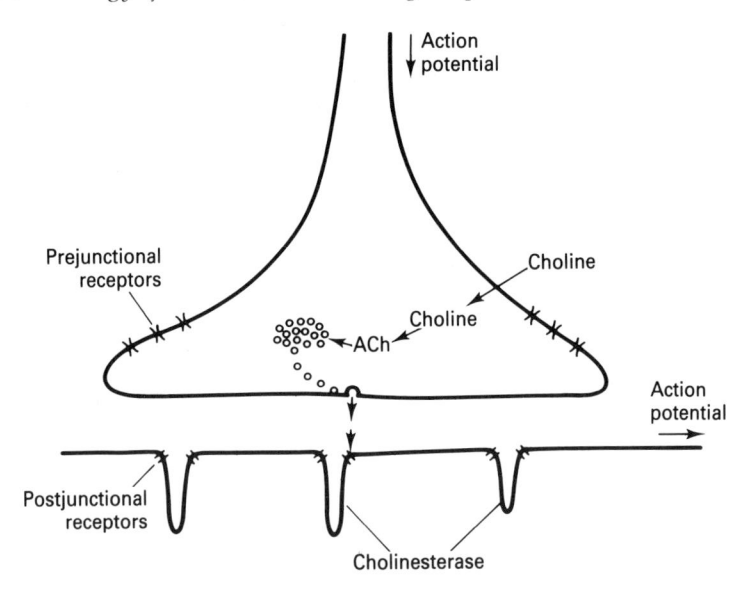

Fig. 2.1 The neuromuscular junction showing potential sites for the modification of transmission.

ACh is manufactured in the nerve terminal by the acetylation of choline, catalysed by the enzyme choline-acetyltransferase. Choline enters the nerve terminal by an active transport system and is supplied both from plasma and from the breakdown products of ACh. Acetate is supplied bound to coenzyme A and ACh synthesis is an energy-dependent process (Fig. 2.2). The ACh is then packed into vesicles in another energy-dependent process. Most of the vesicles are situated within the cell cytoplasm (the reserve) whilst a small number lie close to the inside of the cell membrane (the immediately available store). The movement of vesicles from reserve to immediately available store is transmitter mobilization and is also an energy-dependent process involving calcium ions.

The process of neuromuscular transmission involves the arrival of an action potential at the nerve ending triggering the release of ACh which comes from the immediately available store, rapidly replenished from the

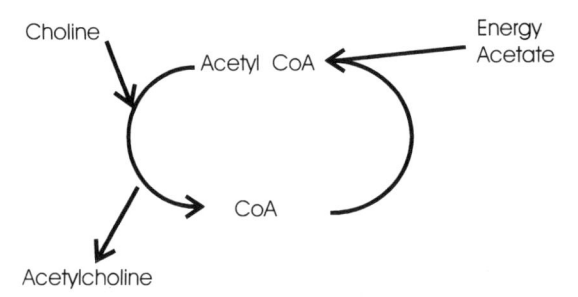

Fig. 2.2 The synthetic process for the manufacture of acetylcholine.

intracellular reserve. The released ACh diffuses across the junctional cleft and combines with the receptors. It is necessary for two molecules of ACh to combine with a receptor system and the binding of the first ACh molecule facilitates the binding of the second which results in activation. The channel opens, permitting small cations (sodium mainly) to travel down their concentration gradients with the end result being depolarization of the muscle membrane if enough channels have been activated.

Receptors also exist on the prejunctional nerve endings. The function of most of these receptors is unknown although it would seem likely that they subserve a modulating function controlled by other physiological mediators which may be either released from neighbouring cells, or blood borne.

There are therefore a number of potential sites of action for agents which may modify neuromuscular transmission and these include the following:

1. Inhibition of the propagation of the action potential along the last section of the nerve (e.g. lignocaine).
2. Inhibition of the release of acetylcholine (e.g. botulinum toxin and certain antibiotics).
3. Prevention of the combination of acetylcholine with the postjunctional receptors (e.g. pancuronium).
4. Inhibition of the uptake of choline (e.g. hemicholinium).
5. Inhibition of the synthesis of acetylcholine (e.g. triethylcholine).
6. Inhibition or activation of prejunctional receptors (e.g. tubocurarine).
7. Inhibition of the activity of acetylcholinesterase thereby increasing the amount of acetylcholine available in the junctional cleft (e.g. neostigmine).

Not all of these proposed mechanisms are clinically relevant although those drugs which do affect neuromuscular transmission may act at one or more of the potential sites. Those drugs which are used clinically to modify neuromuscular transmission exert their effect principally by preventing the combination of ACh with the postjunctional receptors. Although occlusion of postjunctional receptors is the main mechanism of action, there are a number of additional mechanisms. These include actions at the prejunctional receptors and also on the ion channels. These drugs may, in addition, also affect the release of ACh. Drugs which inhibit cholinesterase are used clinically to reverse the effect of the muscle relaxants.

Before considering the muscle relaxants themselves in greater detail, it must be remembered that there are many drugs which have an action at the NMJ. Most of these exert their effects in a manner which is not identical to the mode of action of the muscle relaxants. For example, gentamicin inhibits acetylcholine release. The presence of any such drug is likely to modify the action of a muscle relaxant. Either the relaxant or the other drug may alone have little effect on transmission but, together, may produce quite marked block. For a more complete list of drugs which may interact with the muscle relaxants the reader is referred to Chapter 10.

The postjunctional receptors

It is necessary to examine the postjunctional ACh receptors in slightly greater detail in order to understand better the pharmacology of the muscle relaxants. There are probably about 10 000 receptors per square micrometre located on the shoulders of the junctional folds.[1] Each receptor unit is a pentamer of five glycoprotein subunits which together form a central cation channel inserted through the phospholipid bilayer of the postjunctional membrane (Fig. 2.3).[2] The receptors are linked in pairs by disulphide bridges and it is likely that there is cooperation between the two components of each receptor pair. Two of the glycoprotein subunits in each receptor are identical; these are the alpha subunits. The others are the beta, delta and either epsilon in the adult or gamma in the foetus (and in various other species).[3] The two alpha subunits each carry one ACh recognition site. This site may not only bind ACh, but also other agonists, toxins and reversible antagonists. Although the two alpha subunits have an identical amino acid sequence they are functionally different by virtue of their different environments; one alpha subunit is surrounded by beta and epsilon, whereas the other is surrounded by delta and beta. One might not, therefore, expect them to behave in an identical fashion in their interaction with various antagonists.

There are several possible mechanisms which can take place and which will result in potential inhibition of neuromuscular transmission. These include effects on the ACh receptor sites themselves, effects on the ion channels and extra effects around the channels (Fig. 2.4).

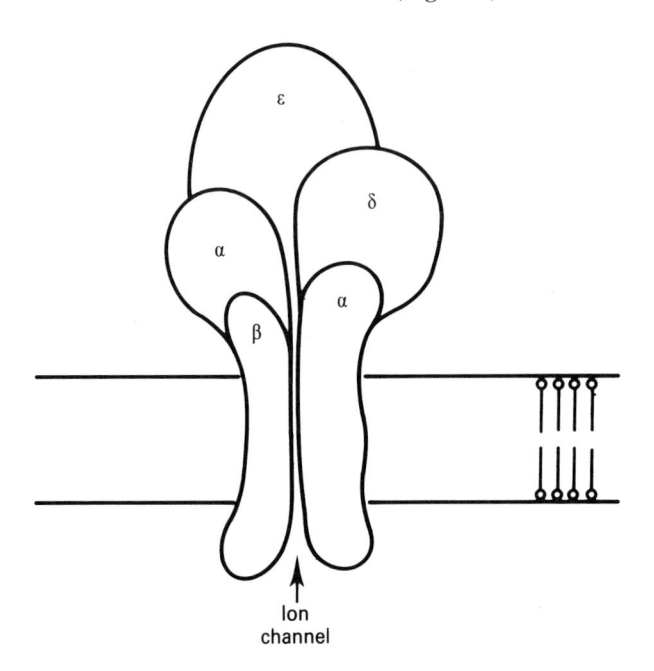

Fig. 2.3 The acetylcholine receptor channel protein macromolecule.

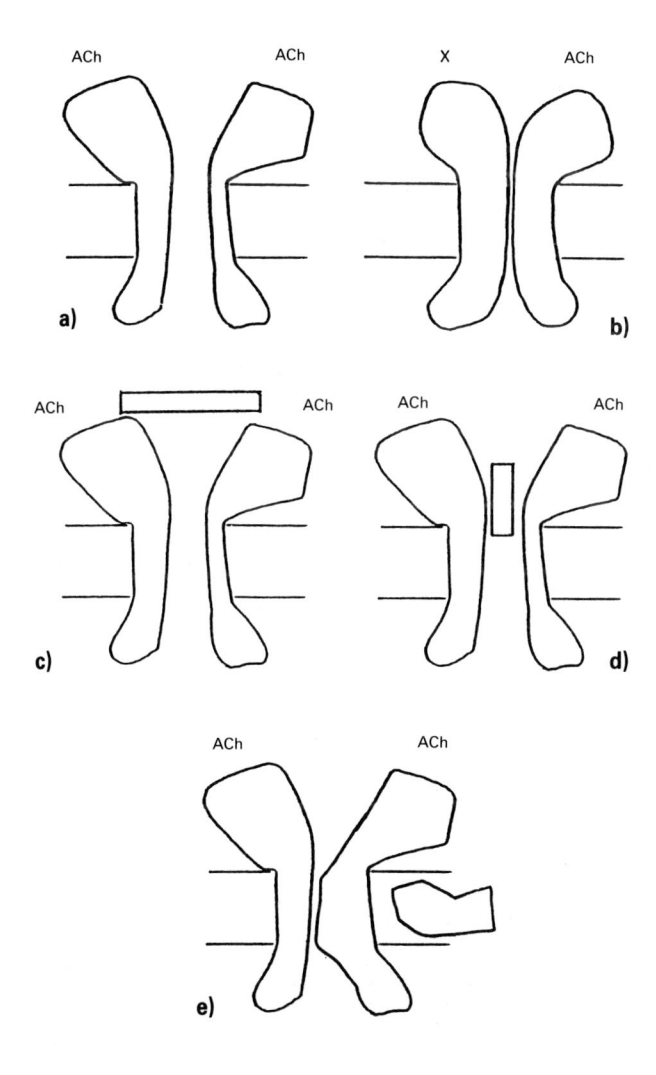

Fig. 2.4 Pharmacological antagonism of neuromuscular transmission. a) *Normal activation*. Both alpha subunits must be occupied simultaneously with agonist molecules for the channel to open. b) *Receptor occlusion*. If either one or both receptor sites in the alpha subunits are occupied by antagonist molecules, the channel cannot open. c) *Closed channel block*. A molecule binding to any part of the receptor complex around the open 'funnel' end can physically obstruct the passage of molecules through the channel. d) *Open channel block*. The channel opens and a molecule passes into the channel, thus restricting the normal flow of ions. e) *Local effects*. A drug binding to a nearby site, or dissolving in the lipid bilayer, may disturb the fluidity or function of the membrane and thus the receptor channel complex.

Receptor occlusion

Any chemical substance which can bind to the ACh recognition site on either alpha subunit of the receptor channel complex has the potential to produce neuromuscular block by physically preventing any other molecule from binding (Fig. 2.4b). The result will be the inability of ACh to bind and the channel will not open. Neuromuscular transmission will be interrupted. In view of the necessity for both alpha subunits to be occupied simultaneously by ACh molecules before the channel can open, it is necessary for an antagonist molecule to occupy only one of the two binding sites to prevent the channel from opening. It is presently believed that the major component of the action of the muscle relaxants takes place through this mechanism.

Closed channel block

The protein macromolecule which forms the receptor system possesses a number of amino acid residues which can function as alternative binding sites. A variety of chemical substances may bind to one or more of these sites. Any bulky molecule which binds to this area may physically obstruct the mouth of the channel like a lid, either partially or totally (Fig. 2.4c). The passage of physiological ions will therefore be impeded. If fewer ions cross during the open time, the transmembrane current will be reduced, so potentiating any other blocker. This binding will take place irrespective of whether the channel is open or closed and is therefore not use dependent. Closed channel block has been proposed as the mechanism underlying the actions of a number of drugs, e.g. the tricyclic antidepressants.[4]

Open channel block

The phenomenon of open channel block occurs when a drug enters the ion channel. The direct result of the presence of this molecule within the lumen of the channel physically obstructs the flow of ions through the channel (Fig. 2.4d). The current flow is therefore less with the result being a failure of transmission. In order for the drug to enter the channel, it must, of course, have been already opened by the presence of an agonist occupying both ACh recognition sites. In addition, the drug requires to be positively charged (the channel is selective for cations) and to be present in a fairly high concentration.[5]

Open channel block was first demonstrated with barbiturates and with local anaesthetic agents. Since that time, a wide range of chemical compounds has been shown to exhibit this phenomenon. Such compounds include volatile anaesthetic agents, alcohols, atropine, phencyclidine derivatives, anti-arrhythmic agents, certain antibiotics (particularly aminoglycosides and polymyxin) and many of the non-depolarizing muscle relaxants.[5]

Channel block differs from receptor occlusion in one fundamental characteristic, namely that whilst receptor interactions in general are highly stereospecific, studies have shown that channel block is almost entirely non-stereospecific. By its very nature, therefore, the phenomenon of channel block is non-competitive and requires a high concentration of antagonist. It is unlikely that channel block is of major importance in the clinical situation; nevertheless, it should be borne in mind when considering the interactions between muscle relaxants and other drugs or when unpredictable effects are observed.

The prejunctional nerve ending – prejunctional receptors

The presence of receptors on the nerve endings at the neuromuscular junction was recognized initially following the observation that adrenaline and noradrenaline increase the end-plate potentials resulting from motor nerve stimulation.[6] Since that time, receptors for many other humoral mediators have been reported. Receptors for the following mediators have been described on the pre-synaptic nerve terminal at cholinergic nerve endings: acetylcholine (nicotinic and muscarinic), noradrenaline (alpha and beta adrenergic), dopamine, gamma amino butyric acid (GABA), opoids, substance-P, adenosine, 5-hydroxytryptamine, prostaglandins, glutamate, benzodiazepines and angiotensin-2.[7] It is likely that the function of these receptors is to allow modulation of transmission by other physiological mediators either released from neighbouring cells, or blood borne. The physiological function of the majority of these receptors is presently unclear. It is likely that the muscle relaxants may also modify neuromuscular transmission by their action on prejunctional receptors.[8]

Prejunctional nicotinic acetylcholine receptors

It is this receptor, of all of the prejunctional receptors, which has attracted the most attention. Early studies reported that ACh depolarized nerve endings, an effect which was antagonized by tubocurarine. This led to the suggestion that these prejunctional receptors might control ACh release. It is now generally accepted that activation of these receptors controls the mobilization and synthesis of ACh rather than its actual release.[8]

Under conditions where there is a rapid rate of firing, a small dose of tubocurarine will result in a failure of transmission. A particular pattern of response is observed where the amplitude of successive stimuli falls from the first, which is the largest, until a constant lower amplitude is reached (tetanic fade).[9] The more rapid is the stimulation (or firing) rate, the more rapid is this fade. Under normal circumstances, the amount of ACh released by a nerve impulse is greatly in excess of that required to just evoke depolarization of the end-plate.[10] The total store and rate of output of ACh declines over the duration of a stimulus train ultimately to reach a constant level lower than at the beginning of the train. In order to sustain ACh release at these higher rates the nerve terminal must be capable of quickly mobilizing the transmitter and this would appear to be mediated through an action of some of the ACh which has been released from the nerve ending acting upon the prejunctional nicotinic receptors. If these receptors are blocked by the introduction of tubocurarine, then mobilization decreases, resulting in fade at higher rates of stimulation.[11] Decamethonium, a substance with cholinomimetic properties, increases ACh mobilization, an effect which is blocked by tubocurarine.[12] ACh thus increases its own mobilization at high rates of stimulation creating a positive feedback system. Such high frequencies of activation are seen in normal voluntary muscle movement and so this mechanism has significant clinical relevance.

The synthesis of ACh in nerve endings is known to be enhanced by nerve stimulation and this appears to be linked to intracellular sodium changes.[13] There are ACh-sensitive sodium channels along the entire axon which are probably involved in the generation or maintenance of the action potential.[14] Similar sodium channels on the nerve endings are probably a continuation of those normally present on the axon but in the nerve ending the end result is an effect on the synthesis of ACh. Choline transport into cells also rises immediately following a nerve impulse.[15]

It appears that the pharmacological profile of the prejunctional nicotinic receptors differs from those on the postjunctional membrane. Hexamethonium (a potent ganglion blocker) produces fade with little depression of transmission, whereas pancuronium (a weak ganglion blocker and potent neuromuscular blocker) produces only modest fade in comparison to its neuromuscular blocking action. The inference is that the prejunctional receptors resemble those nicotinic ACh receptors in autonomic ganglia more than they do the postjunctional receptors.[8] This conclusion is supported by the observation that alpha bungarotoxin, which binds selectively and irreversibly to the ACh recognition sites on the postjunctional membrane,[16] neither binds to the nerve endings nor produces fade.[17]

There may yet be further functions ascribed to the prejunctional nicotinic receptors and the effect of drugs on those receptors may not have yet been fully explained. It is likely that they play a role in the production of tetanic fade but it cannot be concluded with certainty that blockade of prejunctional receptors is the only mechanism underlying tetanic fade.

Prejunctional muscarinic receptors

In 1980, it was reported that neuromuscular junctions in the electric organs of a species of electric eel contained presynaptic muscarinic receptors.[18] Stimulation of those receptors inhibited the release of ACh from the nerve ending and also led to a reduction in MEPP frequency.[19, 20] The proposal was advanced that those prejunctional muscarinic receptors were concerned with a negative feedback mechanism on ACh release. Wessler, however, in 1987,[21] showed that not only can stimulation of prejunctional muscarinic receptors inhibit the release of ACh, but that it can also increase the release of ACh under certain circumstances. It seems likely that these two observations are related to stimulation of two different subtypes of muscarinic receptor on the prejunctional region.[22]

Prejunctional catecholamine receptors

Evidence for the existence of presynaptic alpha receptors is well established. The administration of noradrenaline or adrenaline increases ACh output, an effect which can be antagonized by the use of alpha blocking agents.[23] The function of these catecholamine receptors in unknown, but it has been postulated that they may enhance transmitter output in times of stress.

Prejunctional opioid receptors

These have also been described although their function is unknown. It is possible that they respond to circulating enkephalins at times of stress thus modulating ACh release.

Additional effects

Drugs are selected as muscle relaxants by virtue of their ability to block the receptors on the postjunctional membrane. ACh receptors are present at other sites in the body and it is likely that actions at these additional sites may be responsible for some of the major side-effects of the muscle relaxants. There are nicotinic ACh receptors in autonomic ganglia. Although these are not identical to those on the postjunctional membrane, they seem to be closely similar to those on the prejunctional region of the neuromuscular junction.[8] It would therefore not be surprising to find an effect of certain of the non-depolarizing muscle relaxants at autonomic ganglia (tubocurarine is the most potent).

There may also be an effect of certain of the non-depolarizing muscle relaxants on the muscarinic receptors of the cardiac vagus nerve. An atropine-like effect will be seen with ensuing tachycardia. Gallamine is the most notable relaxant to block the vagus. Indeed, whereas gallamine is here classified as a muscle relaxant with muscarinic side-effects, these side-effects are so prominent that gallamine is regarded by many pharmacologists as a muscarinic blocker which also has an action at the neuromuscular junction. It is possible that the existence of muscarinic receptors on the prejunctional nerve terminal of the neuromuscular junction may be relevant to production of a neuromuscular block.

The receptor–ion channel complex is inserted into the lipid bilayer cell membrane. The normal functioning of the channel will be dependent upon the integrity of the lipid bilayer. Anything which binds elsewhere on the cell membrane in the vicinity of the receptor complex (Fig. 2.4e), or which binds to one of the beta, gamma, delta or epsilon subunits of the receptor may well affect the properties of the channel. A substance which dissolves in the lipid bilayer membrane may disturb the fluidity of the membrane or distort its structure, so altering the function of the channel.[24] Any restrictions on its opening are likely to enhance the neuromuscular blocking action of other drugs.

Muscle relaxants

There are two ways to prevent the normal action of ACh at the neuromuscular junction. Firstly, the binding site can be occupied by an agent which does not produce depolarization but prevents the access of ACh (an antagonist), or secondly the binding site can be occupied by an agent which does produce depolarization but this is prolonged, rendering the receptor mechanism insensitive for a time (an agonist). Although direct evidence is presently lacking, it seems highly likely that the binding site on the alpha subunit is negatively charged.[25] We would therefore expect any blocking agent to possess a positive charge and to have affinity for the negative charge on the receptor. It is interesting to note that all clinically useful muscle relaxants possess at least one positively charged nitrogen. There is an association between possession of two positive charges and increased potency.

Agonists

The agonists include ACh, decamethonium and suxamethonium, the latter being the only depolarizing muscle relaxant in current clinical use. The size of the positively charged nitrogen groups and also the bulk of the interonium supporting structure are important in determining the ability to depolarize the muscle end-plate via the ACh receptors.

The agonists bind to the ACh receptors resulting in activation, opening of the ion channels and depolarization of the postjunctional membrane. Normally the end-plate is depolarized by ACh only for a very short period of time before it becomes repolarized and ready to transmit another impulse. If the agonist persists, depolarization of the end-plate will continue with rapid inactivation of the voltage-gated sodium channels in the muscle membrane, immediately adjacent to the motor end-plate. There is a temporary insulating zone of electrical inexcitability around the end-plates through which impulses cannot pass, which prevents initiation and propagation of a muscle action potential.

In the case of suxamethonium, hydrolysis does not take place by acetylcholinesterase within the junction. There is therefore a persistent depolarization of the end-plate region with neuromuscular block (phase I block). Continued presence of the depolarizing agent may give rise to a form of desensitization of the receptor system which is known as phase II block (dual block). This phenomenon is not fully understood (Chapter 4).

Antagonists

The antagonists form the non-depolarizing family of muscle relaxants, the larger of the two groups of neuromuscular blocking agents. The terminology here is interesting. The agonist group are described by an effect which they possess (depolarizing agents produce depolarization), whereas this group of antagonists are described by an effect which they do not possess (the non-depolarizing agents do not produce depolarization). This latter group was originally known by the alternative name of 'competitive' because their action appeared to be of a competitive nature. Although this may well be true, it is unlikely that direct competition at one site is the sole mechanism of action. The term non-depolarizing was therefore thought to be less incorrect.

The non-depolarizing muscle relaxants bind in a reversible manner to the ACh recognition site on the alpha subunits of the receptor macromolecules and physically prevent binding by ACh. Activation of the receptor is therefore prevented. The binding of the antagonists is readily reversible and a dynamic equilibrium exists, which favours either the agonist (ACh) or the antagonist depending upon their relative concentrations within the junction.

If the generalized molecular structures of the non-depolarizing muscle relaxants are examined, some similarities can be seen. They are all large, bulky molecules (so-called pachycurares) with at least one positively charged nitrogen. It was originally believed that there had to be two nitrogens separated by a distance of approximately 1 nm and the early molecules conformed to this ideal. Some of the more recent molecular structures, however, are very flexible molecules with more variable inter-onium distances.

It has to be remembered that those agents which are in regular clinical use have been discovered and developed in a largely empirical manner. The discovery of the naturally occurring compound tubocurarine stimulated much of the work on muscle relaxants, particularly following its isolation and structural determination by King in 1935.[26] Metocurine is a simple methylated derivative of tubocurarine. The benzylisoquinolinium relaxants (atracurium, mivacurium and doxacurium) are also based loosely on tubocurarine. The design of atracurium was not actually wholly empirical, but centred on its ability for spontaneous breakdown *in vivo*. Alcuronium resulted from a minor modification to toxiferine, another naturally occurring alkaloid. Gallamine was the first truly synthetic muscle relaxant. The steroid muscle relaxants which include pancuronium, vecuronium, pipecuronium and rocuronium followed the observation that a naturally occurring steroid, malouetine possessed a neuromuscular blocking action. It was realized that the steroid nucleus would provide a rigid supporting structure thus allowing accurate prediction and measurement of inter-onium distance. Despite the empirical background, the more recent muscle relaxants have evolved into very precise pharmacological agents, with more specific receptor selectivity and minimal side-effects. We have not yet, however, reached our Holy Grail. There is still a need for an ultra short-acting non-depolarizing agent with a very rapid onset to replace suxamethonium.

Non-relaxant actions

The muscle relaxants in clinical use are all different. The differences are many and are described in greater detail in Chapter 5. In general, those differences fall into four major categories, namely onset, duration, elimination and side-effects. The first two are inherent in the molecular structure of the drug. The third is partly related to the structure and partly to factors unique to that patient at that time. The fourth, however, is a factor which may have a relationship to the principal pharmacological action of the drug.

It is probably true to state that all drugs have effects which are supplementary to their principal action and the muscle relaxants are no exception. It is possible to turn some of these side-effects to the anaesthetist's advantage although the majority of them are unwanted. It is important to be aware of these additional effects in order to enable a more logical choice of drug for each individual patient.

The principal unwanted effects of the muscle relaxants are those on the cardiovascular system and the potential for histamine release. A number of other considerations are important however, which might include effects on the central nervous system, autonomic nervous system, muscles, plasma electrolytes and cholinesterase, as well as potential effects of the metabolic breakdown products of the relaxants.

References

1. Peper K, Dreyer F, Sandri C, Akert K, Moor H. Structure and ultrastructure of the frog motor endplate. A freeze-etching study. *Cell and Tissue Research* 1974; **149**: 437–455.
2. Stroud RM. Acetylcholine receptor structure. *Neuroscience Commentaries* 1983; **1**: 124–138.
3. Mishina M, Takai T, Imoto K, Noda M, Takahashi T, Numa S, et al. Molecular distinction between fetal and adult forms of muscle acetylcholine receptor. *Nature* 1986; **321**: 406–411.
4. Schofield GG, Witkop B, Warwick JE, et al. Differentiation of the open and closed states of the ionic channels of nicotinic acetylcholine receptors by tricyclic antidepressants. *Proceedings of the National Academy of Science of the USA* 1981; **78**: 5240–5244.
5. Lambert JJ, Durant NN, Henderson EG. Drug-induced modification of ionic conductance at the neuromuscular junction. *Annual Reviews of Pharmacology and Toxicology* 1983; **23**: 505–539.
6. Krnjevic K, Miledi R. Some effects produced by adrenaline upon neuromuscular propagation in rats. *Journal of Physiology* 1958; **141**: 291–304.
7. Starke K. Presynaptic receptors. *Annual Reviews of Pharmacology and Toxicology* 1981; **21**: 7–30.
8. Bowman WC. Prejunctional and postjunctional cholinoceptors at the neuromuscular junction. *Anesthesia and Analgesia* 1980; **59**: 935–943.
9. Standaert FG. The action of d-tubocurarine on the motor nerve terminal. *Journal of Pharmacology and Experimental Therapeutics* 1964; **143**: 181–186.
10. Paton WDM, Waud DR. The margin of safety of neuromuscular transmission. *Journal of Physiology* 1967; **191**: 59–90.
11. Glavinovic MI. Presynaptic action of curare. *Journal of Physiology* 1979; **290**: 499–506.
12. Blaber LC. The effect of facilitatory concentrations of decamethonium on the storage and release of transmitter at the neuromuscular junction of the cat. *Journal of Pharmacology and Experimental Therapeutics* 1970; **175**: 664–672.
13. Grewaal DS, Quastel JH. Control of synthesis and release of radioactive acetylcholine in brain slices from the rat. *Biochemical Journal* 1973; **132**: 1–14.
14. Nachmansohn D. *Chemical and Molecular Basis of Nerve Activity* 1975, New York: Academic Press.
15. Bowman WC. The neuromuscular junction: recent developments. *European Journal of Anaesthesia* 1985; **2**: 59–93.
16. Chang CC, Lee CY. Isolation of toxins from the venom of *Bungarus multicinctus* and their modes of neuromuscular blocking action. *Archives Internationales de Pharmacodynamie et de Therapie* 1963; **144**: 241–257.
17. Jones SW, Salpeter MM. Absence of [^{125}I]-alpha-bungarotoxin binding to motor

nerve terminals of frog, lizard and mouse muscle. *Journal of Neuroscience* 1983; **3:** 326–331.
18. Kloog Y, Michaelson DM, Sokolovsky M. Characterization of the presynaptic muscarinic receptor in synaptosomes of *Torpedo* electric organ by means of kinetic and equilibrium binding studies. *Brain Research* 1980; **194:** 97–115.
19. Duncan CJ, Publicover SJ. Inhibitory effects of cholinergic agents on the release of transmitter at the frog neuromuscular junction. *Journal of Physiology* 1979; **294:** 91–103.
20. Abbs ET, Joseph DN. The effects of atropine and oxotremorine on acetylcholine release in rat phrenic nerve-diaphragm preparations. *British Journal of Pharmacology* 1981; **73:** 481–483.
21. Wessler I, Karl M, Mai M, *et al.* Muscarinic receptors on the rat phrenic nerve, evidence for positive and negative feedback mechanisms. *Naunyn Schmeidebergs Archives of Pharmacology* 1987; **335:** 605–612.
22. Wessler, I. Control of transmitter release from the motor nerve by presynaptic nicotinic and muscarinic autoceptors. *Trends in Pharmacological Sciences* 1989; **10:** 110–114.
23. Kuba K. Effects of catecholamines on the neuromuscular junction in the rat diaphragm. *Journal of Physiology* 1970; **211:** 551–570.
24. Forman SA, Miller KW. Molecular sites of anesthetic action in postsynaptic nicotinic membranes. *Trends in Pharmacological Sciences* 1989; **10:** 447–452.
25. Stenlake JB. Some chemical aspects of neuromuscular block. *Progress in Medicinal Chemistry* 1963; **3:** 1–51.
26. King H. Curare alkaloids, Part I. Tubocurarine. *Journal of the Chemical Society* 1935; **2:** 1381–1389.

3

Pharmacokinetics of Neuromuscular Blocking Drugs

CJR Parker

Section 1: General principles

Introduction

Neuromuscular blocking drugs are almost invariably given intravenously; thus absorption is considered to be complete. They are transported to the body tissues, partially bound to plasma proteins in the circulation, where they are distributed in accordance with their physical and chemical properties. Since they are ionized, they are excreted in the urine; some accumulate in the liver, and biotransformation and biliary excretion of both native drug and of metabolites occur to a variable extent. Some undergo metabolism by plasma cholinesterase, and in the case of atracurium, spontaneous degradation may occur.

The importance of pharmacokinetics

Pharmacokinetic information is important for this class of drug, for several reasons. The drugs are intrinsically lethal, and there is no meaningful therapeutic ratio; safe recovery of the patient depends upon the termination of their action at the end of surgery. Although it is usual to antagonize the action of non-depolarizing neuromuscular blocking drugs at the end of surgery with an anticholinesterase, permanent cessation of action demands their removal from the body. The dependence of recovery upon drug removal has been recognized for many years. The consequences of failure to excrete gallamine were reported, in 1963, by Feldman and Levi;[1] they treated a patient with peritonitis who developed renal failure; recovery of neuromuscular function did not occur until after haemodialysis. A similarly prolonged effect of d-tubocurarine was later reported by Riordan and Gilbertson,[2] in a patient suffering from chronic renal failure.

In subsequent years the link between the effect of the non-depolarizing neuromuscular blocking drugs and their disposition has been better defined than for any other class of drug. It has been shown, for example, that in a series of patients the duration of recovery from pancuronium is strongly correlated with the elimination half-time.[3] Furthermore, pharmacokinetic data and measured effect (for example, the depression of evoked twitch tension, or the reduction of the size of the electromyographic signal) have been formally united in a model with a hypothetical effect compartment by Hull and others[4] and Sheiner and others.[5] There is now no doubt that the measured effect of these drugs is determined by the plasma concentration profile.

An early difficulty in pharmacokinetic studies was the lack of methods to measure the drugs at the concentrations found in plasma. An explosion of pharmacokinetic data followed the development and application of methods to assay the concentration of these drugs in the plasma. An early landmark was the spectrophotometric method for tubocurarine,[6] and was followed by other methods such as an radio-immunoassay,[7] high performance liquid chromatography[8] and ion-monitoring mass spectrometry.[9] The disposition of neuromuscular blocking drugs is now known to be affected by age, and body build, as well as by renal and hepatic disease. Such information is essential background to intelligent attempts to provide appropriate therapy. It is the hope of students of pharmacokinetics that such information will allow rational modification of the dose in particular groups of patients.

Approaches to pharmacokinetics

The appreciation of pharmacokinetic data is often hampered by the seeming complexity of the statistical analysis applied to it. It is usual in a pharmacokinetic investigation of a neuromuscular blocking drug to administer a dose, as a bolus or an infusion, and to measure the plasma concentration profile of the drug in the succeeding minutes and hours. Often the total drug concentration is measured in venous plasma samples. If the results of such measurements are to be used to derive information which has a general value, then it must be summarized in a meaningful context. Difficulties are created by the extraction from the raw data of summary measures of drug distribution and excretion.

Definitions

Clearance: The hypothetical volume of plasma from which drug is considered to be completely removed in unit time.

Volume of distribution: The ratio of the amount of drug in the body to its plasma concentration. *For central volume of distribution* it is the ratio of the amount in the hypothetical central compartment to plasma concentration.

Elimination half-time: The time taken for the plasma concentration to fall by half during the elimination phase.

First order: Defines a process in which the rate of change of a variable is directly proportional to the value of that variable.

Zero order: Defines a process in which the rate of change of a variable is independent of the value of that variable.

Rate constant: The fractional change in a variable in unit time.

Time constant: The reciprocal of the rate constant. It is the time taken for a variable to change by a factor e; it is the time which the process would take to reach completion if it continued at its initial rate.

Compartment: A hypothetical space in which drug is considered to be uniformly distributed. It is considered to be well mixed: drug entering the compartment is instantaneously distributed throughout it. A frequent error is to assume that a compartment has a discrete anatomical counterpart.

AUC: The area under the curve of a plot of plasma concentration vs time; when not subscripted, it generally refers to the whole area from time zero (the start of drug administration) to infinity. Thus:

$$\mathrm{AUC} = \int_0^\infty Cp \, dt \tag{1}$$

where: Cp is plasma concentration
t is time.

AUMC: The area under the first moment curve; that is the area under the plot of the product of plasma concentration and time vs time. Thus:

$$\mathrm{AUMC} = \int_0^\infty (Cp \times t) \, dt \tag{2}$$

where: Cp is plasma concentration
t is time.

It should be noted that for the elimination half-time and clearance to be assigned single unique values implies that the rate of decline of plasma concentration is directly proportional to drug concentration; that is the kinetics are first order, and follow an equation of form (3).

$$\frac{dC}{dt} = -kC \tag{3}$$

There are three approaches to the extraction of parameters from a series of measurements of plasma drug concentration.

Compartmental analysis

If the logarithm of the plasma concentration of a neuromuscular blocking drug after an intravenous bolus, is plotted against time, plasma concentration generally falls rapidly initially then more slowly. If the kinetics are first order, the final phase of such a plot should be linear. Often, two or more phases can be identified (Fig. 3.1), and it is possible to interpret the phases of decline of plasma concentration in terms of hypothetical compartments (Fig. 3.2). In principle, this approach can be predictive; an equation of the form (4) could, given the parameters A and B, together with the rate constants for distribution and elimination, usually denoted α and β respectively, allow prediction of the plasma concentration arising from any future dose.

$$C = \text{Dose} \times (Ae^{-\alpha t} + Be^{-\beta t}) \qquad (4)$$

Unfortunately there are many disadvantages. In equation (4), the expression for C is not linear in α or β, and this greatly complicates the derivation of the parameters. When the measured plasma concentrations range over a few orders of magnitude, then a weighting scheme is needed, for the higher the plasma concentration, the greater the absolute error is likely to be in its measurement.[10] Furthermore, the model parameters are estimated with greater percentage uncertainty than are the original data.

There are other problems of principle: the compartments are hypothetical, a criticism that is particularly true of the peripheral compartments. No account is taken of circulatory mixing, and the use of arterial samples

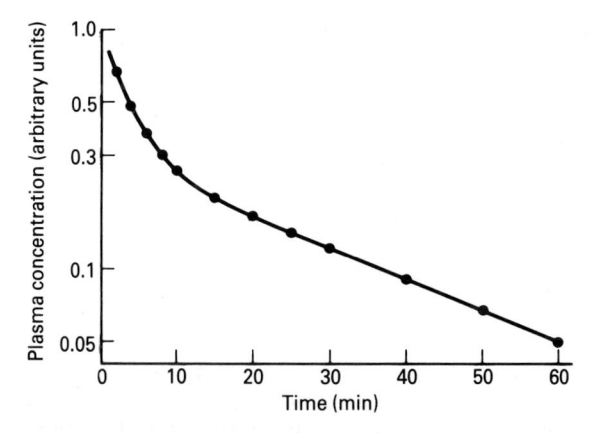

Fig. 3.1 A typical idealized plot of plasma concentration vs time following an intravenous bolus dose of a neuromuscular blocking drug; the plasma concentration is plotted on a logarithmic scale in arbitrary units. Plasma concentration falls rapidly at first and then more slowly, until the rate of decline of the logarithm of plasma concentration is constant. The slope of this final phase is the elimination rate constant, from which the elimination half-time is easily determined. The more rapid initial fall in plasma concentration is envisaged to be due to redistribution.

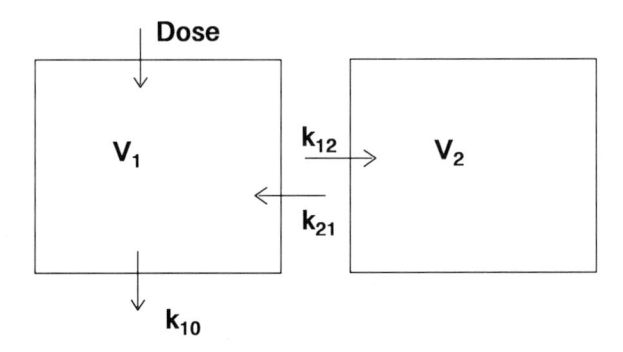

Fig. 3.2 A two-compartment pharmacokinetic model. Two well-stirred compartments, of volumes V_1 and V_2, are connected by a pathway in which drug may be transferred in either direction. The rate of drug transfer from a compartment is proportional to the drug concentration in that compartment, and the rate constant associated with the path. Drug is considered to be administered into the first (central) compartment, and all elimination is usually considered to occur from the first compartment.

gives results quite different from those obtained from venous samples.[11] Finally, the results are not usefully predictive because the individual variability of patients implies that the parameters for a given patient are not known in advance.

Non-compartmental analysis

Many of these problems can be avoided by the use of non-compartmental analysis which avoids a formal model of distribution, but still assumes that the drug is given into a well stirred central compartment from which the drug is eliminated.[12] The method is computationally very simple, and uses just AUC and the AUMC to obtain summary measures of disposition.

Plasma clearance = Dose/AUC   (5)

Mean residence time = AUMC/AUC   (6)

Steady state volume of distribution = Dose × AUMC/AUC²   (7)

These formulae apply to data obtained after a single bolus dose, and with modification, they may be applied to data gathered when the dose is given as a single constant rate infusion. Equations (6) and (7) are not applicable when multiple doses, or infusions at a variable are given, which limits their application in complex situations such as the Intensive Care Unit.

The derivation of the formulae assumes that all drugs may be considered to be eliminated from the central compartment into which the drug is given. For example the application of equation (5) to a drug such as atracurium yields a figure which facilitates comparison between patients, but which does not represent the plasma clearance as defined above.

Population analysis

It has been traditional to attempt to define pharmacokinetic parameters for each individual studied; this process is repeated for several patients, and an estimate is formed of the characteristics of drug disposition in a particular group of subjects. Such studies have often been confined to small homogeneous groups of healthy patients or patients with a specific disease state. The mixed population seen in routine clinical practice is not specifically addressed.

A population approach, as originally suggested for digoxin kinetics,[13] and embodied in the NONMEM programme,[14] eschews the attempt to define the pharmacokinetic parameters for any given individual, but yields answers to such questions as what patient factors, such as age, weight, sex or disease, significantly affect pharmacokinetics, and how variable is disposition within the population. Large numbers of patients are required to be studied, but only a few plasma concentration measurements are needed from each. The major disadvantage is that the statistical theory is complicated, and the necessary calculations are computationally intensive. It seems likely that this approach will be pursued increasingly, and it has already been applied to the pharmacokinetics of rocuronium.[15]

Other approaches to pharmacokinetic analysis

In addition to measurement of drug concentration in the plasma, important information can be derived from analysis of drug in collections of urine and bile.[16,17] In clincial practice, bile collection is unlikely to be complete, but this approach is well suited to the experimental setting.

The analysis of plasma concentration of likely metabolites has been undertaken for some neuromuscular blocking drugs, and can give an impression of the relative importance of particular metabolic routes *in vivo*,[18] or the clinical importance of particular metabolites.[18]

Pitfalls in pharmacokinetic analysis

In addition to the problems of statistical method outlined above, three other pitfalls beset study of the pharmacokinetics of neuromuscular blocking drugs. First, the assays used to measure the parent drug have sometimes been unable to distinguish the parent drug from its metabolites. For pancuronium, two assays have been widely used; the first uses fluorimetric detection of total bisquaternary ammonium compounds,[19] the second uses mass spectrometric ion detection[9] and is able to quantify the parent drug and metabolites directly. Whilst steps can be taken to mitigate the contribution of the metabolites, as was done by Agoston and others,[16] suspicion must hang over the fluorimetric method where its results conflict with those of the more specific assay.

A second, more subtle pitfall arises from the fact that some neuromuscular blocking drugs exist as a mixture of stereoisomers, which may have widely differing pharmacokinetic properties. For example, atracurium comprises three major groups of stereoisomers, ten in all. The *trans-trans* group appears to be rapidly eliminated,[20] yet much published work on the pharmacokinetics of atracurium has used an assay which does not distinguish between the groups of isomers. Likewise mivacurium chloride is also a mixture of three stereoisomers, of which one (*cis-cis*) is much more slowly metabolized than the other two,[21] though it may be less potent. This can lead to an estimate of the elimination half-time which is paradoxically long in relation to the duration of drug action.[22] Clearly the definition of the pharmacokinetics of a drug with multiple stereoisomers requires a stereoselective assay. With recent developments in the production of stereospecific drugs, the issue will continue to be important.

Finally, the validity of the results obtained depends upon the acquisition of a sufficient amount of data across the whole time course of drug disposition. The work of Matteo and others[23] contains a salutary warning in this respect; the serum concentration of d-tubocurarine was measured for up to 96 hours after a single intravenous bolus. When data for serum concentration up to 6 hours were used to fit a model, the pharmacokinetic parameters were unremarkable, and seemed to be estimated with reasonable precision. The drug was still detectable in the serum at 6 hours however, and when the data gathered up to 96 hours were used, a quite different set of pharmacokinetic parameters emerged, with a terminal elimination half-time of 40 hours and a total volume of distribution of over $3 l kg^{-1}$. Whilst the significance of the very low concentration of tubocurarine found several hours after a dose is uncertain, it is clear that adequate definition of the pharmacokinetics of these drugs requires an adequate duration of sampling.

Section 2: Specific neuromuscular blocking drugs

Protein binding

All the non-depolarizing neuromuscular blocking drugs are bound to plasma proteins to some extent, and so, in principle, alterations in plasma protein binding could lead to alterations in drug disposition. In general the bound fraction is modest, and so small changes in protein binding will not lead to large changes in free fraction. Furthermore, although drug binding by plasma proteins has been the focus of much interest, this is at least in part because it is a tractable issue; disposition *in vivo* must also depend upon binding to other sites, both in blood and in the tissues.

Several basic drugs bind chiefly to alpha 1 acid glycoprotein,[24] although Cohen and others have shown that the fractional binding of tubocurarine is similar for the albumin, globulin and fibrinogen fractions,[25] and tritiated dimethyltubocurarine binds chiefly to gamma globulin.[26]

Plasma protein binding has been cited to explain pharmacodynamic differences between individuals; for example Stovner and others[27] showed that dose requirements (judged clinically) for gallamine correlated with plasma albumin concentration. The correlations of course do not prove causation; and results have not always been consistent between studies. Thus, in contrast, in a different series of patients, gallamine requirements were correlated with globulin concentration.[28]

Altered plasma protein binding has also been suggested to explain pharmacodynamic differences in different disease states, for example chronic liver disease due to schistosomiasis.[29] Patients with liver disease had greater requirements of tubocurarine, and also greater concentrations of globulin; a causal relationship was suggested. When the protein binding of tubocurarine by serum from cirrhotic patients was measured directly,[30] it was found to be only moderately bound, and little changed from that in healthy subjects. A similar result was obtained by Ghonheim and others,[31] and by Walker and others[32] who found little change in the plasma protein binding of tubocurarine in either renal or hepatic disease. It seems that altered plasma protein binding is unlikely to account for the marked variability in individual response to these drugs.

A final caveat in the interpretation of plasma protein binding in the context of drug disposition *in vivo*, is that it is important that the measurement be made at physiological pH and temperature, and at a reasonable drug concentration. This imposes a particular difficulty for atracurium, which is unstable under these conditions, and for which there is no entirely satisfactory measurement[33,34] (see Table 3.1).

Table 3.1 The fractional protein binding of some non-depolarizing neuromuscular blocking drugs in healthy human plasma, shown as a percentage, together with the method used and the source

Drug	Protein binding	Method	Ref. Source
Tubocurarine	42–49%	Equilibrium dialysis	25
	56%	Ultracentrifugation	30
	50%	Equilibrium dialysis	32
	51%	Equilibrium dialysis	35
Dimethyltubocurarine	35%	Equilibrium dialysis	35
Pancuronium	29%	Ultracentrifugation	30
	7%	Equilibrium dialysis	36
Vecuronium	91%	Indirect	33
	30%	Ultracentrifugation	30
Atracurium	82%	Indirect	33
	50%	Dialysis (pH 6.2)	34
	37%	Indirect (*in vivo*)	Hunter (unpublished)
Doxacurium	28–34%	Ultrafiltration	37

Pathways of drug elimination

Renal excretion

Renal excretion occurs for all the neuromuscular blocking drugs to some extent, because they are all ionized, water soluble and not excessively protein bound; therefore they may undergo glomerular filtration. The extent of renal excretion differs widely within the group however; it may be the major pathway of excretion, or a minor one amongst several other mechanisms.

At one extreme, the vast majority of a dose of gallamine can be recovered from the urine in the perioperative period,[38] and for this drug other routes of elimination are negligible (see below). On the other hand, suxamethonium, as well as *cis-trans* and *trans-trans* mivacurium chloride, is so rapidly cleared by plasma cholinesterase that the renal route of elimination has a minor role.

Most other neuromuscular blocking drugs lie between these extremes, though atracurium, which has several alternative pathways of elimination, undergoes renal excretion to a minor extent; renal clearance has been estimated to be 10% of total systemic clearance.[17, 39] For other non-depolarizing drugs the renal route accounts for elimination of a higher proportion of the dose. Attempts to measure the proportion of a dose excreted in the urine are likely to result in a underestimate; details of some attempts to measure renal excretion in this way are shown in Table 3.2.

Table 3.2 The proportion of an injected dose of muscle relaxant recovered in the urine in healthy subjects is shown as a percentage

Drug	Percentage recovered	n	Ref. Source
Tubocurarine	38% at 24 h	5	40
	44% at 24 h	9	23
	63% at 48 h	5	35
Dimethyltubocurarine	52% at 48 h	5	35
Gallamine	98% at 48 h	4	38
Pancuronium	43% at 30 h	6	16
	67% at 24 h	8	41
Vecuronium	30% at 24 h	6	42
	(20% as unchanged vecuronium)		
	(10% as 3-desacetyl vecuronium)		
Atracurium	11% at 6 h	17	17
	6% at up to 500 min	3	39
Doxacurium	25% at 6–12 h	9	43
	31% at 24 h	3	44
Pipecuronium	38–43% at 24 h	12	45
	(37–39% as unchanged pipecuronium)		
	(1–4% as 3-desacetyl pipecuronium)		
Rocuronium	1–21% at 24 h	3	15
	33% at 24 h	10	46

Hepatic uptake

These drugs are taken up, metabolized and excreted by the liver to a variable extent. In the extreme case of gallamine, there is hardly any hepatic elimination, with 0.1% recovery in the bile.[38] In contrast, for vecuronium, liver uptake is substantial; in a pig model intraportal drug administration results in much reduced effect, and hepatic exclusion from the circulation leads to prolongation of action.[47]

Lipophilicity appears to be an important determinant of hepatic uptake which is favoured by the presence of just one quaternary nitrogen centre (as in vecuronium and tubocurarine), compared with two (as in pancuronium and dimethyltubocurarine), or three as with gallamine. Thus in the isolated perfused rat liver, biliary excretion is greater and clearance from the plasma is more rapid for tubocurarine (51% biliary excretion in 2 hours) than dimethyltubocurarine (16.5% biliary excretion over the same time).[48] When the same model was applied to a series of steroid neuromuscular blocking drugs, both uptake of drug into the liver and its excretion in the bile were positively correlated with lipophilicity. In particular, vecuronium, with an octanol:water partition coefficient of 2.56 was cleared 14 times more rapidly than pancuronium for which the octanol:water partition coefficient is 0.0033.[49] Studies on subcellular fractions suggest that these drugs distribute passively between plasma and the hepatocyte cytosol in accordance with the electrochemical gradient, but they are bound to intracellular organelles, in particular the mitochondria, and are actively secreted into the bile canaliculus.[50]

A raised plasma concentration of bile salts has been found to inhibit uptake of basic drugs by the liver. The latter effect has been demonstrated both for tubocurarine and for the steroid agent Org 6368 in the isolated perfused rat liver.[51] Moreover, the duration of action of vecuronium was found to be markedly increased in the intact rat during an infusion of the bile salt taurocholate.

Where human studies allow comparison of the hepatic elimination of different drugs, a similar contrast can be seen. Collection of bile samples is likely to be less complete than in animal models but, for example, biliary recovery of metocurine was found to be 2% of the dose, and of tubocurarine 11%.[35] For pancuronium, an early study[16] found biliary recovery of 11% of the dose of pancuronium in the 30 hours after its administration whereas for vecuronium a figure of 40% of the total dose was derived by Bencini and others,[52] after correcting for the probable incompleteness of the biliary collection. Consistent with this picture is the finding that only about 2% of a dose of pipecuronium (which has two quaternary nitrogen centres) was recovered in the bile.[45] The fact that the hepatic excretion of gallamine is negligible presumably reflects the fact that it has three quaternary ammonium groups.

The role of plasma cholinesterase

Some neuromuscular blocking drugs are esters, and are eliminated by ester hydrolysis. They include suxamethonium and mivacurium, for both of which hydrolysis, catalysed by plasma cholinesterase, is the main route of elimination; doxacurium is metabolized by plasma cholinesterase much more slowly. Atracurium is also an ester, and undergoes degradation in plasma to some extent probably by other non-specific esterases.

Plasma cholinesterase is well characterized;[53] its primary structure is known, as is the structure of several important variants. Reduced activity may be due to an inherited abnormality of enzyme structure, or to a reduction in enzyme concentration in plasma (Chapter 4).

Inherited defects in the enzyme

Major variations in the activity of the enzyme arise from inherited alterations in its structure. Several variants of the enzyme, with a reduced activity against suxamethonium, are known. They include the dibucaine and fluoride-resistant variants, as well as the several genetically distinct inactive ('silent') variants.

Causes of altered amount of enzyme

Certain patient conditions affect the synthesis and activity of the enzyme. These have been listed by Whittaker,[54] and include pregnancy, hepatic disease and chronic renal failure. Alterations in enzyme activity also result from several drugs, including pancuroniuim and neostigmine. The effect is generally more minor than the presence of an inherited defect.

Suxamethonium

Suxamethonium is so rapidly degraded by plasma cholinesterase in normal subjects that the profile of its plasma concentration is difficult to characterize. A recent preliminary study of one healthy patient yielded a terminal elimination half-time of 8 minutes.[55] The relationship between the duration of action of suxamethonium and abnormality of plasma cholenesterase is well established.

Mivacurium

This new non-depolarizing drug comprises three distinct optical isomers, two of which, the *cis-trans* and *trans-trans* isomers, are rapidly degraded by plasma cholenesterase. The elimination half-times are 2.9 minutes and 3.6 minutes respectively.[56] The clearance of these isomers *in vivo* is strongly correlated with plasma cholinesterase activity. The third isomer, *cis-cis*, is metabolized *in vitro* much more slowly, and has an elimination half-time *in vivo* of about 35 minutes.[56] Experience in a subject with plasma cholinesterase deficiency indicates a prolonged duration of action may be expected when the enzyme is deficient.[57]

Doxacurium

The rate of hydrolysis of doxacurium by pooled human plasma *in vitro* is reported to be $0.16\,\mu M\,h^{-1}$, or about 6% of the rate of suxamethonium.[58] Comparisons between doxacurium and suxamethonium are of course complicated by the widely differing potency and molecular weights of the two drugs; administration of $30\,\mu g\,kg^{-1}$ of doxacurium bis-cation to a patient weighing 70 kg with a plasma volume of 5 l would be expected to lead to a peak plasma concentration of around $0.4\,\mu M$. It is clear that hydrolysis of doxacurium by plasma cholinesterase cannot be expected to be rapid.

Atracurium

Ester hydrolysis is one of several modes of elimination of atracurium, accounting for up to 27% of the total degradation of atracurium *in vivo*.[17] The *in vitro* inactivation of atracurium is more rapid in plasma than in buffer at the same temperature and pH, a difference which is reduced by organo-phosphorous compounds.[59] Differences in the activity of plasma esterases presumably explain the wide species differences in the *in vitro* half-times of atracurium in plasma. Esterases other than plasma cholinesterase may be responsible, and indeed a normal duration of action has been reported for atracurium in plasma cholinesterase deficiency.[60]

Pharmacokinetics of metabolites of neuromuscular blocking drugs

The study of the pharmacokinetics of the metabolites of neuromuscular blocking drugs has a two-fold importance. First it may shed light on the possible modes of elimination of the parent drug; second, the metabolites may themselves be active, either at the neuromuscular junction or elsewhere. It is difficult to establish the pharmacokinetic profile of a metabolite by study of its plasma concentration following the administration of the parent drug: crucial information on which the calculation of pharmacokinetic parameters is based – in particular, the dose, the time scale of formation of the metabolite and the site of its formation within the body – is not known, even within quite wide limits. It has sometimes been assumed that all of the parent drug is eventually transformed to a particular metabolite, allowing an estimate of the clearance of the metabolite. This assumption was made for the formation of laudanosine from atracurium,[61] but since some 10% of a bolus dose of atracurium is excreted by the renal route,[39] the assumed dose of metabolite is an overestimate, and the derived metabolite clearance must be also be an overestimate. On the other hand, the direct approach to estimating metabolite pharmacokinetics by administering a dose of the metabolite is not usually acceptable especially if the metabolite is toxic.

In the case of pancuronium and vecuronium, the parent drug is acetylated at the 3 and 17 positions; metabolism may occur by hydrolysis at either position to produce the corresponding hydroxyl compound. Both metabolites have a somewhat greater plasma clearance than the parent drug; 3-hydroxy pancuronium is half as potent as the parent drug, whilst the 17-hydroxy metabolite has only one-fiftieth of the potency.[62] The presence of 3-hydroxy pancuronium in human urine and bile was identified semi-quantitatively by Agoston and others;[16] the recovery of 3-hydroxy pancuronium appeared to account for about 20% of the dose of the parent drug. The importance of the 3-hydroxy metabolite of vecuronium is emphasized by the prolonged block which may result in large part from the 3-hydroxy derivative when a large dose given, over a prolonged period, and the renal route of elimination is impaired.[18]

Atracurium is metabolized to several compounds (Fig. 3.3), linked by a complex net of pathways.[39] The two products of Hofmann degradation are laudanosine and an acrylate; the products of ester hydrolysis are a monoquaternary alcohol and a monoquaternary acid. Laudanosine itself may undergo further metabolism to tetrahydropapaverine.

The estimated volume of distribution of laudanosine is much greater than that of atracurium; values of 1.97 l kg^{-1},[61] and 1.7 l kg^{-1},[63] have been estimated in healthy subjects. The elimination half-time of laudanosine is also much longer than that of atracurium, with estimates ranging from 168[61] to 176 minutes.[17] In the critically ill, the elimination half-time of laudanosine may be very long, with (albeit imprecise) estimates of over 33 hours in patients with combined respiratory and renal failure.[64]

The pharmacokinetic profile of the monoquaternary alcohol (which presumably itself undergoes Hofmann elimination) is less well defined than that of laudanosine. There is a danger that other metabolites may not have been separated from it by the chromatographic method, leading to an overestimate of the metabolite concentration. The estimated elimination half-time of the monoquaternary alcohol has been reported to be 27 minutes,[17] and 26 minutes,[39] and it appears to be increased in renal failure to about 42 minutes.[17] Least is known concerning the pharmacokinetics of the monoquaternary acid and acrylate moieties.

Mivacurium is metabolized by ester hydrolysis to an alcohol and an acid; neither the pharmacokinetics of these products nor their clinical import, are, as yet defined.

The effects of advanced age

The function of several organ systems responsible for drug elimination deteriorates as age advances (Chapter 9); both glomerular filtration rate and splanchnic blood flow decrease. It is therefore to be expected that the pharmacokinetics of several neuromuscular blocking drugs are altered in

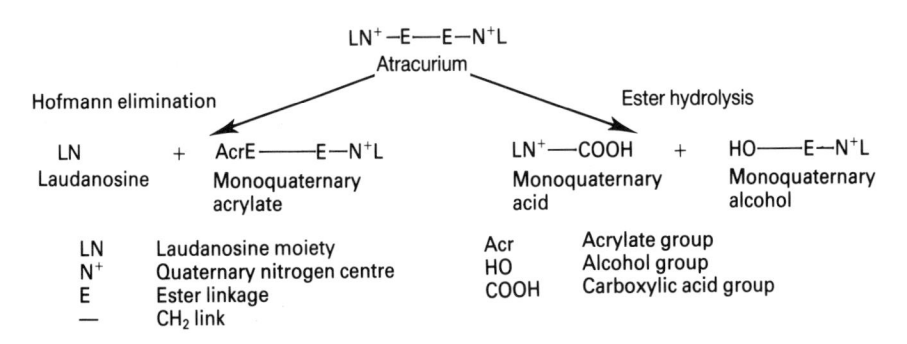

Fig. 3.3 The degradation of atracurium, and its metabolic products; the molecule is symmetrical with two ester linkages, and two sites at which Hofmann elimination may occur. Cleavage at any site destroys the molecule; the monoquaternary alcohol and monoquaternary acrylate are thought to undergo further degradation by cleavage at remaining sites.

the elderly. An early report of this phenomenon is the fall in clearance of pancuronium with age;[65] the fall appeared to be linear, and substantial, from a typical value of clearance around 100 ml min^{-1} at 20 years to around 40 ml min^{-1} at 80 years. In contrast, they found no correlation between age and volume of distribution. These findings were substantially confirmed by Duvaldestin.[41] Later, Rupp and others used a selective mass spectrometric method which measures only parent drug and not metabolite, but studied fewer patients than Duvaldestin, and found only a 20% reduction in clearance in the elderly which did not reach statistical significance.[66]

Subsequently the clearance of both tubocurarine and dimethyltubocurarine was shown to be reduced in the elderly, by about half for tubocurarine and by two-thirds for dimethyltubocurarine; the elimination half-time of both drugs was shown to be increased. In addition both the central and peripheral volumes of distribution were shown to be, if anything, reduced in the elderly and the changes reached statistical significance in the case of dimethyltubocurarine.[67]

The situation for vecuronium and atracurium is less clear. Rupp and others found that both clearance and volume of distribution of vecuronium were reduced in the elderly, with no significant change in the elimination half-time.[66] More recently Lein and others[68] have confirmed the reduction in clearance, and found a correspondingly increased elimination half-time for vecuronium in the elderly.

Atracurium, with its facility for elimination by non-organ dependent routes, might be expected to have pharmacokinetics unchanged by ageing, and indeed any changes are minor. A small increase in elimination half-time has been found with advanced age.[63, 69, 70] Other effects have been less consistent; the largest of these studies found a reduction in clearance with age,[69] whilst Kitts and others, who used a combination of *in vivo* and *in vitro* data, defined a reduction in organ-dependent clearance in the elderly, together with a raised volume of distribution in this group.[70]

The latest generation of neuromuscular blocking drugs has been, as yet, rather less extensively studied. Two reports of the pharmacokinetics of doxacurium in the elderly yield conflicting findings. Thus clearance was found to be reduced by 30%, and elimination half-time increased from 76 to 120 minutes in the elderly.[71] On the other hand, a study using a slightly lower dose of doxacurium (25 µg kg^{-1}) found in increased volume of distribution in the elderly, without change in clearance or elimination half-time.[44]

For pipecuronium, disposition was found to be unchanged in the elderly,[72] whilst for rocuronium, clearance was found to be significantly reduced and elimination half-time more than doubled in a group of elderly patients compared to younger subjects.[73]

The infant

The volume and composition of the body fluid compartments, and the extent of organ function are different in the neonate from the adult (Chapter 11). It is difficult to perform pharmacokinetic studies in this group, as the limited blood volume places an obvious limit on the size and frequency of blood samples. The response to neuromuscular blocking drugs is complicated by relative immaturity of the neuromuscular junction in this age group. Comparison with the adult group is further complicated by the uncertainty of the appropriate normalization of dose, and of pharmacokinetic volume terms for body size.

Perhaps the most illuminating of the pharmacokinetic studies on the older non-depolarizing neuromuscular blocking drugs is that by Fisher and others,[74] who showed that steady state volume of distribution (considered on the basis of body weight) is increased in infancy, being around 704 ml kg^{-1} in the neonate, and 520 ml kg^{-1} in the infant, compared with 300 ml kg^{-1} in the adult. There was little change in plasma clearance (compared in the units of ml min^{-1} kg^{-1}); consequently the elimination half-time was almost doubled in the neonate to 174 minutes (compared with 89 minutes in the adult). The same group has studied the pharmacokinetics of both vecuronium[75] and atracurium[76] in the same paradigm. For both drugs the volume of distribution (measured in the units ml kg^{-1}) is increased in infants. For vecuronium this is associated with a mean residence time of 66 minutes in infants compared with 34 minutes in older children. In the case of atracurium, the volume of distribution at steady state was more than doubled in infants, whilst the elimination half-time was unchanged; this was accompanied by an increase in the estimate of organ-independent clearance, presumably reflecting the spontaneous degradation of atracurium throughout its expanded volume of distribution. Brandom and others[77] also found an increased total volume of distribution and clearance of atracurium in the infant compared with the older child, though the differences disappeared when the comparison of volume terms was made on the basis of ml m^{-2} rather than ml kg^{-1}.

Effects of body size

It is customary to adjust the dose of non-depolarizing neuromuscular blocking drugs drugs according to body weight. This tradition has a long history,[78] but it was also realized very early that administration of tubocurarine to adults solely according to body weight is misguided, and that total muscle mass is likely to be more important.[78]

In subsequent pharmacokinetic investigations it has been usual both to administer the drug, and to scale the resulting volume terms (volumes of distribution and clearance) in proportion to body weight. In 1979, however, it was shown that neither the clearance nor the central or total volume of distribution of pancuronium is correlated with body weight.[65] A similar result was obtained for atracurium[79] in a study of healthy adults, where it was shown that both the absolute central and total volumes of distribution are uncorrelated with body weight, but the 'correction' for body weight, to express the result in units ml kg^{-1} produces a measure of volume of distribution which is strongly negatively correlated with weight.

Since all the neuromuscular blocking drugs are ionized and water soluble, they are not likely to be redistributed into adipose tissue; yet a major reason for the differences between body weight amongst healthy adults is the degree of obesity. The question arises as to the relationship between the pharmacokinetics of the neuromuscular blocking drugs and body size. Knowledge of this relationship should not only provide a rational basis for adjustments of dosage, but also facilitate comparison of pharmacokinetic results from different sources. The issue has been addressed systematically in several recent studies.

Beemer and others[80] showed that within a group of 20 healthy adults, the interindividual variability in both clearance and total volume of distribution was as great if the volume is expressed as such or divided by total body weight; variability was reduced by correction for lean body mass. The same group later extended their work to a study of 80 patients with differing degrees of adiposity,[81] and showed that clearance, estimated during a steady state infusion, is proportional to lean body mass, but that division of clearance by total body weight gives a result which is biased downwards in the obese. A study of the morbidly obese given a single bolus of atracurium demonstrated a similar trend;[82] when expressed as absolute values, both total clearance (ml min^{-1}) and volume of distribution at steady state (l) are comparable with the results in a group of normal body build. When the results were divided by body weight, the values were halved in the morbidly obese relative to those in the lean subjects, leading to a spurious apparent difference from the normal subjects, which reached statistical significance.

For vecuronium, similar findings were obtained by Schwartz and others;[83] the absolute values of volume of distribution and clearance in the normal and obese subject were comparable. Expression of volume terms as a fraction of total body weight resulted in apparently significantly lower values in the obese group. The importance of these findings for appropriate dosage are underlined by the fact that the drug was given as a multiple of body weight (0.1 mg kg^{-1}), and the time to 50% recovery of twitch tension was significantly prolonged in the obese group.

Pregnancy

Pregnancy is associated with a variety of physiological changes which might affect the disposition of neuromuscular blocking drugs. There is an expansion of plasma and extravascular fluid volumes and a rise in glomerular filtration rate, a fall in plasma albumin concentration,[84] and a fall in plasma cholinesterase activity.[54] There is a relative dearth of pharmacokinetic studies of the neuromuscular blocking drugs in the pregnant subject. For pancuronium, a fall in elimination half-time to 114 minutes, compared with 146 minutes in a group of healthy non-pregnant adults studied simultaneously has been found;[85] a rise of plasma clearance was responsible, whilst the total volume of distribution (expressed in units ml kg^{-1}) was little changed. These changes were associated with a rise in the cumulative 24-hour urinary excretion from 50 to 66%. The pharmacokinetics of vecuronium and pancuronium were studied using a mass spectrometric assay by Dailey and others,[86] who found that for each drug the elimination half-time was about half of that obtained in the non-pregnant adult by the same group;[9] thus the elimination half-time of vecuronium was found to be 36 minutes, and for pancuronium to be 72 minutes. The steady state volumes of distribution, expressed (251 ml kg^{-1} for vecuronium and 283 ml kg^{-1} for pancuronium) were similar to those in the non-pregnant subject.[9]

The placental barrier to neuromuscular blocking drugs

Much of the published work on the pharmacokinetics of the neuromuscular blocking drugs in pregnancy has focused on the placental transfer of drug to the infant. The ratio of umbilical venous:maternal venous (UV:MV) concentration of neuromuscular blocking drug at delivery has been documented for a number of non-depolarizing neuromuscular blocking drugs. It is clear that the placental barrier is relative rather than absolute, and all of the non-depolarizing drugs studied cross the placenta to some extent. Indeed when maternal exposure is prolonged and substantial, significant quantities of d-tubocurarine can be transferred to the infant.[87]

A common paradigm for such studies is to relate the ratio of drug concentration in umbilical and maternal veins at delivery to the time between drug administration and delivery. With atracurium for example, the UV:MV ratio rises linearly with time between the atracurium bolus and delivery reaching 0.21 with a dose–delivery interval of 18 minutes.[88] Similar findings were reported for pancuronium[85] where a typical value of UV:MV drug concentration of 0.22 was found to increase with time, and for dimethyltubocurarine.[89] For vecuronium, in contrast, a typical value of 0.11 for the UV:MV concentration ratio was found not to increase with time.[90]

Such a measure as the UV:MV drug concentration ratio is complicated, and must depend not only upon the properties of the placenta, but on the history of maternal drug concentration between administration and delivery. A high UV:MV ratio does not necessarily reflect a high absolute foetal concentration, and the fact that UV:MV ratio rises with time must reflect the fact that maternal plasma drug concentration is falling as a consequence of redistribution within the mother. In the study of Shearer and others, for example, the patient with the shortest dose–delivery interval had an absolute umbilical vein atracurium concentration which was close to the mean for all the patients studied.[88]

The extent of drug redistribution to the foetus can be affected by the mode of drug administration. Thus in an early study of the disposition of alcuronium, in which a toad bioassay was used, drug was detectable in cord blood in ten of twelve infants when the alcuronium had been given as a rapid bolus, but only two of seven infants where the drug had been given as a slow injection.[91]

Hypothermia

Profound reductions in body temperature lead to changes in organ function which affect the metabolism and excretion of drugs. Assessment of the effect of hypothermia on the disposition of neuromuscular blocking drugs is confounded by two facts. First, the induction of hypothermia in humans is usually part of a technique of cardiopulmonary bypass, which itself affects organ function and body fluid compartment volumes. Secondly the response of the neuromuscular junction to non-depolarizing neuromuscular blocking drugs changes in hypothermia; the potency of tubocurarine decreases in mild hypothermia,[92] for pancuronium the reverse may be true.[93]

The interpretation of the effects of hypothermia on the pharmacokinetics of neuromuscular blocking drugs is most clearly shown in two studies in the cat,[92, 93] in which the clearance of both pancuronium and tubocurarine is reduced on cooling to 28–29°C. The reduction is substantial, with a 60% decrease for both drugs, and is associated with an increase in the elimination half-time of around 40%.

In the human undergoing cardiopulmonary bypass, a reduction in pancuronium clearance is suggested by the results of d'Hollander and others,[94] who showed that although the plasma concentration of pancuronium required to sustain a steady 90% depression of twitch was reduced on average by 30% during hypothermic phase of bypass, the infusion rate required to maintain this drug concentration was reduced by 60%. For atracurium, the infusion requirement is decreased.[95] To what extent this had a pharmacokinetic basis is unclear, but in addition to the likely reduction in organ-dependent clearance the rate of Hofmann elimination is temperature dependent and a reduction in organ-independent clearance is to be expected in hypothermia.[96]

Effects of renal failure

The patient with renal failure presents several features which are likely to affect the pharmacokinetics of neuromuscular blocking drugs in addition simply to the loss of an excretory route (Chapter 9). Normal body fluid regulation is impaired; the patient is likely to be overhydrated, but immediately following haemodialysis the reverse may be true. The activity of other pathways may also be altered; thus plasma cholinesterase activity is lowered in chronic renal failure and this accounts quantitatively for the somewhat reduced infusion requirements for mivacurium chloride.[97] Most patients with renal failure take multiple medications, and those with renal transplants require immunosuppressive drugs, leading to the possibility of drug interactions. Cyclosporin for example inhibits the hepatic cytochrome P450 activity,[98] and the multiple medications almost universally required by such patients provide a fertile ground for pharmacokinetic drug interactions.

The influence of renal failure on the pharmacokinetics of several non-depolarizing neuromuscular blocking drugs is shown in Table 3.3. In several instances, the elimination half-time is increased in renal failure, and the plasma clearance is decreased; there are exceptions however, and in the case of rocuronium, a significant increase in the elimination half-time is attributable to an increase in the volume of distribution in renal failure.[15] Perhaps the most noteworthy feature of Table 3.3 is that there are three reports on the pharmacokinetics of atracurium, each failing to find a significant alteration in renal failure; the pharmacokinetic basis for preference of this drug in renal failure would seem to be sound.

Effects of hepatic disease

As with renal disease, the crucial issue for the clinician is the overall effect of a disease state on drug disposition. Hepatic diseases are heterogeneous and a distinction should be drawn between at least three manifestations. First the patient with acute and severe hepatic failure, where the function of other organ systems may also be deranged; second the patient with hepatic cirrhosis; and, finally, the patient with biliary obstruction of previously healthy liver (Chapter 9).

Hepatic failure

In the patient with fulminant hepatic failure, the pharmacokinetics of a single bolus of atracurium were studied by Ward and Neill;[106] the volume of distribution was found to be raised, but the elimination half-life was almost unchanged at 22 minutes compared to 21 minutes in a group of healthy subjects. Since several of the patients in this study were suffering from acute renal failure in addition to fulminant hepatic failure, these data provide powerful evidence of the robustness of organ-independent elimination of atracurium.

The pharmacokinetics of both mivacurium and doxacurium have been studied in patients with hepatic failure sufficient to warrant hepatic transplantation by Cook and others.[22, 43] For doxacurium, no significant change was found in the pharmacokinetic parameters; for mivacurium the clearance was halved and the mean residence time more than doubled in hepatic failure, a difference which presumably reflected the lower activity of plasma cholinesterase in this group.

Hepatic cirrhosis

Hepatic cirrhosis is a chronic disease, with a histological definition, in which a series of physiological changes ensues, including fluid retention, a hyperdynamic circulation and alteration in the pattern of plasma protein concentrations, with increased levels of gamma globulin and reduced albumin levels. There is evidence of resistance to non-depolarizing neuromuscular blocking drugs.

Table 3.3 The effect of renal failure on the pharmacokinetics of some neuromuscular blocking drugs. The results for the comparator group of healthy patients are shown below those for the patients in renal failure

Drug		Pharmacokinetic parameter Clearance	Vd_{ss}	$t_{1/2}\beta$	n	Ref. Source
Dimethyltubocurarine	R	0.38 ml kg^{-1} min^{-1}*	353 ml kg^{-1}*	684 min*	5	99
	H	1.20	472	360		
Gallamine	R	0.24 ml kg^{-1} min^{-1}*	285 ml kg^{-1}*	752 min*	8	100
	H	1.20	207	131		
Pancuronium	R	53 ml min^{-1}*	296 ml kg^{-1}	257 min*	10	101
	H	123	261	133		
Vecuronium	R	4.5 ml kg^{-1} min^{-1}	347 ml kg^{-1a}	68 min	10	102
	H	3.6	242	50		
	R	3.1 ml kg^{-1} min^{-1}*	241 ml kg^{-1}	83 min*	12	103
	H	5.3	199	53		
Atracurium	R	6.7 ml kg^{-1} min^{-1}	224 ml kg^{-1b}	23.7 min	9	104
	H	6.1	182	20.6		
	R	5.8 ml kg^{-1} min^{-1}	141 ml kg^{-1a}	20.1 min	6	39
	H	5.5	153	19.3		
	R	7.8 ml kg^{-1} min^{-1}	165 ml kg^{-1}	19.7 min	6	17
	H	10.8	280	17.3		
Mivacurium ('active[c]')	R	76.6 ml kg^{-1} min^{-1d}	150 ml kg^{-1d}	1.9 min[e]	9	22
	H	70.4	112	1.5		
(cis-cis[c])	R			34.3 min	9	22
	H			18.4		
Doxacurium	R	1.2 ml kg^{-1} min^{-1}*	270 ml kg^{-1}	221 min	8	43
	H	2.7	220	99		
Pipecuronium	R	1.6 ml kg^{-1} min^{-1}*	426 ml kg^{-1}*	247 min*	20	105
	H	2.4	307	118		
Rocuronium	R	2.89 ml kg^{-1} min^{-1}	264 ml kg^{-1}*	97 min*	10	15
	H	[f]	207	71		

R renal failure; H healthy patients; * Significantly different from the group of healthy patients.
[a] Vβ
[b] VdAREA
[c] The assay did not separate the three stereoisomers whose renal excretion may be different.
[d] Derived values.
[e] Mean residence time.
[f] No significant difference between renal failure and healthy patients; population analysis did not give separate values for each group.

The pharmacokinetic alterations in cirrhosis are shown for some neuromuscular blocking drugs in Table 3.4. The most nearly consistent feature is an increased volume of distribution, which has been demonstrated for pancuronium, fazadinium, atracurium and rocuronium, though not for vecuronium. For pancuronium, fazadinium and rocuronium this leads to a prolongation of the elimination half-time.

Extrahepatic biliary obstruction

The complete obstruction of the biliary tract eliminates the biliary route of excretion, as well as resulting in a greatly raised concentration of bile salts (see above). The disposition of pancuronium in patients with extrahepatic biliary obstruction has been studied by Westra and others[38] and by Somogyi and others.[112] Both groups found a substantial and significant prolongation of the elimination half-time, which was more than doubled in one of the studies,[112] but the mechanism differed between the two studies. Thus in one study plasma clearance was halved,[112] whereas in the other, a greater than 50% increase in steady state volume of distribution was responsible.[38]

The same two groups also examined the pharmacokinetics of gallamine in patients with extrahepatic cholestasis, and the alterations in disposition were much less striking. Thus one study found no significant alteration in any pharmacokinetic parameter, whereas Ramzan and others[113] found only a 20% increase in steady state volume of distribution, a change which did not significantly affect the elimination half-time.

Thermal injury

Patients with burns are resistant to non-depolarizing neuromuscular blocking drugs. The pharmacokinetics of atracurium,[34] tubocurarine[114] and dimethyltubocurarine[115] have been studied in these patients. In general the pharmacokinetics are little altered in patients with thermal injury, though the estimated plasma protein binding of atracurium was increased from around 50% to 63%.[34] Resistance is not a result of changes in drug disposition. The subject is considered in greater detail in Chapter 9.

The new non-depolarizing neuromuscular blocking drugs

Much of our knowledge of the pharmacokinetics of neuromuscular blocking drugs pertains to drugs which will be used to a decreasing extent in the future, as their place is taken by newer agents. Where pharmacokinetic data are available for these drugs, they have been discussed above. It is appropriate, however, to attempt to summarize the main pharmacokinetic features of four of the more recently evaluated non-depolarizing neuromuscular blocking drugs.

The two new benzylisoquinolinium drugs, mivacurium and doxacurium, have a quite different pharmacokinetic profile. The two most abundant stereoisomers of mivacurium are rapidly eliminated by ester

Table 3.4 The effect of hepatic cirrhosis upon the pharmacokinetics of some neuromuscular blocking drugs. The results for the comparator group of healthy patients are shown below those for the cirrhotic patients

Drug		Pharmacokinetic parameter Clearance	Vd_{ss}	$t_{1/2}\beta$	n	Ref. Source
Pancuronium	C	1.45 ml min^{-1} kg^{-1}*	416 ml kg^{-1}*	208 min*	14	107
	H	1.86	279	114		
Fazadinium	C	75 ml min^{-1} m^2	448 ml kg^{-1}*	153 min*	8	108
	H	96	287	82		
Atracurium	C	8.0 ml min^{-1} kg^{-1}*	282 ml kg^{-1a}	24.5 min	7	61
	H	6.6	202	20.9		
Vecuronium	C	2.73 ml min^{-1} kg^{-1}*	253 ml kg^{-1}	84 min*	12	109
	H	4.26	246	58		
	C	4.4 ml min^{-1} kg^{-1}	220 ml kg^{-1}	51.4 min	10	110
	H	4.5	180	57.7		
Rocuronium	C	3.0 ml min^{-1} kg^{-1}	322 ml kg^{-1}*	173 min*	6	111
	H	3.4	174	79		

C hepatic cirrhosis; H healthy patients
* Significantly different from the group of healthy patients.
^{a} Vβ

hydrolysis. The drug is thus the most rapidly eliminated of all the currently available non-depolarizing drugs.[21] The mode of elimination of the minority *cis-cis* isomer and its importance are still uncertain, though cholinesterase and renal function may both play a role.[116] Doxacurium has a longer elimination half-time, with ester hydrolysis probably accounting for only a small fraction of elimination; its pharmacokinetic profile is typical of a conventional bisquaternary compound.

The two new aminosteroid muscle relaxants, rocuronium and pipecuronium, are both entirely organ dependent for their elimination. The use of either drug in patients with renal or hepatic disease will require consideration of this fact. It might be expected that rocuronium, with its structural similarity to vecuronium (in particular its single quaternary nitrogen group), should undergo considerable hepatic uptake. Indeed, whilst the elimination of rocuronium is dependent on the renal route to some extent,[15] there is preliminary evidence of prolonged elimination in patients with hepatic dysfunction.[111] Pipecuronium, in contrast, with its similarity to pancuronium, and two quaternary nitrogen centres, seems to be more dependent upon renal excretion.[45]

References

1. Feldman SA, Levi JA. Prolonged paresis following gallamine: a case report. *British Journal of Anaesthesia* 1963; **35:** 804–806.
2. Riordan DD, Gilbertson AA. Prolonged curarization in a patient with renal failure: case report. *British Journal of Anaesthesia* 1971; **43:** 506–508.
3. Shanks CA, Somogyi AA, Triggs EJ. Dose-response and plasma concentration response relationships of pancuronium in man. *Anesthesiology* 1979; **51:** 111–118.
4. Hull CJ, VanBeem HBH, McLeod K, Sibbald A, Watson MJ. A pharmacodynamic model for pancuronium. *British Journal of Anaesthesia* 1978; **50:** 1113–1122.
5. Sheiner LB, Stanski DR, Vozeh S, Miller RD, Ham J. Simultaneous modelling of pharmacokinetic and pharmacodynamics: application to d-tubocurarine. *Clinical Pharmacology and Therapeutics* 1979; **25:** 358–371.
6. Cohan EN, Paulsen WJ, Elert E. Studies of d-tubocurarine with measurements of concentration in human blood. *Anesthesiology* 1957; **18:** 300–309.
7. Horowitz PE, Spector S. Determination of serum d-tubocurarine concentration by radio-immunoassay. *Journal of Pharmacology and Experimental Therapeutics* 1973; **185:** 94–100.
8. Simmonds RJ. Determination of atracurium, laudanosine and related compounds in plasma by high-performance liquid chromatography. *Journal of Chromatography* 1985; **343:** 431–436.
9. Cronnelly R, Fisher DM, Miller RD, Gencarelli P, Nguyen-Gruenke L, Castagnoli N. Pharmacokinetics and pharmacodynamics of vecuronium (ORG NC45) and pancuronium in anesthetized humans. *Anesthesiology* 1983; **58:** 405–408.
10. Boxenbaum H, Riegelman S, Elashoff R. Statistical estimation in pharmacokinetics. *Journal of Pharmacokinetics and Biopharmaceutics* 1974; **2:** 123–148.
11. Donati F, Varin F, Ducharme J, Gill SS, Theoret Y, Bevan DR. Pharmacokinetics and pharmacodynamics of atracurium obtained with arterial and venous blood samples. *Clinical Pharmacology and Therapeutics* 1991; **49:** 515–522.

12. Gibaldi M, Perrier D. *Pharmacokinetics* 2nd edn, 1982, New York: Marcel-Dekker, pp. 409–417.
13. Sheiner LB, Rosenberg B, Marathe VV. Estimation of population characteristics of pharmacokinetic parameters from routine clinical data. *Journal of Pharmacokinetics and Biopharmaceutics* 1977; **5:** 455–479.
14. Boeckmann AJ, Sheiner LB, Beal SL. *NONMEM Users' Guide, Part V Introductory Guide*, NONMEM Project Group, University of California, San Francisco, November 1992, pp. 32–40.
15. Szenohradszky J, Fisher DM, Segredo V, Caldwell JE, Bragg P, Sharma M, *et al*. Pharmacokinetics of rocuronium bromide (ORG 9426) in patients with normal renal function of patients undergoing cadaver renal transplantation. *Anesthesiology* 1992; **77:** 899–904.
16. Agoston S, Vermeer GA, Kersten UW, Meijer DKF. The fate of pancuronium bromide in man. *Acta Anaesthesiologica Scandinavica* 1973; **17:** 267–275.
17. Vandenbrom RHG, Wierda JMKH, Agoston S. Pharmacokinetics and neuromuscular blocking effects of atracurium besylate and two of its metabolites in patients with normal and impaired renal function. *Clinical Pharmacokinetics* 1991; **19:** 230–240.
18. Segredo V, Caldwell JE, Matthay MA, Sharma ML, Gruenke LD, Miller RD. Persistent paralysis in critically ill patients after long-term administration of vecuronium. *New England Journal of Medicine* 1992; **327:** 524–528.
19. Kersten UW, Meijer DKF, Agoston S. Fluorimetric and chromatographic determination of pancuronium bromide and its metabolites. *Clinica Chimica Acta* 1973; **44:** 59–66.
20. Tsui D, Graham GG, Torda T. The pharmacokinetics of atracurium isomers *in vitro* and in humans. *Anesthesiology* 1987; **67:** 722–728.
21. Head-Rapson AG, Devlin JC, Lovell GG, Parker CJR, Hunter JM. Pharmacokinetics of the isomers of mivacurium chloride in the healthy adult. *British Journal of Anaesthesia* 1993; **70:** 487P.
22. Cook DR, Freeman JA, Lai AA, Kang Y, Stiller RL, Aggarwal S, *et al*. Pharmacokinetics of mivacurium in normal patients and in those with hepatic or renal failure. *British Journal of Anaesthesia* 1992; **69:** 580–585.
23. Matteo RS, Nishitateno K, Pua EK, Spector S. Pharmacokinetics of d-tubocurarine in man: effect of an osmotic diuretic on urinary excretion. *Anesthesiology* 1980; **52:** 335–338.
24. Wood M. Plasma drug binding: implications for anesthesiologists. *Anesthesia and Analgesia* 1986; **65:** 786–804.
25. Cohen EN, Corbascio A, Fleischli G. The distribution and fate of d-tubocurarine. *Journal of Pharmacology and Experimental Therapeutics* 1965; **147:** 120–129.
26. Skivington MA. Protein binding of three tritiated muscle relaxants. *British Journal of Anaesthesia* 1972; **44:** 1030–1034.
27. Stovner J, Theodorsen L, Bjelke E. Sensitivity to gallamine and pancuronium with special reference to plasma proteins. *British Journal of Anaesthesia* 1971; **43:** 953–958.
28. Stout RJ. Some clinical evidence for a quantitative relationship between dose of relaxant and plasma protein. *West Indian Medical Journal* 1963; **12:** 256–264.
29. Baraka A, Gabali F. Correlation between tubocurarine requirements and plasma protein pattern. *British Journal of Anaesthesia* 1968; **40:** 89–93.
30. Duvaldestin P, Henzel D. Binding of tubocurarine, fazadinium, pancuronium and ORG NC45 to serum proteins in normal man and in patients with cirrhosis. *British Journal of Anaesthesia* 1982; **54:** 513–516.
31. Ghonheim MM, Kramer E, Bannow R, Pandya H, Routh JI. Binding of d-tubocurarine to plasma proteins in normal man and in patients with hepatic or renal disease. *Anesthesiology* 1973; **49:** 410–415.

32. Walker JS, Shanks CA, Brown KF. Determinants of d-tubocurarine plasma protein binding in health and disease. *Anesthesia and Analgesia* 1983; **62:** 870–874.
33. Foldes FF, Deery A. Protein binding of atracurium and other short-acting neuromuscular blocking agents and their interaction with human cholinesterases. *British Journal of Anaesthesia* 1983; **55:** 31S–34S.
34. Marathe PH, Dwersteg JF, Pavlin EG, Haschke RH, Heimbach DM, Slattery JT. Effect of thermal injury on the pharmacokinetics and pharmacodynamics of atracurium in humans. *Anesthesiology* 1989; **70:** 752–755.
35. Meijer DKF, Weitering JG, Vermeer GA, Scaf AHJ. Comparative pharmacokinetic of d-tubocurarine and metocurine in man. *Anesthesiology* 1979; **51:** 402–407.
36. Wood M, Stone WJ, Wood AJJ. Plasma binding of pancuronium – effect of age, sex and disease. *Anesthesiology* 1980; **53:** S286.
37. de Angelis R, Loebs P, Maehr R. High-performance liquid chromatographic analysis of doxacurium, a new long-acting neuromuscular blocker. *Journal of Chromatography* 1990; **525:** 389–400.
38. Westra P, Vermeer GA, De Lange AR, Scaf AHJ, Meijer DKF, Wesseling H. Hepatic and renal disposition of pancuronium and gallamine in patients with extrahepatic cholestasis. *British Journal of Anaesthesia* 1981; **53:** 331–337.
39. Ward S, Boheimer N, Weatherley BC, Simmonds RJ, Dopson TA. Pharmacokinetics of atracurium and its metabolites in patients with normal renal function, and in renal failure. *British Journal of Anaesthesia* 1987; **59:** 697–706.
40. Miller RD, Matteo RS, Benet LZ, Sohn YJ. The pharmacokinetics of d-tubocurarine in man with and without renal failure. *Journal of Pharmacology and Experimental Therapeutics* 1977; **202:** 1–7.
41. Duvaldestin P, Saada J, Berger JL, D'Hollander A, Desmonts JM. Pharmacokinetics, pharmacodynamics and dose-response relationships of pancuronium in control and elderly subjects. *Anesthesiology* 1982; **56:** 36–40.
42. Bencini AF, Scaf AHJ, Sohn YJ, Meistelman C, Lienhart A, Kersten UW, et al. Diposition and urinary excretion of vecuronium bromide in anesthetized patients with normal renal function or renal failure. *Anesthesia and Analgesia* 1986; **65:** 245–251.
43. Cook DR, Freeman JA, Lai AA, Robertson KA, Kang Y, Stiller RL, et al. Pharmacokinetics and pharmacodynamics of doxacurium in normal patients and in those with hepatic or renal failure. *Anesthesia and Analgesia* 1991; **72:** 145–150.
44. Dresner DL, Basta SJ, Ali HH, Schwartz AF, Embree PB, Wargin WA, et al. Pharmacokinetics and pharmacodynamics of doxacurium in young and elderly patients during isoflurane anesthesia. *Anesthesia and Analgesia* 1990; **71:** 498–502.
45. Wierda JMKH, Szenohradsky J, De Wit APM, Zentai G, Agoston S, Kakas M, et al. The pharmacokinetics, urinary and biliary excretion of pipecuronium bromide. *European Journal of Anesthesiology* 1991; **8:** 451–457.
46. Wierda JMKH, Kleef UW, Lambalk LM, Kloppenburg WD, Agoston S. The pharmacokinetics and pharmacodynamics of Org 9426, a new nondepolarizing blocking agent, in patients anaesthetized with nitrous oxide, halothane and fentanyl. *Canadian Journal of Anaesthesia* 1991; **38:** 430–435.
47. Motsch J, Hennis PJ, Zimmerman FA, Agoston S. A model for determining the influence of hepatic uptake of non-depolarizing muscle relaxants. *Anesthesiology* 1989; **70:** 128–133.
48. Meijer DKF, Weitering JG, Vonk RJ. Hepatic uptake and biliary excretion of d-tubocurarine and trimethyltubocurarine in the rat *in vivo* and in isolated perfused rat livers. *Journal of Pharmacology and Experimental Therapeutics* 1976; **198:** 229–239.
49. Mol WEM, Rombout F, Paanakker JE, Oosting R, Scaf AHJ, Meijer DKF. Pharmacokinetics of steroidal muscle relaxants in isolated perfused rat liver. *Biochemical Pharmacology* 1992; **44:** 1453–1459.

50. Mol WEM, Meijer DKF. Hepatic transport mechanisms for bivalent organic cations: subcellular distribution and hepato-biliary concentration gradients of some steroidal muscle relaxants. *Biochemical Pharmacology* 1990; **39:** 383–390.
51. Westra P, Keulemans GTP, Houwertjes MC, Hardonk MJ, Meijer DKF. Mechanisms underlying the prolonged duration of action of muscle relaxants caused by extrahepatic cholestasis. *British Journal of Anaesthesia* 1981; **53:** 217–227.
52. Bencini AF, Scaf AHJ, Sohn YJ, Kersten-Kleef UW, Agoston S. Hepatobiliary disposition of vecuroniuim bromide in man. *British Journal of Anaesthesia* 1986; **58:** 988–995.
53. Pantuck EJ. Plasma cholinesterase: gene and variations. *Anesthesia and Analgesia* 1993; **77:** 380–386.
54. Whittaker M. Plasma cholinesterase and the anaesthetist. *Anaesthesia* 1980; **35:** 174–197.
55. Lagerwerf AJ, Vanlinthout LEH, Vree TB. Rapid determination of succinylcholine in human plasma by high-performance liquid chromatography with fluorescence detection. *Journal of Chromatography* 1991; **570:** 390–395.
56. Head-Rapson AG, Devlin JC, Lovell GG, Parker CJR, Hunter JM. Pharmacokinetics of the isomers of mivacurium chloride in healthy patients and in renal failure. *British Journal of Anaesthesia* 1993; **71:** 312P–313P.
57. Mangar D, Kirchoff GT, Rose PL, Castellano FC. Prolonged neuromuscular block after mivacurium in a patient with end-stage renal disease. *Anesthesia and Analgesia* 1993; **76:** 866–867.
58. Basta SJ, Savarese JJ, Ali HH, Embree PB, Schwartz AF, Rudd GD, et al. Clinical pharmacology of doxacurium chloride: a new long-acting non-depolarizing muscle relaxant. *Anesthesiology* 1988; **69:** 478–486.
59. Stiller RL, Cook DR, Chakravorti S. In vitro degradation of atracurium in human plasma. *British Journal of Anaesthesia* 1985; **57:** 1085–1088.
60. Donen N. Use of atracurium in a patient with plasma cholinesterase deficiency. *Canadian Journal of Anaesthesia* 1987; **34:** 64–66.
61. Parker CJR, Hunter JM. Pharmacokinetics of atracurium and laudanosine in patients with hepatic cirrhosis. *British Journal of Anaesthesia* 1989; **62:** 177–183.
62. Miller RD, Agoston S, Booij LHDJ, Kersten UW, Crul JF, Ham J. The comparative potency and pharmacokinetics of pancuronium and its metabolites in anaesthetized man. *Journal of Pharmacology and Experimental Therapeutics* 1978; **207:** 539–543.
63. Kent AP, Parker CJR, Hunter JM. Pharmacokinetics of atracurium and laudanosine in the elderly. *British Journal of Anaesthesia* 1989; **63:** 661–666.
64. Parker CJR, Jones JE, Hunter JM. Disposition of infusions of atracurium and its metabolite, laudanosine, in patients in renal and respiratory failure in an ITU. *British Journal of Anaesthesia* 1988; **61:** 531–540.
65. McLeod K, Hull CJ, Watson MJ. Effects of ageing on the pharmacokinetics of pancuronium. *British Journal of Anaesthesia* 1979; **51:** 435–438.
66. Rupp SM, Castagnoli KP, Fisher DM, Miller RD. Pancuronium and vecuronium pharmacokinetics and pharmacodynamics in younger and elderly adults. *Anesthesiology* 1987; **67:** 45–49.
67. Matteo RS, Backus WW, McDaniel DD, Brotherton WP, Abraham R, Diaz J. Pharmacokinetics and pharmacodynamics of d-tubocurarine and metocurine in the elderly. *Anesthesia and Analgesia* 1985; **64:** 23–29.
68. Lein CA, Matteo RS, Ornstein E, Schwartz AE, Diaz J. Distribution, elimination and action of vecuronium in the elderly. *Anesthesia and Analgesia* 1991; **73:** 39–42.
69. Parker CJR, Hunter JM, Snowdon SL. Effect of age, sex and anaesthetic technique on the pharmacokinetics of atracurium. *British Journal of Anaesthesia* 1992; **69:** 439–443.

70. Kitts JB, Fisher DM, Canfell PC, Spellman MJ, Caldwell JE, Heier T, et al. Pharmacokinetics and pharmacodynamics of atracurium in the elderly. *Anesthesiology* 1990; **72:** 272–275.
71. Gariepy LP, Varin F, Donati F, Salib Y, Bevan DR. Influence of aging on the pharmacokinetics and pharmacodynamics of doxacurium. *Clinical Pharmacology and Therapeutics* 1993; **53:** 340–347.
72. Ornstein E, Matteo RS, Schwartz AE, Jamdar SC, Diaz J. Pharmacokinetics and pharmacodynamics of pipecuronium bromide (Arduan) in elderly surgical patients. *Anesthesia and Analgesia* 1992; **74:** 841–844.
73. Matteo RS, Ornstein E, Schwartz AE, Stone JG, Ostapkovich N, Spencer HK. Pharmacokinetics and pharmacodynamics of ORG 9426 in elderly surgical patients. *Anesthesiology* 1991; **75:** A1065.
74. Fisher DM, O'Keefe C, Stanski DR, Cronnelly R, Miller RD, Gregory GA. Pharmacokinetics and pharmacodynamics of d-tubocurarine in infants, children and adults. *Anesthesiology* 1982; **57:** 203–208.
75. Fisher DM, Castagnoli K, Miller RD. Vecuronium kinetics and dynamics in anaesthetised infants and children. *Clinical Pharmacology and Therapeutics* 1985; **37:** 402–406.
76. Fisher DM, Canfell PC, Spellman MJ, Miller RD. Pharmacokinetics and pharmacodynamics of atracurium in infants and children. *Anesthesiology* 1990; **73:** 33–37.
77. Brandom BW, Stiller RL, Cook DR, Woelfel SK, Chakravorti S, Lai A. Pharmacokinetics of atracurium in anaesthetized infants and children. *British Journal of Anaesthesia* 1986; **58:** 1210–1213.
78. Prescott F, Organe G, Rowbotham S. Tubocurarine chloride as an adjunct to anaesthesia: report on 180 cases. *Lancet* 1946; **ii:** 80–84.
79. Parker CJR, Hunter JM. Relationship between volume of distribution of atracurium and body weight. *British Journal of Anaesthesia* 1993; **70:** 443–445.
80. Beemer GH, Bjorksten AR, Crankshaw DP. Pharmacokinetics of atracurium during continuous infusion. *British Journal of Anaesthesia* 1990; **65:** 668–674.
81. Beemer GH, Bjorksten AR, Crankshaw DP. Effect of body build on the clearance of atracurium: implication of drug dosing. *Anesthesia and Analgesia* 1993; **76:** 1296–1303.
82. Varin F, Ducharme J, Theoret Y, Besner J-C, Bevan DR, Donati F. Influence of extreme obesity on the body disposition and neuromuscular blocking effect of atracurium. *Clinical Pharmacology and Therapeutics* 1990; **48:** 18–25.
83. Schwartz AE, Matteo RS, Ornstein E, Halevy JD, Diaz J. Pharmacokinetics and pharmacodynamics of vecuronium in the obese surgical patient. *Anesthesia and Analgesia* 1992; **74:** 515–518.
84. Shnider SM, Levinson G. Anesthesia for obstetrics. In Miller RD ed. *Anesthesia* 3rd edn, 1990, New York: Churchill Livingstone.
85. Duvaldestin P, Demetriou M, Henzel D, Desmonts JM. Placental transfer of pancuronium and its pharmacokinetics during caesarean section. *Acta Anesthesiologica Scandinavica* 1978; **22:** 327–333.
86. Dailey PA, Fisher DM, Schnider SM, Baysinger CL, Shinohara Y, Miller RD, et al. Pharmacokinetics, placental transfer and neonatal effects of vecuronium and pancuronium for caesarean section. *Anesthesiology* 1984; **60:** 569–574.
87. Older PO, Harris JM. Placental transfer of tubocurarine: a case report. *British Journal of Anaesthesia* 1968; **40:** 459–463.
88. Shearer ES, Fahy LT, O'Sullivan EP, Hunter JM. Transplacental distribution of atracurium, laudanosine and monoquaternary alcohol during elective caesarean section. *British Journal of Anaesthesia* 1991; **66:** 551–556.
89. Kivalo I, Saarikara S. Placental transfer of ^{14}C-dimethyltubocurarine during caesarean section. *British Journal of Anaesthesia* 1976; **48:** 239–241.

90. Demtriou M, Depoix J-P, Diakite B, Fromentin M, Duvaldestin P. Placental transfer of ORG NC45 in women undergoing caesarean section. *British Journal of Anaesthesia* 1982; **54:** 643–645.
91. Thomas J, Climie CR, Mather LE. The placental transfer of alcuronium. *British Journal of Anaesthesia* 1969; **41:** 297–301.
92. Ham J, Miller RD, Benet LZ, Matteo RS, Roderick LL. Pharmacokinetics and pharmacodynamics of d-tubocurarine during hypothermia in the cat. *Anesthesiology* 1978; **49:** 324–329.
93. Miller RD, Agoston S, van der Pol F, Booij LHDJ, Crul JF, Ham J. Hypothermia and the pharmacokinetics and pharmacodynamics of pancuronium in the cat. *Journal of Pharmacology and Experimental Therapeutics* 1978; **207:** 532–538.
94. d'Hollander AA, Duvaldestin P, Henzel D, Nevelsteen M, Bomblet JP. Variations in pancuronium requirement, plasma concentration, and urinary excretion induced by cardiopulmonary bypass with hypothermia. *Anesthesiology* 1983; **58:** 505–509.
95. Flynn PJ, Hughes R, Walton B. The use of atracurium in cardiopulmonary bypass with induced hypothermia. *Anesthesiology* 1983; **59:** A262.
96. Merrett RA, Thompson CW, Webb FW. *In vitro* degradation of atracurium in human plasma. *British Journal of Anaesthesia* 1982; **55:** 61–66.
97. Phillips BJ, Hunter JM. Use of mivacurium chloride by constant infusion in the anephric patient. *British Journal of Anaesthesia* 1992; **68:** 492–498.
98. Cunningham C, Burke MD, Whiting PH, Simpson JG, Wheatley DN. Ketoconazole, cyclosporin and the kidney. *Lancet* 1982; **ii:** 1464.
99. Brotherton WP, Matteo RS. Pharmacokinetics and pharmacodynamics of metocurine in humans with and without renal failure. *Anesthesiology* 1981; **55:** 273–276.
100. Ramzan MJ, Shanks CA, Triggs EJ. Gallamine disposition in surgical patients with chronic renal failure. *British Journal of Clinical Pharmacology* 1981; **12:** 141–147.
101. Somogyi AA, Shanks CA, Triggs EJ. The effect of renal failure on the disposition and neuromuscular blocking action of pancuronium. *European Journal of Clinical Pharmacology* 1977; **12:** 23–29.
102. Meistelman C, Leinhart A, Leveque C, Bitker MO, Pigot B, Viars P. Pharmacology of vecuronium in patients with end-stage renal failure. *European Journal of Anesthesiology* 1986; **3:** 153–158.
103. Lynam DP, Cronnelly R, Castagnoli KP, Canfell PC, Caldwell J, Arden J, *et al*. The pharmacokinetics and pharmacodynamics of vecuronium in patients anesthetized with isoflurane with normal renal function or with renal failure. *Anesthesiology* 1988; **69:** 227–231.
104. Fahey MR, Rupp SM, Fisher DM, Miller RD, Sharma M, Canfell PC, *et al*. The pharmacokinetics and pharmacodynamics of atracurium in patients with and without renal failure. *Anesthesiology* 1984; **61:** 699–702.
105. Caldwell JE, Canfell PC, Castagnoli KP, Lynam DP, Fahey MR, Fisher DM, *et al*. The influence of renal failure on the pharmacokinetics and duration of action of pipecuronium bromide in patients anesthetized with halothane and nitrous oxide. *Anesthesiology* 1989; **70:** 7–12.
106. Ward S, Neill EAM. Pharmacokinetics of atracurium in acute hepatic failure (with acute renal failure). *British Journal of Anaesthesia* 1983; **55:** 1169–1172.
107. Duvaldestin P, Agoston S, Henzel D, Kersten UW, Desmonts JM. Pancuronium pharmacokinetics in patients with liver cirrhosis. *British Journal of Anaesthesia* 1978; **50:** 1131–1135.
108. Duvaldestin P, Saada J, Henzel D, Daumon G. Fazadinium pharmacokinetics in patients with liver disease. *British Journal of Anaesthesia* 1980; **52:** 789–793.

109. Lebrault C, Berger JL, d'Hollander AA, Gomeni R, Henzel D, Duvaldestin P. Pharmacokinetics and pharmacodynamics of vecuronium (ORG NC45) in patients with cirrhosis. *Anesthesiology* 1985; **62:** 601–605.
110. Arden JR, Lynam DP, Castagnoli KP, Canfell PC, Cannon JC, Miller RD. Vecuronium in alcoholic liver disease: a pharmacokinetic and pharmacodynamic analysis. *Anesthesiology* 1988; **68:** 771–776.
111. Magorian T, Wood P, Caldwell JE, Szenohradsky J, Sharma H, Gruenke LD, *et al*. Pharmacokinetics, onset, and duration of action of rocuronium in humans: normal vs hepatic dysfunction. *Anesthesiology* 1991; **75:** A1069.
112. Somogyi AA, Shanks CA, Triggs EJ. Disposition kinetics of pancuronium bromide in patients with total biliary obstruction. *British Journal of Anaesthesia* 1977; **49:** 1103–1107.
113. Ramzan IM, Shanks CA, Triggs EJ. Pharmacokinetics and pharmacodynamics of gallamine triethiodide in patients with total biliary obstruction. *Anesthesia and Analgesia* 1981; **60:** 289–296.
114. Martyn JAJ, Matteo RS, Greenblatt DJ, Lebowitz PW, Savarese JJ. Pharmacokinetics of d-tubocurarine in patients with thermal injury. *Anesthesia and Analgesia* 1982; **61:** 241–246.
115. Martyn JAJ, Goudsouzian NG, Matteo RS, Liu LMP, Szyfelbein SK, Kaplan RF. Metocurine requirements and plasma concentrations in burned paediatric patients. *British Journal of Anaesthesia* 1983; **55:** 263–268.
116. Parker CJR, Head-Rapson AG, Devlin JC, Hunter JM. Pharmacokinetics of *cis-cis* mivacurium: a population approach. *British Journal of Anaesthesia* 1993; **71:** 766P–767P.

4

Suxamethonium

NJN Harper

Pharmacology: mechanism of neuromuscular blockade

Succinylcholine is a bisquaternary ammonium compound; a feature in common with many non-depolarizing muscle relaxants. Several factors influence the neuromuscular blocking actions of quaternary ammonium compounds and the distance between the cationic quaternary nitrogen groups appears to be critical. It is these groups which form an electrostatic attraction with the acetylcholine receptors. Many of the actions of suxamethonium can be explained by the similarity of its chemical structure to two molecules of acetylcholine.

Depolarizing blockade

Under normal conditions, acetylcholine is removed rapidly from the neuromuscular junction by acetylcholinesterase and the resting membrane potential is restored rapidly with the result that acetylcholine-induced neuromuscular blockade is not seen. Depolarizing agents increase membrane permeability to sodium and potassium simultaneously by holding open the appropriate membrane channels. Suxamethonium is removed from the neuromuscular junction a thousand times more slowly than acetylcholine and the muscle membrane is maintained in a depolarized state for a period of several minutes. Partial depolarization of the muscle membrane might be expected to increase the chance of a nerve impulse resulting in an action potential in the adjacent electrically excitable muscle membrane. However, the continuing presence of suxamethonium increases the threshold of the post-synaptic membrane. In the end-plate region an action potential may be generated but in the adjacent muscle membrane the increased permeability to sodium is only transient whereas the increased permeability to potassium is persistent. Consequently, the surrounding membrane becomes unexcitable (a process known as accommodation). It has been suggested that 'accommodation blockade' might be a more appropriate description of the actions of suxamethonium than depolarizing blockade.

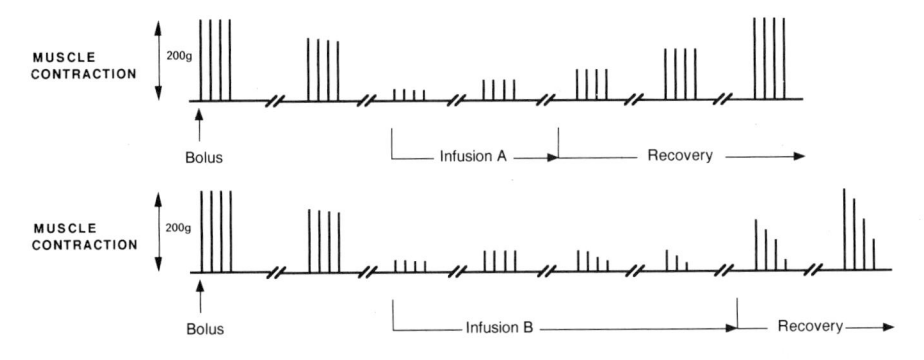

Fig. 4.1 Short (A) and long (B) infusions of suxamethonium demonstrating the development of Train-of-Four fade with time.

Phase II blockade

Repeated bolus administration or prolonged continuous infusion of suxamethonium is associated with the gradual development of phase II blockade in which the characteristics of a pure depolarizing block are apparently replaced by those of a non-depolarizing block. Phase II block is characterized by tetanic and Train-of-Four fade (Fig. 4.1) and post-tetanic facilitation. Pure depolarizing block due to suxamethonium (phase I block) is enhanced by neostigmine or edrophonium but phase II block may be reversed by edrophonium providing that there is considerable fade and the Train-of-Four ratio is less than 0.4.[1] Neostigmine may reverse phase II block even when the Train-of-Four ratio exceeds 0.4.[2] Phase II block may be less reversible with anticholinesterases if it is caused by plasma cholinesterase deficiency.[3] Unfortunately, in clinical practice the tactile or visual assessment of Train-of-Four fade is so inaccurate that reversal of phase II blockade cannot be recommended unless the muscle twitch is measured using a force or acceleration transducer or electromyographically (Chapter 6) and it is preferable to simply await the full return of neuromuscular function as assessed by Double Burst stimulation.

The mechanisms responsible for the development of phase II block are still the subject of debate. It is for this reason that the descriptive term Phase II block is preferred to the mechanistic terms Depolarization or Dual block. The term Desensitization is a general description of a receptor state in which an agonist may bind to the receptor in an apparently normal manner but the receptor channel does not open. It is not specific to the administration of suxamethonium although desensitization may be partially responsible for phase II block. In addition to cholinergic agonists, many drugs including local anaesthetics, phenothiazines, cholinesterase inhibitors and alcohols will induce desensitization *in vitro*. After repeated doses of suxamethonium the degree of depolarization associated with a given depth of neuromuscular blockade decreases, suggesting that a mechanism other than depolarization is increasingly important.

It is now generally accepted that fade in the responses to repetitive stimuli is predominantly a prejunctional phenomenon and it is attractive to speculate that the onset of phase II block coincides with increasing blockade of the prejunctional receptors that normally mediate the enhancement of acetylcholine mobilization in the nerve terminal.

The probability of the development of phase II block depends on the total dose and the duration of administration of suxamethonium. There is wide variation between patients. Inhalational anaesthetic agents accelerate the onset of phase II block by 33–50% during a continuous infusion of suxamethonium.[4,5] During a continuous infusion of suxamethonium at a rate sufficient to produce 90% twitch depression the Train-of-Four ratio will have decreased to approximately 0.5 after a period of 30–40 minutes. Phase II block occurs earlier in patients with plasma cholinesterase deficiency because the neuromuscular junction is exposed to persistently greater concentrations of suxamethonium.

Tachyphylaxis and bradyphylaxis

The onset of phase II block is associated with the clinical phenomenon of tachyphylaxis: i.e., a greater dose is needed to produce the same effect.[6] This phenomenon has been attributed to mutual antagonism between the depolarizing and non-depolarizing processes at the neuromuscular junction. Conversely, if a continuous infusion of suxamethonium is continued for 90 minutes or more, the infusion requirements begin to decrease dramatically, particularly in the presence of volatile agents.[7] This phenomenon is known as bradyphylaxis.

Clinical use of suxamethonium

Indications

Rapid sequence induction

Suxamethonium remains the muscle relaxant of choice when the interval between the induction of anaesthesia and endotracheal intubation needs to be as short as possible. In situations where the airway is precarious it is wise to avoid any muscle relaxant until local anaesthesia or deep general anaesthesia has permitted laryngoscopy to take place. In addition, the short duration of suxamethonium is a potential safety factor in the hazardous situation that arises when both endotracheal intubation and ventilation of the lungs are impossible. The failed-intubation procedure for caesarean section relies on this attribute of suxamethonium. Suxamethonium is not without side-effects (Table 4.1), some of which are related to its depolarizing action at the neuromuscular junction. In current practice the choice of suxamethonium is usually based on a particular clinical requirement for rapid, straightforward endotracheal intubation. A dose of 1 mg kg^{-1} almost invariably produces complete ablation of the twitch response of the thumb within 90 seconds and optimum conditions for endotracheal intubation within 60 seconds. It is frequently possible to intubate with ease before maximum blockade has occurred at the adductor pollicis muscle[8] because neuromuscular blockade becomes maximal at the masseter

58 Suxamethonium

Table 4.1 Complications of the administration of suxamethonium

Fasciculations
Myalgia
Masseter spasm
Bradycardia
Hyperkalaemia
Raised intraocular pressure
Raised intragastric pressure
Prolonged neuromuscular blockade
Myoglobinuria
Release of catecholamines
Malignant hyperthermia
Anaphylaxis

muscles and the vocal cords before the adductor pollicis.[9,10] Despite the rapid onset of suxamethonium at the larynx and masseter muscles, it is likely that difficulties in endotracheal intubation are sometimes due to impatience rather than individual resistance to neuromuscular blockade or the administration of a partially degraded drug as the result of storage at too high a temperature.

It is not surprising that a great deal of attention has been given to developing a non-depolarizing agent with a rapid onset. The onset of non-depolarizing muscle relaxant drugs can be accelerated by using the priming technique but this manoeuvre cannot be recommended when there is a risk of regurgitation of gastric contents because the competence of the airway-protective reflexes and the ability of the patient to maintain oxygenation[11] may be compromised by the priming dose.

Many factors affect the rate of onset of muscle relaxant drugs and the view that non-depolarizing agents necessarily have a slower onset of action than depolarizing agents is no longer tenable. Rocuronium exhibits the most rapid onset among the currently available non-depolarizing drugs. A dose of twice the ED95 produces maximum neuromuscular blockade at the larynx in a mean time of 1.4 minutes;[12] but it is unable to match the rapidity of action of suxamethonium which develops its maximum effect in approximately 1 minute. Muscle relaxants in general appear to block neuromuscular transmission more rapidly at the larynx than at peripheral muscles but this phenomenon appears to be particularly marked after suxamethonium,[10] a factor which mitigates against the use of muscle relaxants other than suxamethonium during rapid sequence induction.

Non-depolarizing muscle relaxants following suxamethonium

It is usual practice to delay the administration of a non-depolarizing muscle relaxant drug until some signs of returning neuromuscular function have been observed after an intubating dose of suxamethonium. After suxamethonium in a dose of 1 mg kg^{-1} recovery of the twitch response of the thumb to 80% of control takes approximately 5 minutes at which time some signs of returning neuromuscular transmission could be expected.[13] The pattern of suxamethonium blockade appears to be subject to geographical or ethnic influence and the duration is shorter by approximately 35% in London than in New York.[14] The simultaneous administration of suxamethonium and a non-depolarizing agent cannot be recommended because the duration of the depolarizing drug may be reduced to such an extent that the conditions for endotracheal intubation might be jeopardized if anatomical difficulties were encountered.[13]

Electroconvulsive therapy

Tubocurarine (without general anaesthesia) was the first muscle relaxant drug to be used to prevent the traumatic complications of electroconvulsive therapy (ECT) in 1941. Suxamethonium rapidly became the favoured relaxant in combination with general anaesthesia. In addition to its short duration, the rapid onset of suxamethonium at the masseter muscles[9] may be responsible for its popularity. A 25 mg dose of suxamethonium appears to offer the optimum balance between the attenuation of muscle contractions and the duration of action.[15] It must be borne in mind that small doses of suxamethonium are associated with exaggerated inter-patient variability in neuromuscular blockade: a dose of 0.3 mg kg^{-1} results in a range of blockade from 4 to 90% at the adductor pollicis.[16]

Maintenance of neuromuscular blockade: continuous infusion

The popularity and appropriateness of suxamethonium as a continuous infusion has waned dramatically over recent years as first, intermediate-duration and, then, short-duration non-depolarizing muscle relaxants have been introduced. Discussion of this subject will, therefore, be brief. Tachyphylaxis, bradyphylaxis and phase ll block, have been discussed above. It appears to be no less easy to adjust the depth of neuromuscular blockade during a relatively short suxamethonium infusion than during a mivacurium infusion.[17] Following an intubating dose of 1 mg kg^{-1} the infusion requirement for suxamethonium to produce 95% twitch depression is approximately 90 μg kg^{-1} min^{-1} during the first hour of nitrous oxide and opioid anaesthesia. Under these conditions, the recovery indices (recovery from 75 to 25% twitch depression) for suxamethonium and mivacurium are identical (approximately 6 minutes). Phase II block is likely to develop after approximately 30–60 minutes and a continuous infusion of suxamethonium lasting greater than 1 hour should be embarked upon with caution.

Duration of suxamethonium and plasma cholinesterase

The action of suxamethonium is terminated by reduction of the concentration at the neuromuscular junction consequent on rapid plasma clearance as a result of hydrolysis by the enzyme plasma cholinesterase (PCHE). The products of hydrolysis are succinylmonocholine, a weak agonist, and choline. Succinylmonocholine is further hydrolysed to succinate and choline. Approximately 70% of a 100 mg bolus of suxamethonium is hydrolysed within 60 seconds.[18] There is a significant variation in the concentration of plasma cholinesterase in a group of normal individuals but the concentration is rarely so low as to prolong the action of suxamethonium. The half-time of plasma cholinesterase *in vivo* is approximately 5 days.[19] The inherited disorders of plasma cholinesterase synthesis are discussed below. The first trimester of pregnancy is associated with a reduction of approximately 25–30%[20] before being restored to normal by approximately 6 weeks postpartum.[21]

Acquired plasma cholinesterase deficiency

Plasma cholinesterase may be reduced in several chronic diseases, but rarely to an extent that prolongs significantly the effect of suxamethonium (Table 4.2). There is conflicting information concerning the activity of plasma cholinesterase in patients with chronic renal failure[22, 23] but it is unlikely that the action of suxamethonium will be prolonged to a clinically significant extent. Severe hepatic disease may be associated with reduction in plasma cholinesterase sufficient to prolong the clinical duration of suxamethonium.

Haemodilution,[24] including cardiopulmonary bypass[25] and plasmaphoresis[26] reduce the concentration of plasma cholinesterase but clinical problems are uncommon. Interactions between suxamethonium and other drugs are described in Chapter 10.

Table 4.2 Causes of decreased plasma cholinesterase activity

Inherited	Cholinesterase variants (abnormal enzyme)
Physiological	Last trimester of pregnancy
	Neonates and infants
Acquired	Hepatic dysfunction
	Cachexia
	Widespread carcinoma
	Uraemia
	Myxoedema
	Severe burns
Iatrogenic	Pancuronium
	Mivacurium
	Neostigmine
	Organophosphorous agents
	Esmolol
	Propranolol
	Plasmaphoresis
	Cardiopulmonary bypass
	Haemodilution
	MAOIs
	Oestrogens

Inherited plasma cholinesterase deficiency

The clinical duration of suxamethonium blockade is approximately inversely proportional to the activity of plasma cholinesterase.[27] Plasma cholinesterase is a large molecule comprising four polypeptide chains of molecular weight 80 000 daltons. Each chain contains approximately 400 amino acids. It is important to distinguish between an abnormally low concentration of the normal enzyme, and the presence of low cholinesterase activity as a result of an abnormal plasma cholinesterase variant. The chemical structure of these variants is such that suxamethonium is hydrolysed more slowly or not at all. Small changes in the amino acid sequence of the cholinesterase molecule have considerable functional consequences.

Synthesis of the normal (usual) enzyme is dependent on the two, normal allelic genes at the E1 locus which correspond to the genotype E1u E1u. The more commonly occurring variants are E1u E1a (heterozygous atypical or dibucaine resistant); E1u E1f (heterozygous fluoride resistant) and E1u E1s (heterozygous silent) (Table 4.3). Each genotype is associated with a considerable variation in sensitivity to suxamethonium. For example, although the frequency of the E1u E1a variant in the population is approximately 1 in 15, only 1 in 500 will demonstrate prolongation of the action of suxamethonium. Heterozygotes for the atypical, fluoride-resistant and silent genes are clearly more frequent than clinical experience might indicate and it is probable that the action of suxamethonium is clinically prolonged in only a small proportion of individuals[28]. The E1a E1a genotype is the most frequent homozygous variant (1 in 2000) and suxamethonium neuromuscular blockade may be prolonged for up to 2 hours. There are considerable racial differences in the frequency of plasma cholinesterase variants: the atypical variant is more common in Caucasians and less common in Negro races.

Table 4.3 Plasma cholinesterase variants: mean enzyme activity, dibucaine number, fluoride number, frequency in the UK population and sensitivity to suxamethonium

Genotype	Activity	Dubicaine number Mean	Range	Fluoride number Mean	Range	Frequency	Sensitivity
E1u E1u	100	80	77–83	61	56–68	96.0%	1:2500 +
E1u E1s	50	80	77–83	61	56–68	1:190	1:1902 +
E1u E1f	86	74	70–83	52	46–54	1:200	1:200 +
E1u E1a	77	62	48–69	50	44–54	1:25	1:500 +
E1a E1f	59	53	45–59	33	28–39	1:20 000	All +
E1f E1f	74	67	64–69	36	34–43	1:154 000	All +
E1f E1s	37	67	64–69	36	34–43	1:150 000	All +
E1a E1a	43	21	8–28	19	10–28	1:2000	All ++
E1a E1s	22	21	8–28	19	10–28	1:29 000	All ++
E1s E1s	Enzymic activity too low to measure					1:100 000	All ++

After Whittaker (1980)[29]

The dibucaine number (DN) is the percentage inhibition of the rate of hydrolysis of benzoylcholine by the plasma sample when a fixed concentration of dibucaine is added. In a large population the DN exhibits a tri-modal distribution corresponding to the genotypes E1u E1u (DN approximately 80), E1u E1a (DN approximately 60) and E1a E1a (DN approximately 20). The gene E1s is associated with an extremely low plasma cholinesterase activity: heterozygotes for this gene (E1u E1s) are equally sensitive to suxamethonium as homozygotes for the atypical gene; and homozygotes may have no measurable plasma cholinesterase activity. The fluoride-resistant gene occurs with a similar frequency to the silent gene but the total plasma cholinesterase activity is greater.

Many of the incidents of apparent 'suxamethonium apnoea' are due to factors other than the prolongation of action of suxamethonium. Respiratory depression consequent on a combination of hyperventilation with an inhalational agent, with or without an opioid, is a more common event. Clearly, the diagnosis can be made only with a peripheral nerve stimulator which should be available whenever suxamethonium is administered.

Although the administration of purified plasma cholinesterase derived from pooled plasma donation has been advocated in the past, the risk of transmitting viral disease, poor availability and cost preclude its use in most centres. It is possible that future developments in biological engineering may make available entirely synthetic plasma cholinesterase without these disadvantages. The administration of synthetic plasma cholinesterase may prove to be an advantage in managing 'suxamethonium apnoea' or 'mivacurium apnoea' in the future. The current management of 'suxamethonium apnoea' simply comprises continued anaesthesia and mechanical ventilation until adequate return of neuromuscular function has been demonstrated with Double Burst stimulation (Chapter 6). It is probably inadvisable to attempt to reverse prolonged suxamethonium blockade with neostigmine or edrophonium even if appreciable Train-of-Four fade is present unless a method of measuring the muscle response accurately is available.

Investigation and counselling

It has been emphasized that elucidation of the various genotypes may necessitate the investigation of the proband's entire family.[29] Cooperation among families is usually forthcoming when it is explained that approximately 1 in 8 of the relatives who are investigated will be sensitive to suxamethonium. If a patient is suspected of having a plasma cholinesterase abnormality the anaesthetist should contact one of the reference centres which benefit from many years of accumulated experience in interpreting the various genotypes. A hazard card should be issued to affected individuals who must be given the opportunity of discussing the implications with an anaesthetist, preferably in an anaesthetic outpatient clinic. There is some debate concerning the necessity of issuing a hazard card to heterozygotes for the atypical enzyme (E1u E1a) because the prolongation of suxamethonium blockade is rarely of clinical significance unless the concentration of the enzyme is also very low. The physiological decrease in plasma cholinesterase concentration associated with pregnancy may result in a clinically important sensitivity to suxamethonium in this group of patients and many reference centres will issue a hazard card to heterozygotes for the atypical enzyme only if they are pre-menopausal women.

If a patient with a known plasma cholinesterase deficiency requires anaesthesia with a rapid sequence induction, several options are available to the anaesthetist. Awake intubation is one possible solution, but this technique should be practised only by experienced anaesthetists and is not favoured in many countries. A normal dose of suxamethonium may be given (with neuromuscular monitoring) in the expectation that there will be a requirement for prolonged mechanical ventilation. This is probably the optimum approach for the known heterozygote. A large dose of a non-depolarizing agent, for example, rocuronium, in the well-preoxygenated patient will enable endotracheal intubation to be performed within 90 seconds. It is possible that induction of anaesthesia with propofol and alfentanil might facilitate sufficiently rapid endotracheal intubation but the reliability of this technique in the 'full stomach' situation is open to question.

Complications of suxamethonium

Fasciculations

The fasciculations induced by suxamethonium are frequently, but not invariably, associated with many of the other complications of this agent. The mechanism is unclear: repeated fasciculations can occur in the same group of muscles and fasciculations may be grouped on a motor unit basis. The administration of suxamethonium is associated with a transient increase in the force of a muscle contraction evoked by stimulation of the motor nerve. This is separate from fasciculation and there is no increase in the evoked electromyographic activity of the muscle suggesting that suxamethonium causes a brief increase in muscle contractility. It has been postulated that the fasciculations are mediated by the combination of suxamethonium with prejunctional receptors and the activation of local axon reflexes.[30] The involvement of prejunctional receptors is supported by the observation that non-depolarizing relaxants reduce the incidence and severity of fasciculation (but not myalgia) in proportion to their predilection for prejunctional receptors.[31] Thus, gallamine is more effective than pancuronium in preventing fasciculations due to suxamethonium.

Fasciculations occur in approximately 40–75% of adult patients receiving suxamethonium 1.5 mg kg^{-1}.[32] Both fasciculation and myalgia are more common in women and less common in children. It has been suggested that the myofibrillar damage caused by suxamethonium-induced fasciculations may be the result of an increase in the intracellular calcium concentration.

Myalgia

Postoperative myalgia remains a significant problem despite a considerable amount of research into possible methods of prevention. It has been suggested that suxamethonium-induced myalgia is the product of shearing

64 Suxamethonium

Table 4.4 Possible strategies for reducing the incidence of suxamethonium-induced myalgia

---

Preoperative stretch exercises
Pre-curarization with a non-depolarizing muscle relaxant
Pretreatment with suxamethonium (self-taming)
Benzodiazepines
Opioids
Local anaesthetic agents
Calcium gluconate
Calcium chloride
Magnesium sulphate
Prostaglandin synthase inhibitors
Anti-oxidants (including vitamin C)
Dantrolene

---

forces between muscle and fascia.[33] There is no significant correlation between the severity of suxamethonium-induced fasciculations and the severity of postoperative myalgia in usual clinical practice.[32, 34, 35] The reported frequency of suxamethonium-induced myalgia varies considerably. The incidence in obstetric practice is as low as 1.5%.[36] A frequency of 40–50% might be expected in a general population.[32, 37] Suxamethonium myalgia is more common in patients who are active within a short time after their surgical procedure[38] and it is a common complication after day case surgery. Interestingly, a programme of stretch exercises 1 hour before gynaecological laparoscopy is associated with a four-fold decrease in the incidence of postoperative suxamethonium-induced myalgia[37] and, in these circumstances, there is a significant association between the severity of fasciculations and the frequency of myalgia.

Numerous strategies have been suggested for reducing the frequency and severity of suxamethonium-induced myalgia (Table 4.4). Postoperative myalgia may occur in the absence of suxamethonium; approximately 10% of patients undergoing gynaecological laparoscopy experience significant myalgia on the first postoperative day.[39]

Pre-curarization with a non-depolarizing neuromuscular blocking drug remains the generally favoured technique. It is necessary to administer the pre-curarizing agent 2–3 minutes before suxamethonium: the optimum dose appears to be approximately 20% of a usual bolus dose suitable for intubation. There is some debate concerning the best choice of non-depolarizing agent for this purpose. Gallamine and tubocurarine may be more effective than pancuronium.[34] Other workers have reported that tubocurarine, gallamine, pancuronium and vecuronium are equally effective[32] in reducing the incidence of myalgia from approximately 40 to 20%. Atracurium appears to be as effective as tubocurarine.[40]

Pre-curarization is not without potential problems. Because there is a wide inter-patient variation in sensitivity to non-depolarizing muscle relaxants, a significant proportion of patients will suffer diplopia, ptosis and a sensation of suffocation after a small dose of a non-depolarizing agent given either to prevent suxamethonium myalgia or to act as a priming dose to accelerate non-depolarizing blockade (Chapter 10).[41] In one study 40% of patients experienced difficulty in swallowing and developed a significant reduction in static lung volumes and forced ventilatory measurements[42] and this may be associated with a fall in oxygen saturation in the elderly.[43] Pulmonary aspiration of gastric contents has been described after a priming dose of vecuronium.[44] After pre-curarization, a larger than normal dose of suxamethonium is required if intubating conditions are to be acceptable.[45]

It has been suggested that pretreatment with a benzodiazepine may reduce the incidence of suxamethonium myalgia. However, in 1990 Mingus and colleagues reported that midazolam 0.025 mg kg^{-1} given 3 minutes before suxamethonium 1.5 mg kg^{-1} was ineffective.[39]

If the intubating dose of suxamethonium is preceded by a smaller dose (10 mg), the incidence of fasciculations is reduced[46] but this technique of 'self-taming' does not prevent postoperative myalgia.[47] Self-taming may be useful for reducing the increase in serum potassium associated with suxamethonium[48] (see below).

The non-steroidal anti-inflammatory drugs appear to offer valuable prophylaxis against suxamethonium myalgia. Aspirin 600 mg administered orally 1 hour before anaesthesia reduces the incidence of suxamethonium myalgia by approximately 50% and is equally successful as pretreatment with tubocurarine 0.05 mg kg^{-1}, 3 minutes before induction of anaesthesia. It has been suggested that inhibition of prostaglandin synthesis is a possible mechanism.[45,49] Aspirin reduces the incidence of suxamethonium myalgia without attenuating the release of Serum Creatine Kinase (CK) from injured muscle. Diclofenac 75 mg administered i.m. 20 minutes preoperatively significantly reduces postoperative myalgia.[50]

Chlorpromazine inhibits phospholipase A2 (PLA2) and the effectiveness of chlorpromazine 0.1 mg kg^{-1} i.v. 3 minutes before induction of anaesthesia in reducing the frequency of suxamethonium myalgia and attenuating the release of CK has been attributed to reduction of the PLA2-induced release of arachidonic acid metabolites.[45]

In summary, tubocurarine 0.05 mg kg^{-1} appears to provide the best protection against suxamethonium myalgia, but the technique may be associated with the dual hazards of poor intubating conditions and pre-induction muscle weakness. A non-steroidal anti-inflammatory drug or chlorpromazine may be acceptable alternatives.

Bradycardia

Suxamethonium emulates two linked acetylcholine molecules and it inevitably acts as an agonist at muscarinic receptors in the sinus node in addition to nicotinic receptors at sympathetic and parasympathetic ganglia. The net cardiac response is determined by the balance of vagal and sympathetic blockade. In the majority of patients muscarinic stimulation and suxamethonium-induced catecholamine release are mutually antagonistic; but, in a small proportion of individuals, the muscarinic response predominates. Suxamethonium-induced bradycardia was first described by Phillips in 1954[51] and occurs most commonly after repeated doses. Nonetheless, a single dose may precipitate severe bradycardia[52] or even asystole.[53] A continuous infusion of suxamethonium is less likely to provoke bradycardia than repeated boluses. In addition to sinus bradycardia, junctional rhythm or ventricular escape beats are sometimes observed. Hyperkalaemia, hypoxia and hypercarbia are said to predispose to suxamethonium-induced arrhythmias. Bradyarrhythmias are more common in children than in adults. The prior administration of atropine or glycopyrrolate is effective in preventing the adverse cardiac effects of suxamethonium[54] and one of these drugs should always proceed a second dose of suxamethonium.

Masseter spasm

A transient increase in the tone of the masseter muscles is a normal accompaniment of the generalized increase in tone that precedes flaccid paralysis due to suxamethonium neuromuscular blockade. Van der Spek has demonstrated that mouth opening may be limited in approximately 50% of children for up to 10 minutes after suxamethonium.[55] It is often difficult to open the patient's mouth if insufficient time has been permitted to elapse after administration of suxamethonium, or if an inadequate dose has been given. A separate entity has been described in which, despite correct technique, masseter tone is markedly increased for several minutes, sometimes in association with generalized muscle spasm and severe reduction in chest-wall compliance. In this hazardous situation it is impossible to protect the airway against aspiration of gastric contents and it may be difficult to ventilate the lungs adequately. However, this phenomenon is most feared because of its association with malignant hyperthermia (MH). Approximately 65–68% of MH probands develop muscle spasm in response to suxamethonium.[56] Some 5% of such patients present with masseter spasm as the only abnormal sign. Muscle spasm in response to suxamethonium is seen in patients with myotonia congenita which may be associated with MH susceptibility. Ellis[56] has stated that, 'Whenever muscle spasm develops with suxamethonium a cause should be sought and the patient should be assumed to be potentially at risk from MH.' Careful attention should be paid during anaesthesia to confirming or excluding MH in these patients and the end-tidal carbon dioxide concentration, ECG and core temperature should be monitored assiduously. In the past it has been strongly advised that the observation of masseter spasm necessitates discontinuing anaesthesia and considering the administration of dantrolene. More recently, it has been suggested that anaesthesia should be continued with the appropriate monitoring and with dantrolene at hand should the early signs of MH appear. The rationale for this protocol is the extreme infrequency with which masseter spasm proceeds to MH and the thesis that masseter spasm represents an extreme manifestation of the normal, agonist effects of suxamethonium. Nonetheless, if masseter spasm is observed, possible trigger agents for MH should be avoided. It is current practice in the UK to investigate all patients who develop severe masseter spasm for susceptibility to MH.

Hyperkalaemia

It should be remembered that the serum potassium is normally 0.5 mmol l^{-1} higher than the plasma potassium. The administration of suxamethonium is normally followed by a rapid, transient increase in plasma potassium as a result of prolonged depolarization of the muscle membrane. This increase does not usually exceed 0.5 mmol l^{-1} and appears to be harmless.[57] In several disease states characterized by physical or pharmacological denervation of muscle, the rapid increase in plasma potassium may be magnified ten-fold, leading to ventricular fibrillation or asystole.

Lesions of upper or lower motor neurones are commonly implicated (Table 4.5) and prolonged immobility *per se* is sufficient to introduce this hazard to anaesthesia.[58] The common mechanism appears to be an abnormal proliferation of cholinoceptors which spread out to coat the entire muscle fibre outside the neuromuscular junctions. The number of ion channels in the muscle membrane is thereby increased vastly and the leakage of potassium produced by suxamethonium-induced depolarization of the muscle membrane is increased in proportion. Muscle trauma may also permit leakage of potassium during fasciculations. Opinion is divided as to how rapidly after the injury the hyperkalaemic response to suxamethonium becomes exaggerated. After trauma, the dangerous period may begin after 1 week and, after denervation, may persist for 6 months.[59] The hyperkalaemic response develops more rapidly in experimental animals than in man. The 'window of safety' for suxamethonium appears to extend reliably to only 24 hours after the injury in man.

The increase in serum potassium is not exaggerated in chronic renal failure[60] but the rise in potassium may be hazardous if there is pre-existing hyperkalaemia. It is generally accepted that the use of suxamethonium should be avoided in chronic renal failure if the serum potassium exceeds 5.0 mmol l^{-1}. Numerous pharmacological strategies have been suggested for

Table 4.5 Disorders associated with excessive hyperkalaemia in response to suxamethonium

Neurological
  Encephalitis
  Stroke
  Cerebral aneurysm
  Head injury
  Spinal cord injury
  Multiple sclerosis
  Major peripheral nerve injury
  Polyneuropathy

Other
  Immobility
  Muscle trauma
  Cold injury
  Burns

mitigating the exaggerated rise in serum potassium in susceptible individuals. It is advisable to avoid suxamethonium altogether if possible in this situation and the introduction of rapid-onset non-depolarizing muscle relaxant drugs, of which rocuronium is in the vanguard, may remove this particular hazard completely.

Myoglobin release

Myoglobin is normally released from muscle during exercise. Excessive release is seen occasionally after suxamethonium, especially in children or after repeated doses. The associated rise in CK has been used as an indicator of muscle damage and massive CK release has been recorded in patients with a myopathy. However, there is no clear relation between the rise in CK and the severity of suxamethonium myalgia.[45, 61]

Intraocular pressure

Intraocular pressure may be increased by laryngoscopy and endotracheal intubation *per se*;[62] an effect that is attenuated by topical anaesthesia. Suxamethonium, with or without endotracheal intubation, increases intraocular pressure by 5–10 mmHg for up to 10 minutes. Much effort has been expended in investigating possible strategies for obtunding the rise in intraocular pressure which may be hazardous in a penetrating eye injury or in severe glaucoma. Deep general anaesthesia offers some protection. Unfortunately, pre-curarization with a non-depolarizing muscle relaxant is not sufficiently reliable in attenuating the rise in intraocular pressure to use this technique with equanimity in a patient with a penetrating eye injury.[63] The truly rapidly acting non-depolarizing drug, of which rocuronium is the prototype, may solve this problem if a sufficiently large dose is given.

Intragastric pressure

The increase in intragastric pressure following suxamethonium is extremely variable. Miller and Way[64] demonstrated an increase of up to 40 cm water although they were unable to demonstrate any elevation in pressure in one-third of individuals. An intragastric pressure in excess of 30 cm water is sufficient to cause gastro-oesophageal reflux in cadaveric experiments and, when pre-curarization had been demonstrated to obtund the suxamethonium-induced elevation of the intragastric pressure,[64] this technique gained popularity as a method of attempting to increase the safety of rapid sequence induction. However, it is now generally accepted that the elevation of intragastric pressure is more than balanced by a simultaneous, greater increase of the lower oesophageal pressure such that the risk of regurgitation and aspiration of gastric contents is not increased by the administration of suxamethonium, and may actually be decreased.[65]

Intracranial pressure

Clinical studies of the effect of suxamethonium on the intracranial pressure have been unable to distinguish between the consequences of giving suxamethonium and those due to laryngoscopy or the initiation of mechanical ventilation.[66] It is important to pay meticulous attention to all those factors that can elevate intracranial pressure in those patients who are at risk and it appears that suxamethonium is of minor significance.

Malignant hyperthermia

Malignant hyperthermia (MH) was first recognized by Denborough in 1960 who described a family in which several, anaesthesia-related deaths had occurred.[67] The prevalence varies on a geographical basis between 1:10 000 and 1:50 000. Despite its rarity, this syndrome is important because the mortality is extremely high unless the diagnosis is made early and the correct management is instituted immediately. Autosomal dominant inheritance can be demonstrated in 50% of cases and, in 20% of individuals, inheritance is either recessive or multifactorial. MH is more common in the younger age groups (10–20 years). The male to female ratio is approximately 2:1. Triggering agents include suxamethonium and the anaesthetic vapours. A method of measuring core temperature, ECG and end-tidal carbon dioxide concentration should be available wherever suxamethonium is given.

Clinical features

Muscle rigidity is not invariably present, may be transient, and may be generalized or localized to the masseter muscles. The implications of masseter spasm are discussed above. Clinical features may follow immediately after the administration of the trigger agent, or may be delayed for several hours. In some patients suxamethonium may be the trigger agent; in others an inhalational agent is the precipitating factor, either alone or in combination with suxamethonium. The earliest signs are likely to include tachycardia, tachyarrhythmias, tachypnoea, elevated end-tidal carbon dioxide concentration and cyanosis (Table 4.6). A dramatic rise in end-tidal carbon dioxide

Table 4.6 Clinical features of MH

Specific
    Hyperthermia increasing at $> 2°C\,h^{-1}$
    Elevated end-tidal carbon dioxide concentration
    Mixed metabolic and respiratory acidosis
    Muscle rigidity
    Rhabdomyolysis

Non-specific
    Hyperthermia increasing at $< 2°C\,h^{-1}$
    Hyperkalaemia
    Increased oxygen consumption
    Mottled cyanosis
    Impaired coagulation
    Hypertension
    Tachycardia

concentration may occur irrespective of whether the patient is mechanically ventilated or breathing spontaneously and this is probably the most useful single diagnostic feature.[68] In the early stages it may be difficult to differentiate MH from the effects of elevated circulating catecholamine concentrations as a result of excessively light anaesthesia, thyrotoxicosis or phaeochromocytoma. Hyperpyrexia may be a late sign. Peripheral vasoconstriction in the early phase of the syndrome may even reduce the skin temperature[69] and the measurement of core temperature is mandatory. The rate of rise of core temperature may be dramatic and $4\,°C\,h^{-1}$ is not unusual. Mottling of the skin is common. Metabolic acidosis and hyperkalaemia are invariably present. Late signs include myoglobinuria, acute renal failure and disseminated intravascular coagulation. If rapid laboratory facilities are available, measurement of arterial blood pH, PCO_2, PO_2, and potassium will clinch the diagnosis. A blood sample should be saved for estimation of CK and lactate as soon as possible. Repeated arterial pH and blood gas estimations will be required to monitor treatment.

Pathogenesis

Malignant hypothermia occurs in patients with abnormal muscle physiology. Morphological muscle abnormalities are common in individuals who are susceptible to MH but the exact pathological changes are variable, even between siblings, and do not correlate with the severity of the MH reaction.[69] There appears to be some overlap with the morphological features of myotonia congenita with which MH may be associated, in common with Duchenne muscular dystrophy and central core disease. The genetic aspects of MH are currently the subject of much research and an association has been established in a few families between MH and the q12–13.2 region of chromosome 19.[70] The calcium release channel in the sarcoplasmic reticulum, the ryanodine receptor, is coded in this region and certain DNA markers from the ryanodine gene have been connected with MH susceptibility in several families. It is possible that eventually a genetic test for MH susceptibility will be developed when the pathogenic alleles of the ryanodine receptor become known.

The primary event in MH is a massive rise in calcium ion concentration in the cytosol of the muscle cells. Muscle cell contraction then becomes persistent and uncontrolled with the generation of large amounts of heat and cellular destruction. It has been postulated that trigger agents for MH interact with the abnormal ryanodine receptor to unleash this outpouring of calcium ions from the sarcoplasmic reticulum which normally releases calcium in a highly controlled fashion to initiate muscle contraction.

Management of an MH crisis

Management of masseter spasm is discussed above. The early administration of dantrolene is central to the management of MH. Immediately MH is suspected, dantrolene $1\,mg\,kg^{-1}$ should be given i.v. This corresponds to three or four 20 mg vials, the contents of which require to be dissolved in 60 ml sterile water. Dantrolene, in at least this dose, should be available within a few minutes at all sites where anaesthetics are given. It may be necessary to continue the administration of dantrolene at 1 mg kg^{-1} to a total dose of 10 mg kg^{-1} until it is certain that the clinical features are responding. It is probably wise to continue administration for 24 hours following the resolution of the acute MH reaction with 0.25–0.5 mg kg^{-1} i.v. at intervals of 6 hours; or, if appropriate, 1–2 mg kg^{-1} can be given p.o. at similar intervals. The place of surface cooling is uncertain and much time may be wasted in sending for ice and cold intravenous fluids. The immediate administration of dantrolene is more efficacious in restoring a normal temperature. In addition, surface cooling may encourage shivering which is detrimental. If shivering is pronounced, it should be treated with i.v. chlorpromazine. Volatile agents should be withdrawn immediately and the patient should be hyperventilated with oxygen. If it is imperative to continue anaesthesia because interruption of surgery would be life threatening, a continuous i.v infusion of propofol,[71] and bolus doses of fentanyl would appear to be safe. If muscle relaxation is necessary, vecuronium[72] and atracurium[73] have been used safely. Reversal of residual neuromuscular blockade with neostigmine and glycopyrrolate appears to be safe in MH susceptible patients[73] although, in established MH, it is preferable to permit spontaneous recovery of neuromuscular transmission. Arrhythmias may require specific treatment with a beta blocking drug. Hyperkalaemia should be treated with i.v. dextrose and insulin and sodium bicarbonate should be given (via a central venous line) to correct the metabolic acidosis providing that mechanical ventilation has been instituted with large minute volume. A urinary catheter should be inserted and urine output encouraged with intravenous fluids and a diuretic if necessary. Each 20 mg vial of dantrolene contains 3 g of mannitol and sodium hydroxide is added to achieve a pH of 9.5. The initial dose of dantrolene will contain approximately 12 g of mannitol which is less than the 0.5–1.0 g kg^{-1} which may be needed for renal protection. Urine should be examined for myoglobin which may be present in sufficient quantity to colour the urine red or brown.

Laboratory investigation of an individual with possible MH susceptibility

Any patient who has exhibited MH or a clinical picture suspicious of MH should be investigated by muscle biopsy, both to confirm the diagnosis of MH and to diagnose any coexisting muscle disorder. The muscle biopsy is taken from the quadriceps muscle in the region which includes the motor point. The European MH Group distinguishes three diagnostic categories: MH susceptible individuals (MHS), individuals undoubtedly not susceptible to MH (MHN) and individuals with equivocal laboratory results (MHE). A muscle biopsy specimen from an MHS patient develops a strong contracture when exposed to halothane ($\leq 2\%$ v/v), caffeine (≤ 2 mmol l), or a combination of these agents. This is known as the *in vitro* contracture test (IVCT). The halothane/caffeine test has a very high sensitivity (100%) and specificity (80%) for MHS.[74] CK estimation in isolation is not sufficiently specific. An unexplained elevated CK found incidentally in the absence of muscle disease may be associated with MH susceptibility.[75]

MH susceptible patients may develop muscle stiffness and myalgia when exposed to severe stress or halogenated hydrocarbons found in the workplace, and treatment with oral dantrolene may be necessary. It is advisable for all MH susceptible patients to wear a Medic Alert bracelet and to carry an Anaesthetic Hazard Warning card. An MH support group is to be found in most countries. First, second and third degree relatives should be counselled and advised to undergo screening. If screening is not taken up, the relative should carry MH warnings as if he or she were MH susceptible.

Anaesthesia in the known MH susceptible patient

Surgery should be performed under regional blockade if possible. If emergency general anaesthesia is needed and there is insufficient time to permit the stomach to empty, the choice of technique lies between awake intubation or the administration of a large dose of rocuronium immediately after the induction of anaesthesia. If difficult intubation is suspected, awake intubation is recommended. If triggering agents are assiduously avoided, general anaesthesia should not be associated with excess risk. Suxamethonium and the anaesthetic vapours are absolutely contraindicated. The provision of a 'halothane-free' anaesthesic machine is now considered to be unnecessary provided that the vaporizers are removed, the breathing circuit is replaced with new, disposable plastic tubing and the circuit is flushed with oxygen for 30 minutes. Used sodalime should be replaced because inhalational agents are adsorbed onto the granules. The use of prophylactic oral dantrolene is considered to be unnecessary and it is not without side-effects. The intravenous preparation should be available in the operating theatre. Halsall and Ellis[76] have compiled a list of anaesthetic drugs that are acceptable in MH susceptible patients (Table 4.7). Anaesthesia should not be started until the necessary monitoring (see above) is in place and monitoring should continue for 6–8 hours postoperatively.

Table 4.7 Drugs that are safe to use in the MH susceptible patient

Benzodiazepines
Metoclopramide
Droperidol
Thiopentone
Propofol
All analgesics
Nitrous oxide
All local anaesthetics
Atracurium
Vecuronium
Neostigmine + atropine or glycopyrrolate

Table 4.8 Contraindications to the use of suxamethonium

Previous anaphylactic response
MH susceptibility (or family history)
Prolonged immobility
Spinal injury
Burns
Pre-existing hyperkalaemia
Penetrating eye injury
Plasma cholinesterase deficiency (relative contraindication)

Contraindications to suxamethonium

The contraindications to the use of suxamethonium are dominated by those situations in which there is a risk of hyperkalaemia or MH (Table 4.8). Cardiac arrest has been observed when suxamethonium has been given to facilitate endotracheal intubation in the Critical Care Unit after several days of immobility. To these contraindications must be added a history of a previous anaphylactic reaction and the presence of a penetrating eye injury. Plasma cholinesterase deficiency should not be considered an absolute contraindication to suxamethonium, for example in general anaesthesia for caesarean section, although it will be necessary for both the anaesthetist and the patient to be prepared for a prolonged period of mechanical ventilation.

References

1. Lee C. Train-of-Four fade and edrophonium antagonism of neuromuscular block by succinylcholine in man. *Anesthesia and Analgesia* 1976; **55:** 663–667.
2. Donati F, Bevan DR. Intensity of Phase ll succinylcholine block and its antagonism with neostigmine. *Canadian Anaesthetists Society Journal* 1984; **31:** 588–589.
3. James MFM, Howe HC. Prolonged paralysis following suxamethonium. *British Journal of Anaesthesia* 1990; **65:** 430–432.
4. Donati F, Bevan DR. Long-term succinylcholine infusion during isoflurane anaesthesia. *Anesthesiology* 1983; **58:** 6–10.
5. Donati F, Bevan DR. Effect of enflurane and fentanyl on the clinical characteristics of long-term succinylcholine infusion. *Canadian Anaesthetists Society Journal* 1982; **29:** 59–64.
6. Lee C, Barnes A, Katz RL. Magnitude, dose requirement and mode of development of tachyphylaxis to suxamethonium in man. *British Journal of Anaesthesia* 1978; **50:** 189–194.
7. Futter ME, Donati F, Bevan DR. Prolonged suxamethonium infusion during nitrous oxide anaesthesia supplemented with halothane or fentanyl. *British Journal of Anaesthesia* 1983; **55:** 947–953.
8. Stirt JA, Katz RL, Schehl DL, Lee C. Atracurium for intubation in man. A clinical and electromyographic study. *Anaesthesia* 1984; **39:** 1214–1221.
9. Smith CE, Donati F, Bevan DR. Effects of succinylcholine at the masseter and adductor pollicis muscles in adults. *Anesthesia and Analgesia* 1989; **69:** 158–162.
10. Meistelman C, Plaud B, Donati F. Neuromuscular effects of succinylcholine on the vocal cords and adductor pollicis muscles. *Anesthesia and Analgesia* 1991; **73:** 278–282.

11. Mahajan RP, Hennessy N, Aitkenhead AR. Effect of priming dose of vecuronium on lung function in elderly patients. *British Journal of Anaesthesia* 1993; **77:** 1198–1202.
12. Meistelman C, Plaud B, Donati F. Rocuronium ORG 9426 neuromuscular blockade at the adductor muscles of the larynx and adductor pollicis in humans. *Canadian Journal of Anaesthesia* 1992; **39:** 665–669.
13. Harper NJN, Chadwick IS, Linsley A. Suxamethonium and atracurium: sequential and simultaneous administration. *European Journal of Anaesthesiology* 1993; **10:** 13–17.
14. Katz RL, Norman J, Seed RF, *et al*. A comparison of the effects of suxamethonium and d-tubocurarine in patients in London and New York. *British Journal of Anaesthesia* 1969; **41:** 1041–1047.
15. Konarzewski WH, Milosavljevic D, Robinson M, *et al*. Suxamethonium dosage in electroconvulsive therapy. *Anaesthesia* 1988; **43:** 474–476.
16. Chestnut RJ, Harper NJN, Healy TEJ, Farragher BE. Suxamethonium, the relation between dose and response. *Anaesthesia* 1989; **44:** 14–18.
17. Brandom BW, Woelfel SK, Cook DR, *et al*. Comparison of mivacurium and suxamethonium administered by bolus and infusion. *British Journal of Anaesthesia* 1989; **62:** 488–493.
18. Kalow W. The distribution, destruction and elimination of muscle relaxants. *Anesthesiology* 1959; **20:** 505–518.
19. Hall GM, Wood GJ, Paterson JL. Half-life of plasma cholinesterase. *British Journal of Anaesthesia* 1984; **56:** 903–904.
20. Evans RT, Wroe JM. Plasma cholinesterase changes during pregnancy. *Anaesthesia* 1980; **35:** 651–654.
21. Robson N, Robertson I, Whittaker M. Plasma cholinesterase changes during the puerperium. *Anaesthesia* 1986; **41:** 243–249.
22. Ryan DW. Preoperative serum cholinesterase concentration in chronic renal failure. *British Journal of Anaesthesia* 1977; **49:** 945–949.
23. Robertson GS. Serum cholinesterase deficiency 1: disease and inherited. *British Journal of Anaesthesia* 1966; **38:** 355–360.
24. Schuh FT. Influence of haemodilution on the potency of neuromuscular blocking drugs. *British Journal of Anaesthesia* 1981; **53:** 263–265.
25. Jackson SH, Bailey GWH, Stevens G. Reduced plasma cholinesterase following haemodilutional cardiopulmonary bypass. *Anaesthesia* 1982; **37:** 319–320.
26. Wood GJ, Hall GM. Plasmaphoresis and plasma cholinesterase. *British Journal of Anaesthesia* 1978; **50:** 945–947.
27. Viby-Mogensen J. Correlation of succinylcholine duration of action with plasma cholinesterase activity in subjects with the genotypically normal enzyme. *Anesthesiology* 1980; **53:** 517–520.
28. Viby-Mogensen J. Succinylcholine neuromuscular blockade in subjects hererozygous for atypical plasma cholinesterase. *Anesthesiology* 1981; **55:** 231–235.
29. Whittaker M. Plasma cholinesterase variants and the anaesthetist. *Anaesthesia* 1980; **35:** 174–197.
30. Kitamura S, Yoshiya I, Tashir C, *et al*. Succinylcholine causes fasciculation by prejunctional mechanism. *Anesthesiology* 1981; **55:** A22.
31. O'Sullivan EP, Williams NE, Calvey TN. Differential effects of neuromuscular blocking agents on suxamethonium-induced fasciculations and myalgia. *British Journal of Anaesthesia* 1988; **60:** 367–371.
32. Ferres CJ, Mirakhur RK, Craig HJL, *et al*. Pre-treatment with vecuronium as a prophylactic against post-suxamethonium muscle pain. *British Journal of Anaesthesia* 1983; **55:** 735–741.
33. Wates DJ, Mapleson WW. Suxamethonium pains: hypothesis and observation. *Anaesthesia* 1971; **26:** 127–141.

34. O'Sullivan EP, Williams NE, Calvey TN. Differential effects of neuromuscular blocking agents on suxamethonium-induced fasciculations and myalgia. *British Journal of Anaesthesia* 1988; **60:** 367–371.
35. Newman PTF, London JM. Muscle pain following administration of suxamethonium: the aetiological role of muscular fitness. *British Journal of Anaesthesia* 1966; **38:** 533–540.
36. Crawford JS. Suxamethonium pains and pregnancy. *British Journal of Anaesthesia* 1971; **43:** 677–680.
37. Magee DA, Robinson RJS. Effect of stretch exercises on suxamethonium induced fasciculations and myalgia. *British Journal of Anaesthesia* 1987; **59:** 596–601.
38. Oxorn DC, Whatley GS, Knox JWD, Hooper J. The importance of activity and pretreatment in the prevention of suxamethonium myalgias. *British Journal of Anaesthesia* 1992; **69:** 200–201.
39. Mingus ML, Herlich A, Eisenkraft JB. Attenuation of suxamethonium myalgias. *Anaesthesia* 1990; **45:** 834–837.
40. Sosis M, Broad T, Larijani GE, Marr AT. Comparison of atracurium and d-tubocurarine for prevention of succinylcholine myalgia. *Anesthesia and Analgesia* 1987; **66:** 657–659.
41. Engbaek J, Viby-Mogensen J. Precurarization – a hazard to the patient? *Acta Anaesthesiologica Scandinavica* 1984; **28:** 61–62.
42. Mahanjan RP, Laverty J. Lung function after vecuronium pretreatment in young, healthy adults. *British Journal of Anaesthesia* 1992; **69:** 318–319.
43. Mahajan RP, Hennessey N, Aitkenhead AR. Effect of priming dose of vecuronium on lung function in elderly patients. *Anesthesia and Analgesia* 1993; **77:** 1198–1202.
44. Musich J, Walt LF. Pulmonary aspiration after a priming dose of vecuronium. *Anesthesiology* 1986; **64:** 517–519.
45. McLoughlin C, Elliott P, McCarthy G, Mirakhur RK. Muscle pains and biochemical changes following suxamethonium administration after six pretreatment regimens. *Anaesthesia* 1992; **47:** 202–206.
46. Baraka A. Self-taming of succinylcholine induced fasciculations. *Anesthesiology* 1977; **46:** 292–293.
47. Brodsky JB, Brock-Utne JG. Does self-taming with succinylcholine prevent postoperative myalgia? *Anesthesiology* 1979; **50:** 265–267.
48. Magee DA, Gallagher EG. Self-taming of suxamethonium and serum potassium concentration. *British Journal of Anaesthesia* 1984; **56:** 977–980.
49. McLoughlin C, Nesbitt GA, Howe JP. Suxamethonium induced myalgia and the effect of pre-operative administration of aspirin. *British Journal of Anaesthesia* 1988; **43:** 565–567.
50. Kahraman S, Ercan S, Aypar U, Erden K. Effect of preoperative i.m. administration of diclofenac on suxamethonium-induced myalgia. *British Journal of Anaesthesia* 1993; **71:** 238–241.
51. Phillips HS. Physiologic changes noted with the use of succinylcholine chloride, as a muscle relaxant during endotracheal intubation. *Anesthesia and Analgesia* 1954; **33:** 165–177.
52. Baraka A. Severe bradycardia following propofol-suxamethonium sequence. *British Journal of Anaesthesia* 1988; **61:** 482–483.
53. Sorensen M, Engbaek J, Viby Mogensen J, *et al.* Bradycardia following a single injection of suxamethonium. *Acta Anaesthesiologica Scandinavica* 1984; **28:** 232–235.
54. Sorensen O, Eriksen S, Hommelgaard P, *et al.* Thiopental-nitrous oxide-halothane anesthesia and repeated succinylcholine: comparison of preoperative glycopyrrolate and atropine administration. *Anesthesia and Analgesia* 1940; **59:** 686–689.

55. Van der Spek AFL, Reynolds PI, Fang WB, et al. Changes in resistance to mouth opening induced by depolarizing neuromuscular blockade. *British Journal of Anaesthesia* 1990; **64**: 21–27.
56. Ellis FR, Halsall PJ. Suxamethonium spasm. A differential diagnostic conundrum. *British Journal of Anaesthesia* 1984; **56**: 381–384.
57. Weintraub HD, Heisterkamp DV, Cooperman LH. Changes in plasma potassium concentration after depolarizing blockers in anaesthetized man. *British Journal of Anaesthesia* 1969; **41**: 1048–1052.
58. Gronert GA, Theye RA. Effect of succinylcholine on skeletal muscle with immobilization atrophy. *Anesthesiology* 1974; **40**: 268–271.
59. Birch AA, Mitchell GD, Playford GA, Long CA. Changes in serum potassium response to succinylcholine following trauma. *Journal of the American Medical Association* 1969; **210**: 490–493.
60. Koide M, Waud BE. Serum potassium concentrations after succinylcholine in patients with chronic renal failure. *Anesthesiology* 1972; **36**: 142–145.
61. Laurence AS. Myalgia and biochemical changes following intermittent suxamethonium administration. *Anaesthesia* 1987; **42**: 503–510.
62. Joshi C, Bruce DL. Thiopental and succinylcholine action on intraocular pressure. *Anesthesia and Analgesia* 1975; **54**: 471–475.
63. Cook JH. The effect of suxamethonium on intraocular pressure. *Anaesthesia* 1981; **36**: 359–365.
64. Miller RD, Way WL. Inhibition of succinylcholine-induced increased intragastric pressure by non-depolarizing relaxant and lidocaine. *Anesthesiology* 1971; **34**: 185–188.
65. Smith G, Dalling R, Williams TIR. Gastro-oesophageal pressure gradient changes produced by induction of anaesthesia and suxamethonium. *British Journal of Anaesthesia* 1978; **50**: 1137–1143.
66. Marsh ML, Dunlop BJ, Shapiro HM, et al. Succinylcholine-intracranial pressure effects in neurosurgical patients. *Anesthesia and Analgesia* 1980; **59**: 550–551.
67. Denborough MA, Lovell RRH. Anaesthetic deaths in a family. *Lancet* 1960; **ii**: 45.
68. Baudendistel L, Goudsouzian N, Coté C, Strafford M. End-tidal CO2 monitoring. Its use in the diagnosis and management of malignant hyperthermia. *Anaesthesia* 1984; **39**: 1000–1003.
69. Ellis FR. Inherited muscle disease. *British Journal of Anaesthesia* 1980; **52**: 153–164.
70. MacLennan DH, Duff C, Zorzato F, et al. Ryanodine receptor gene is a candidate for predisposition to malignant hyperthermia. *Nature* 1990; **343**: 559–561.
71. Gallen JS. Propofol does not trigger malignant hyperthermia. *Anesthesia and Analgesia* 1991; **72**: 413.
72. Buzello W, Williams CH, Chandra P, et al. Vecuronium and porcine malignant hyperthermia. *Anesthesia and Analgesia* 1985; **64**: 515–519.
73. Ørding H, Nielsen VG. Atracurium and its antagonism by neostigmine plus glycopyrrolate in patients susceptible to malignant hyperthermia. *British Journal of Anaesthesia* 1986; **58**: 1001–1004.
74. Ellis FR, the European MH Group. Laboratory diagnosis of malignant hyperpyrexia susceptibility MHS. *British Journal of Anaesthesia* 1985; **57**: 1038.
75. Larach MG, Landis JR, Bunn JS, Diaz M, the North American Malignant Hyperthermia Registry. Prediction of malignant hyperthermia susceptibility in low-risk subjects: an epidemiologic investigation of caffeine halothane contracture responses. *Anesthesiology* 1992; **76**: 16–27.
76. Halsall PJ, Ellis FR. Malignant hyperthermia. In Fisher M. McD ed. *The Anaesthetic Crisis. Ballière's Clinical Anaesthesiology* Vol 7 No 2, 1993, London: Ballière Tindall, pp. 343–356.

5

Non-depolarizing Muscle Relaxants

Brian J Pollard

Since the introduction of the muscle relaxants into clinical practice in 1942,[1] many potential non-depolarizing relaxants have been synthesized. Only about a dozen have been introduced into clinical practice of which some have been subsequently withdrawn (e.g. fazadinium). This chapter will be confined to those which are commonly available at present. A summary of the basic pharmacokinetic and pharmacodynamic data is given in Table 5.1 (see also Chapter 3).

Table 5.1 Pharmacokinetics and pharmacodynamics of the muscle relaxants in normal adults

Relaxant	Volume of distribution ($l\ kg^{-1}$)	Plasma clearance ($l\ kg^{-1}\ h^{-1}$)	Elimination half-life (min)	ED95 ($mg\ kg^{-1}$)
Alcuronium	0.32	0.08	200	0.25
Atracurium	0.16	0.33	20	0.23
Doxacurium	0.2	0.12	130	0.03
Gallamine	0.21	0.07	135	2.4
Metocurine	0.44	0.07	220	0.28
Mivacurium	0.2	3.3	17	0.08
Pancuronium	0.21	0.1	130	0.06
Pipecuronium	0.31	0.14	138	0.05
Rocuronium	0.23	0.2	120	0.3
Tubocurarine	0.45	0.1	200	0.5
Vecuronium	0.26	0.28	60	0.04

The data in this table are overall means calculated from a number of sources. It should be remembered that the figures have a wide range in most cases. These are mainly related to differences in the methodology between the various studies.

Alcuronium

This muscle relaxant is synthesized from the naturally occurring alkaloid, C-toxiferine and was introduced into clinical practice in 1961 as an alternative to tubocurarine. It is a bisquaternary compound (Fig. 5.1). Its popularity has declined considerably in the last 10 years with the introduction of the newer intermediate-acting relaxants.

Pharmacokinetics

The distribution volume is approximately 0.33 lkg^{-1} with a plasma clearance of 14 $ml\,kg^{-1}\,min^{-1}$. Alcuronium is almost entirely renally excreted as the unchanged molecule and the clearance is therefore prolonged in patients with renal impairment. The clearance is halved in the elderly age group, which is probably related to changes in renal function. Elimination half-life is approximately 200 minutes in normal patients, rising to twice that value or more in the elderly.[2,3] Alcuronium is about 40% bound to plasma proteins, principally to albumin.[4]

Pharmacodynamics

The ED50 of alcuronium is approximately 0.13 mg kg^{-1} and the ED95 0.25 mg kg^{-1}. Alcuronium is potentiated by volatile anaesthetic agents to the extent that the ED50 is reduced by between 10 and 20% in the presence of halothane, enflurane or isoflurane.

The onset of action of alcuronium is relatively slow. An ED95 dose would be expected to reach maximum block in between 5 and 6 minutes, although this can be accelerated by the use of an increased bolus dose or the technique of priming.

The duration of action of alcuronium is similar to that of pancuronium and a little shorter than that of tubocurarine. An intubating dose can be expected

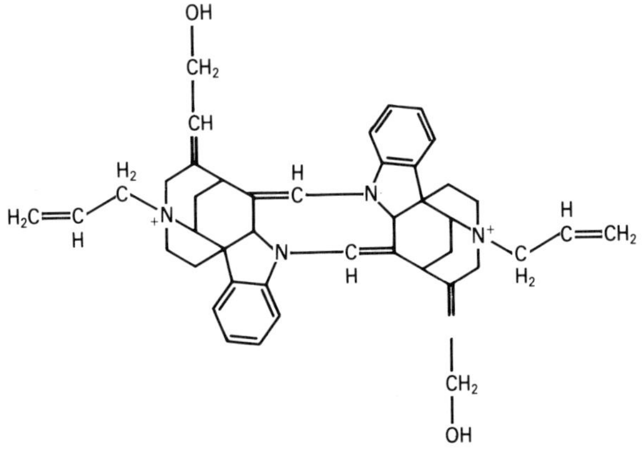

Fig. 5.1 Structural formula of alcuronium.

to have reached greater than 90% recovery after approximately 60 minutes. The spontaneous recovery index is about 30 minutes.[5] Recovery can be satisfactorily accelerated by the use of an anticholinesterase in a routine fashion (Chapter 8).

Side-effects

Alcuronium is a weak blocker at autonomic ganglia. It may therefore produce hypotension in hypovolaemic patients although not as marked as that with tubocurarine.[6] Alcuronium also possesses weak vagolytic action and the heart rate can be expected to rise by between 5 and 10% following its administration.[7]

Histamine release is not a problem with alcuronium, although it has been implicated in a number of anaphylactic reactions[8] (Chapter 13).

Atracurium

Atracurium is a completely synthetic relaxant, although its origins do lie in the substance petaline which is a plant derivative. The molecule is a bulky diester with two positively charged nitrogen atoms (Fig. 5.2). It was developed because of its novel method of metabolism by the Hofmann elimination reaction.[9] The original Hofmann reaction involved the conversion of an amide to an amine under alkaline conditions at high temperature and in the molecule of atracurium this reaction takes place at a pH of 7.4 and a temperature of 37°C. It is thus unstable at body temperature and pH and is stored at 4°C buffered to a pH of 3. Breakdown products of atracurium include a number of substances, in particular laudanosine, acrylates and quaternary organic acids.

Interest has focused on the breakdown products of atracurium for several years. Laudanosine will produce convulsions in certain laboratory animals,[10] but has never been implicated in any similar way in the human. The acrylates are highly reactive substances which may be hepatotoxic,[11] although once again no problems have hitherto been reported in the human. The interest has grown because atracurium may be administered by continuous infusion for a prolonged period of time in Intensive Care patients and laudanosine has been shown to accumulate in these patients (Chapter 12).

Hofmann elimination is not the only method for breakdown of atracurium. It is also metabolized by esterases in both the plasma and the liver. It is not attacked by plasma cholinesterase. The ratio of ester hydrolysis to Hofmann elimination differs from species to species, but in the human up to about 60% probably undergoes ester hydrolysis in the liver, although there is not uniform agreement.[12]

Atracurium is a complex molecule which has four asymmetric centres, giving a total of sixteen possible stereoisomers. Recent research has focused on the possibility of selecting out one isomer and one such substance, known at present as 51W89, is currently undergoing clinical trials.

Pharmacokinetics

The rapid spontaneous degradation of atracurium at body temperature and pH has made detailed pharmacokinetic studies difficult. Data are, however, available for not only normal patients but also those with various organ dysfunctions. The volume of distribution in normal patients is approximately 0.16 l kg^{-1}, and clearance 5.5 ml^{-1} kg^{-1} min^{-1}. The elimination half-life is remarkably constant throughout the studies at approximately 20 minutes. The interesting observation concerning atracurium is that these figures are very little changed by marked renal or hepatic dysfunction (Chapter 9). The elimination half-life is also about 20 minutes in the anephric patient and in patients with severely impaired liver function. The clearance and volume of distribution are elevated to a small extent.[13]

Pharmacodynamics

The ED50 of atracurium is approximately 0.13 mg kg^{-1}, and the ED95 approximately 0.23 mg kg^{-1}.

The onset of action of an ED95 dose is about 3–5 minutes. This can be reduced by increasing the intubating dose although there is a lower limit of just over 2 minutes.[14] The priming principle is also effective and reduces the onset time to about 2 minutes.

The duration of action depends upon the dose administered. The duration of action is prolonged by volatile anaesthetic agents and this ranges from an approximately 15% increase with halothane to a 40–50% increase with isoflurane.

Spontaneous recovery occurs reliably from an atracurium neuromuscular block such that an intubating dose can be expected to provide muscular relaxation for between 25 and 40 minutes in the normal healthy patient.[15] In view of this predictable recovery profile, reversal may not always be necessary. Most anaesthetists do, however, still administer a reversal agent (usually neostigmine) in order to ensure a rapid and guaranteed return of muscular power at the end of the procedure.

Atracurium is very little affected by the age of the patient (Chapter 9). The onset time is more rapid and the duration of action slightly shorter in infants and children (Chapter 11). The elderly patient behaves no differently to the fit young adult, probably because atracurium is affected to a negligible extent by changes in renal function. The breakdown of atracurium is decreased by hypothermia and this is manifest in a reduction in requirements during hypothermic cardiopulmonary bypass.

The reliable and predictable action of atracurium has made it the drug of choice for short to intermediate procedures for many anaesthetists. It has found a regular use in day cases in particular. It is presently regarded as the relaxant of choice for myasthenic patients (Chapter 9), but should be given in small doses of between 5 and 10 mg at a time while monitoring the response.

Side-effects

One of the key features of atracurium is its lack of cardiovascular effects. This was well demonstrated by Healy and Palmer in 1982[16] who showed that the dose ratio for neuromuscular block to ganglion block was almost 50:1 whereas that for tubocurarine was 10:1. The dose to produce vagal block is about 25 times that to produce neuromuscular block[17] and this factor may underlie the common observation that bradycardia may occur during surgery when atracurium is the muscle relaxant. This is particularly noticeable when using a high-dose narcotic technique.

Histamine release may be a problem with atracurium when administered at doses in excess of about 0.5–0.6 mg kg^{-1}. It is seldom a problem with normal clinical doses. A pink flush is commonly seen spreading up the arm used for injection and across the face and neck. It is rarely associated with any systemic disturbance and rapidly fades. Histamine reactions appear to be more common when atracurium immediately follows thiopentone and it has been suggested that the precipitate formed by the mixing of the two drugs in the infusion tubing might be to blame (Chapter 13). The line should always be flushed with saline between drugs.

Doxacurium

This bisquaternary benzylisoquinolinium relaxant has recently been introduced into the USA. It is unlikely, however, to reach clinical use in Europe. Its chemical structure, as might be expected, bears a close similarity to atracurium (Fig. 5.2).

Pharmacokinetics

Doxacurium is excreted principally by the kidneys and its duration of action is therefore prolonged in patients with impaired renal function.[18] It is affected only very little by decreases in liver function. The duration of action is prolonged slightly in the elderly age group (Chapter 9).

Pharmacodynamics

The ED95 of doxacurium is 0.03 mg kg^{-1}. It is thus the most potent muscle relaxant presently available.

The onset of action of doxacurium is very slow. An ED95 dose may take up to 10 minutes to reach maximum effect. This can be shortened to about 3–4 minutes by increasing the intubating dose to 0.08 mg kg^{-1}, but the administration of such a high dose will produce a block which may last up to 2.5 hours.[19]

Doxacurium has a very long duration of action. One ED95 dose can be expected to produce muscular relaxation for up to 60 minutes and increasing the dose may considerably extend the duration of action. There is a degree of unpredictability about the action duration with wide individual variations and it may be necessary to delay reversal until adequate spontaneous recovery is present.

Although complete spontaneous recovery will ultimately ensue, it is recommended that a doxacurium block be always reversed. Reversal from profound levels of block may be prolonged (Chapter 8).[20]

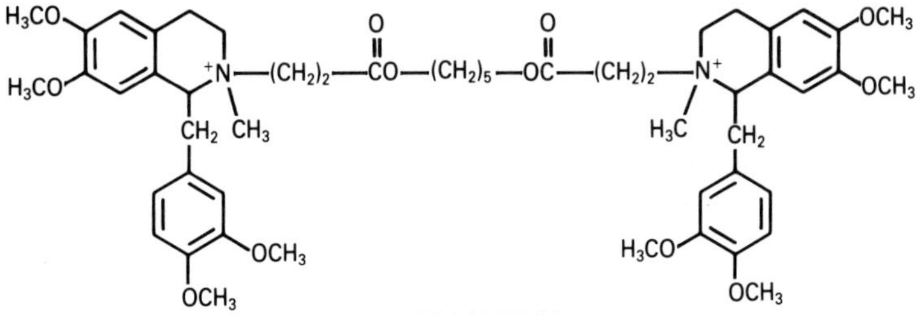

ATRACURIUM

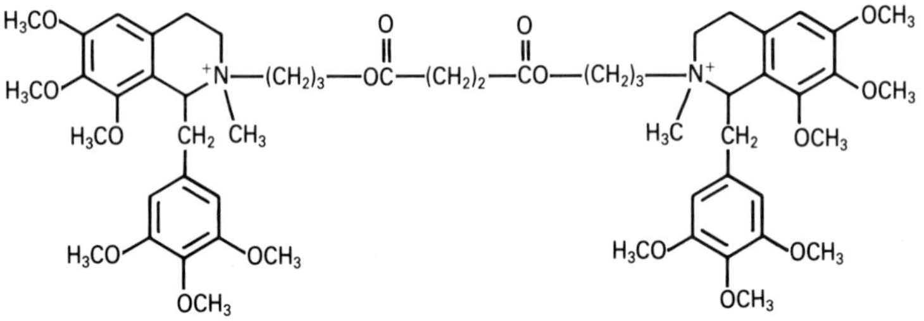

DOXACURIUM

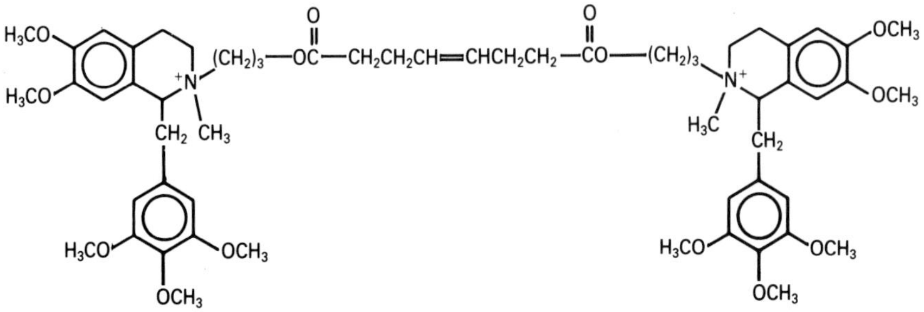

MIVACURIUM

Fig. 5.2 Structural formulae of the benzylisoquinolinium relaxants atracurium, doxacurium and mivacurium.

Side-effects

A striking feature of doxacurium is its clean cardiovascular profile. There are no significant effects on heart rate, blood pressure or cardiac output from doxacurium.[21] Doxacurium does not release histamine.

Gallamine

Gallamine was the first truly synthetic muscle relaxant. It is the only one with three positively charged nitrogen atoms in the molecule (Fig. 5.3). Gallamine enjoyed a wide use during the 1950s and 1960s, but its use has declined considerably in the last 10 years. It is still used in some institutions in doses of 10–20 mg for pre-curarization to reduce some of the unwanted effects of suxamethonium for which purpose it appears to be better than many of the other relaxants (Chapter 4).[22]

Pharmacokinetics

Gallamine is not metabolized in the body, but excreted unchanged in the urine. Negligible amounts appear in the bile. Gallamine is therefore an unwise choice in the patient with impaired renal function because a grossly prolonged block may ensue (Chapter 9).[23] The duration of action of gallamine is prolonged in the elderly and this appears to be due to a reduced plasma clearance secondary to a decrease in renal function.

Pharmacodynamics

Gallamine is the least potent of the muscle relaxants in current clinical use. The ED50 is approximately 1.0 mg kg^{-1} and the ED95 is approximately 2.4 mg kg^{-1}.

The onset of action of gallamine is relatively slow, with maximum block being reached in about 4–7 minutes. The duration of action is similar to that of pancuronium, in that a clinical duration of action of about 60 minutes can be expected from an intubating dose.[24]

The spontaneous recovery of gallamine is slow, the recovery index being about 45 minutes. Gallamine is rapidly and effectively reversed by an anticholinesterase although no more so than any of the other longer-acting relaxants.

Side-effects

The most notable side-effect seen with gallamine is a dose-related tachycardia which results principally from its vagal blocking action.[25] So marked is the action on the vagus that many pharmacologists regard gallamine as an

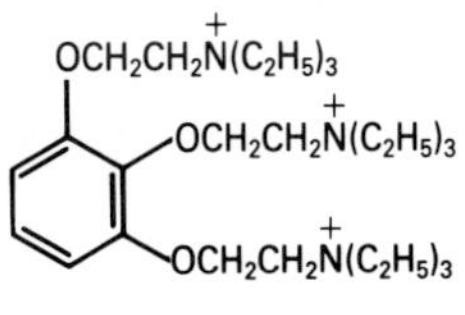

Fig. 5.3 Structural formula of gallamine.

84 Non-depolarizing muscle relaxants

antimuscarinic drug which happens also to possess activity at the neuromuscular junction instead of a neuromuscular blocking drug which is an antimuscarinic. The tachycardia is accompanied by a small increase in cardiac output and blood pressure. Gallamine does not produce maximum vagal block because a dose of atropine will produce a further increase in heart rate. This side-effect has been put to use by anaesthetists because it is powerful enough to obtund the oculocardiac reflex during ophthalmic surgery.

In the normal clinical dose range, gallamine is not a ganglion blocker. It does not often cause histamine release. Anaphylactoid reactions have, however, been described.

Metocurine

This relaxant was produced by the methylation of the hydroxyl groups of tubocurarine (Fig. 5.4). It is therefore a synthetic derivative of a naturally occurring agent. Although initially named dimethyltubocurarine, it has since been realized that there are actually three hydroxyl groups which are methylated and so the name trimethyltubocurarine is a better description.

Pharmacokinetics

Metocurine does not undergo metabolism in the human. The majority is excreted in the urine with a small amount being secreted into the bile. The drug therefore accumulates in patients with impaired renal function and should be used with care in these patients.[26] Its action is also prolonged in the elderly secondary to the decrease in renal function observed in this age group.

Pharmacodynamics

The potency of metocurine is approximately twice that of tubocurarine. The ED50 is about 0.13 mg kg^{-1} and the ED95 about 0.28 mg kg^{-1}.

The rate of onset of metocurine is similar to that of tubocurarine. An intubating dose of 0.3 mg kg^{-1} will produce maximum block after about 5–6 minutes.[27] The onset can be further accelerated by increasing the dose.

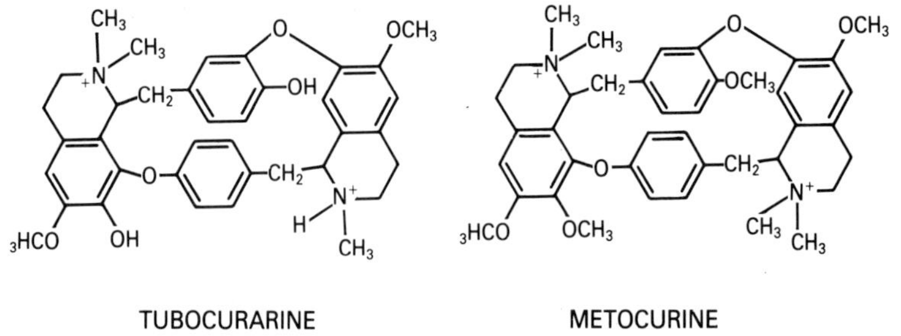

Fig. 5.4 Structural formulae of tubocurarine and metocurine.

The duration of action is very similar to that of tubocurarine. An intubating dose of 0.3 mg kg^{-1} can be expected to provide muscular relaxation for about 60–80 minutes.[27] Satisfactory reversal can be achieved with the use of neostigmine in normal doses of 0.05 mg kg^{-1}.

Side-effects

The reason why metocurine was developed and marketed was because it offered a cleaner alternative to tubocurarine. Although not devoid of side-effects, these are far fewer than tubocurarine at clinical doses. At higher doses, autonomic block does become manifest producing a reduction in blood pressure.[27] The potential for histamine release is also much less than that of tubocurarine. Since the advent of newer drugs with even less potential for side-effects, the use of metocurine has declined.

Mivacurium

Mivacurium is a relatively new addition to the muscle relaxant family. It is a benzylisoquinolinium structure (the atracurium family – Fig. 5.2).

Pharmacokinetics

Mivacurium is the only non-depolarizing muscle relaxant to undergo metabolism by plasma cholinesterase. It is broken down at about 75% of the rate of suxamethonium. This mechanism of metabolism means that mivacurium is cleared from the plasma very rapidly, and makes for a relaxant with a short duration of action.

This mechanism of metabolism confers an interesting property on mivacurium. The interaction of an enzyme and its substrate conforms to the Law of Mass Action and it follows from this law that the rate of the reaction is proportional to the active mass (concentration) of each of the reacting components. The rate of reaction is therefore proportional to both the plasma concentration of mivacurium and to the concentration of plasma cholinesterase. Under normal circumstances, plasma cholinesterase is present in great excess. This results in the rate of breakdown being approximately proportional to the concentration of mivacurium – i.e. the larger the dose, the higher the concentration, the more rapid the breakdown.

Under circumstances where the activity or quantity of plasma cholinesterase is reduced, there might be expected to be an effect on the breakdown of mivacurium. This has been confirmed. The duration of action of mivacurium is prolonged in patients who are homozygous for the atypical cholinesterase gene.[28]

In patients with impaired renal function, the duration of action of mivacurium is prolonged (Chapter 9).[29] The duration of action of mivacurium does not appear to be different in the elderly age group when compared to younger adult patients (Chapter 9).[30]

Pharmacodynamics

The ED95 of mivacurium is about 0.075 mg kg^{-1} in adult patients.

The administration of one ED95 dose will result in a maximum block being reached after about 4–5 minutes. The onset time can be shortened by increasing the intubating dose such that 0.25 mg kg^{-1} has an onset time of about 2.5 minutes and increasing the dose further results in little improvement. The onset time cannot be further shortened by priming.

Mivacurium is a short acting agent. An intubating dose of 0.15 mg kg^{-1} can be expected to take about 10–12 minutes to recover to 25% block and a further 5–10 minutes to reach 95% recovery. Because of its metabolism by plasma cholinesterase, the duration of action is little affected by increasing the dose, unlike the other non-depolarizing agents. In one typical study, increasing the dose from 0.15 to 0.25 mg kg^{-1} only increased the duration of action from 15 to 20 minutes.[31]

Recovery, once it has begun, is rapid and complete. The recovery index is approximately 6–10 minutes and seems to be unaffected by the total dose of mivacurium which has been administered to the patient. It is not often necessary to use an anticholinesterase agent to antagonize a mivacurium block, but if clinically indicated, reversal is rapidly secured with a routine dose of 0.05 mg kg^{-1} of neostigmine (Chapter 8).

Side-effects

Studies have shown no significant changes in cardiovascular variables following mivacurium in bolus doses up to 0.25 mg kg^{-1} which could be attributed to a direct effect of the drug on autonomic systems. A transient fall in the blood pressure may sometimes be seen, particularly with larger doses due to the release of histamine.[31] Flushing may also be seen in the arm used for venepuncture and this may spread onto the upper body. These phenomena can be reduced by administering the mivacurium bolus more slowly over about 15–30 seconds.

Pancuronium

Pancuronium was the first steroid-based muscle relaxant to reach clinical use. Although a steroid, it has no hormonal activity. The rigid steroid structure holds the two quaternary nitrogen atoms a constant fixed distance apart, the separation between them being close to the optimum for neuromuscular blocking action. Inspection of its structure (Fig. 5.5) shows that it has two acetylcholine moieties embedded within the molecule.

Pharmacokinetics

Pancuronium is metabolized in the liver to the 3-OH, 17-OH and 3,17-DiOH derivatives although only the 3-OH compound has been detected in

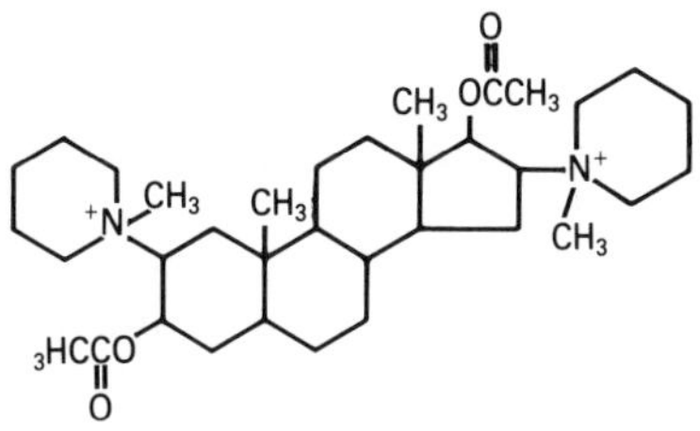

PANCURONIUM

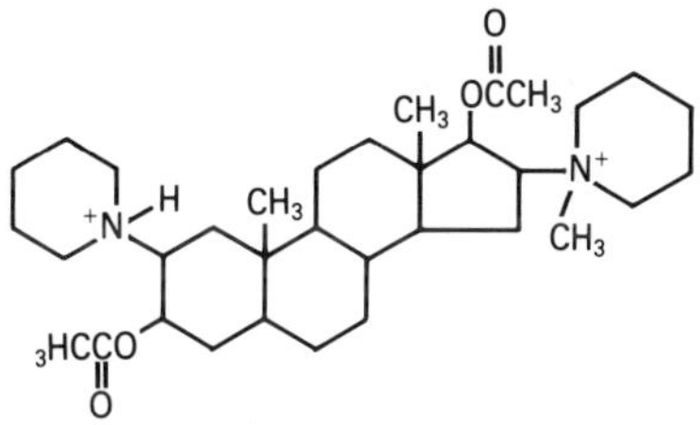

VECURONIUM

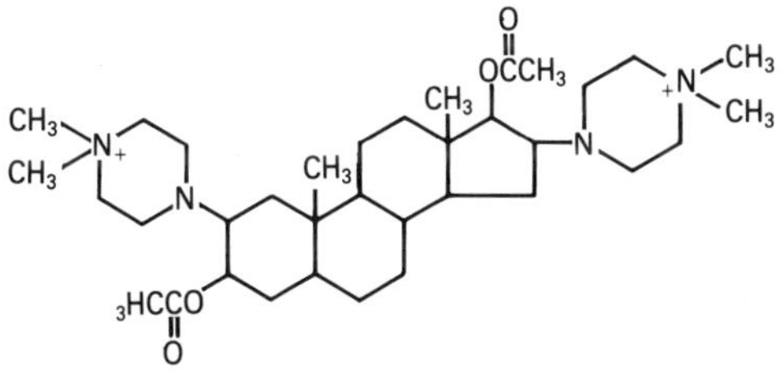

PIPECURONIUM

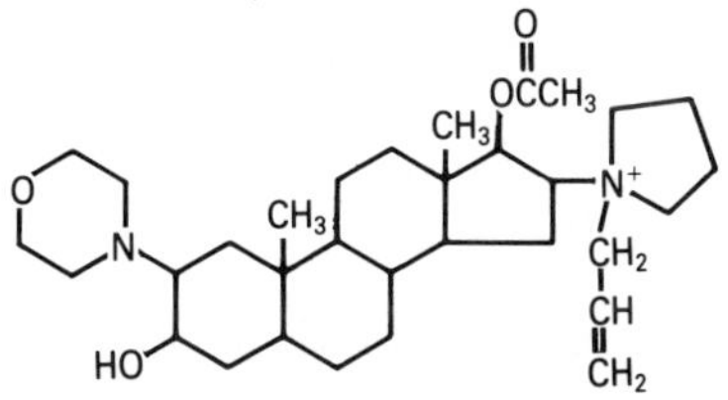

ROCURONIUM

Fig. 5.5 Structural formulae of the steroid relaxants pancuronium, vecuronium, pipecuronium and rocuronium.

man. This metabolite possesses neuromuscular blocking activity with a potency about 40% that of the parent compound. The metabolites together with the parent compound are excreted by the kidneys.

In cases where there is impairment of renal or hepatic function the duration of action of pancuronium may be prolonged (Chapter 9). The volume of distribution is also increased in renal failure. The decrease in renal function with advancing age leads to a small prolongation of effect in the elderly population.

Pharmacodynamics

Pancuronium is a popular relaxant and a great many studies have been performed to investigate its properties. The ED50 is about 0.03–0.04 mg kg^{-1} and the ED95 about 0.06 mg kg^{-1}.

Following the administration of one ED95 dose, maximum block can be expected to be reached in between 3 and 4 minutes. Increasing the dose to 0.1 mg kg^{-1} reduces the onset time to about 2 minutes but this cannot be further improved upon. Priming will also accelerate the onset of block by a similar amount.

An ED95 dose will last approximately 60 minutes. Higher doses last proportionately longer, and cumulation can be observed following repeated doses.

The spontaneous recovery index is about 25–30 minutes.[24] A block can be reliably antagonized with neostigmine in normal clinical doses (Chapter 8).

Side-effects

One of the stimuli to introduce pancuronium into clinical practice was its favourable cardiovascular profile. The blood pressure is well maintained and the hypotension seen with several other relaxants is not a feature of pancuronium. Indeed, a tachycardia, hypertension and an increased cardiac output may be seen.[32]

Pancuronium does not appear to have any activity at autonomic ganglia. The tachycardia which is often seen following the administration of pancuronium is not usually of any significance. The mechanisms behind the cardiovascular side-effects of pancuronium include block of cardiac vagal receptors and an increased release of noradrenaline together with a decreased re-uptake of noradrenaline from sympathetic nerve fibres.

These side-effects have led to the popularity of pancuronium for use in cardiac surgery. Although newer relaxants exist which have a cleaner cardiovascular profile, the tachycardia and elevation of cardiac output following the use of pancuronium have helped it to retain its place in cardiac surgery.

Pancuronium appears to be devoid of any propensity to release histamine.

Pipecuronium

This muscle relaxant belongs to the steroid family and is a bisquaternary molecule where the inter-nitrogen distance is greater than the other steroids (Fig. 5.4). It was originally developed in Hungary in 1980.[33] It is now available in the USA and will soon be introduced into the UK and other parts of Europe.

Pharmacokinetics

Pipecuronium does not appear to undergo very much metabolism in the body. Most of an administered dose can be recovered from the urine. One might therefore expect its action to be extended in renal failure and this has indeed been reported.[34] The elimination half-life, which is about 137 minutes in the normal patient, increases to double that value in patients with absent renal function. Care should therefore be taken with this drug in patients with impaired renal function (Chapter 9).

Animal studies suggest that the liver may also be involved in either the inactivation or excretion of pipecuronium.[35] This would be logical because although the duration is prolonged in the anephric patient, it is not prolonged indefinitely.

The dose requirements of pipecuronium are slightly less in the elderly age group (Chapter 9). Although this observation may reflect reduced renal clearance in this age group, Matteo *et al*. (1991)[36] were unable to detect significant pharmacokinetic changes.

Pharmacodynamics

The ED95 of pipecuronium is about 0.05 mg kg^{-1}. The onset of action of an ED95 dose is about 4–6 minutes and this can be shortened to about 2.5–3 minutes by the administration of a larger bolus dose.

Pipecuronium has a long duration of action. An ED95 dose can be expected to last between about 40 and 60 minutes. This is proportionately increased with increases in the intubating dose.

The spontaneous recovery from a pipecuronium block is slow; the recovery index is about 30 minutes.[37] Antagonism is therefore recommended at the end of surgery using neostigmine. Reversal may take up to 8–10 minutes and it may occasionally be necessary to administer a further dose of neostigmine in order to be certain that adequate muscle power has returned.

Side-effects

The principal feature of pipecuronium is its almost complete lack of cardiovascular side-effects.[38] There is either no, or an extremely small, change in heart rate, blood pressure and cardiac output following the administration of pipecuronium. Histamine release does not appear to be a problem.

Rocuronium

This is another steroid-based relaxant (Fig. 5.4). It has only just been introduced into clinical practice. It seems likely that it will secure a place in anaesthesia.

Pharmacokinetics

The largest proportion of a dose of rocuronium is taken up by the liver and secreted into the bile. A significant amount is excreted in the urine in normal patients; the liver thus appears to be the principal route of elimination for rocuronium. It is eliminated unchanged and undergoes minimal breakdown in the body.

In patients with impaired renal function, the onset and duration of action of rocuronium are little changed (Chapter 9).[39] In patients with marked hepatic dysfunction, however, the duration of action is increased (Chapter 9).[40]

In the elderly patient, the onset of action is slightly slower and the duration of action prolonged (Chapter 9).[41]

Pharmacodynamics

With the exception of gallamine, this drug is the least potent of the non-depolarizing relaxants. The ED95 of rocuronium is about 0.3 mg kg^{-1} in adults.

The important feature of rocuronium is that there is a rapid onset of block. One ED95 will produce maximum block in approximately 4 minutes and this can be reduced to 2–2.5 minutes with doses of twice the ED95. Following the administration of larger doses, between three and four times the ED95, the maximum block is reached in about 1.7 minutes.[42] Intubating conditions, however, are not dependent upon there being a maximum block with rocuronium. The onset of rocuronium demonstrates a very fast initial phase up to about 80% block which then slows down.[43] Intubation is thus possible after only 60 seconds with more than 0.45 mg kg^{-1} of rocuronium making it a truly rapid-onset drug. Use of priming does not further accelerate the onset.

The duration of action is similar to that of vecuronium. A bolus dose of 0.6 mg kg^{-1} (twice the ED95 and a dose which will give good intubating conditions in about 60–75 seconds) can be expected to last about 30–40 minutes. Cumulation has not so far been reported following repeated doses or use by infusion.

Rapid and satisfactory reversal can be accomplished by the use of routine clinical doses of neostigmine (Chapter 8).

Side-effects

Rocuronium is a very clean relaxant. There are negligible ganglion blocking actions and minimal effect on the vagus in doses larger than those likely to be used clinically.[44] Histamine release has not been observed in doses up to four times the ED95.[45]

Tubocurarine

Tubocurarine was the first muscle relaxant to be discovered and also the first to find its way into clinical practice.[1] It is a monoquaternary compound, although it was originally thought to be bisquaternary (Fig. 5.4); the second nitrogen becomes protonated at body pH. It is still derived from its natural source, *Chondodendron tomentosum*. Although its use has declined in recent years following the introduction of newer agents with a shorter duration of action and negligible side-effects, it still remains the yardstick against which new relaxants are compared.

Pharmacokinetics

Very little tubocurarine is metabolized in the body.[46] The drug is excreted unchanged in both the urine and the bile. The possession of two routes of elimination have led to tubocurarine being used safely for patients with either impaired liver function or impaired renal function in the past. The clearance is nevertheless reduced and the duration of action consequently prolonged in renal failure (Chapter 9).

Recovery from a tubocurarine neuromuscular block is slower in the elderly age group, presumably due to the decreased renal function in those patients (Chapter 9).[47]

Pharmacodynamics

Tubocurarine, like the other non-depolarizing relaxants has an action on both the postjunctional and prejunctional acetylcholine receptors. The action of tubocurarine on the prejunctional nicotinic receptors is thought to be more potent than that of the other non-depolarizing relaxants because of the pronounced fade to a repetitive train of stimuli with tubocurarine.[48] The ED50 of tubocurarine is about 0.26 mg kg^{-1} and the ED95 about 0.5 mg kg^{-1}.

The onset of block when using tubocurarine is slow, an ED95 dose taking between 4 and 7 minutes to achieve maximum effect. The onset can be accelerated by priming. The use of a larger dose to improve onset characteristics is not feasible with tubocurarine because troublesome side-effects appear at doses not much greater than the ED95.

The duration of action is similar to that of pancuronium, lasting about 50–70 minutes. Being a long-acting drug, it accumulates if given in repeated doses leading to a prolonged block. The action of tubocurarine appears to be affected by acid-base changes to a greater extent than any of the other relaxants (Chapter 9).

The spontaneous recovery index is about 30 minutes. The routine use of an anticholinesterase to hasten recovery is recommended (Chapter 8).

Side-effects

Tubocurarine is a potent antagonist at autonomic ganglia. Hypotension commonly follows the administration of tubocurarine and it is therefore an unwise choice in the patient who already has a low blood pressure, has any degree of hypovolaemia or has coronary artery disease. There is a reduction in both the cardiac output and the systemic vascular resistance. Interestingly, the reduction in blood pressure is accompanied by only a limited rise in heart rate.[49] This may be a reflection of a significant block of sympathetic ganglia.

Tubocurarine may also release histamine at clinical doses. Skin flushing is common following the administration of tubocurarine and bronchospasm may be seen in susceptible patients. The release of histamine may exaggerate the hypotension caused by block of autonomic ganglia.

The propensity of tubocurarine to lower the blood pressure made it a favourite for use when deliberate hypotension was being practised.

Vecuronium

Vecuronium is a synthetic steroid-based molecule which came from the demethylation of pancuronium (Fig. 5.5). It is a monoquaternary compound. Vecuronium is unstable in aqueous solution and is therefore supplied as a freeze-dried powder which has to be dissolved in water before use. Once dissolved, it will retain its potency for 24 hours.

Pharmacokinetics

Vecuronium, like pancuronium, breaks down to 3-OH, 17-OH and 3,17-DiOH compounds.[50] This process takes place principally by spontaneous deacetylation although liver metabolic pathways are also involved. The 3-OH derivative has muscle relaxant activity with a potency about 50–70% of that of the parent compound. Vecuronium is eliminated through the kidneys both as the parent compound and as its breakdown products.

The elimination half-life of vecuronium is fairly long, at about 60 minutes. Its relatively short duration of action appears to be due to rapid redistribution rather than rapid metabolism (Chapter 3). The implication from this is that cumulation may occur following larger doses, especially in the presence of renal failure and this has indeed been reported.[51, 52] In cases of severe impairment of hepatic function, the duration of action of vecuronium is considerably prolonged (Chapter 9).[53]

Pharmacodynamics

The ED50 of vecuronium is about 0.025 mg kg^{-1} and the ED95 about 0.04 mg kg^{-1}. The potency is therefore not much different from that of pancuronium.

The onset of action of one ED95 dose is between 5 and 6 minutes. This can be reduced considerably by increasing the intubating dose or by priming. A dose of vecuronium of over 0.2 mg kg^{-1} will produce maximum block in approximately 2 minutes.[14]

Vecuronium is an intermediate-acting muscle relaxant with the duration of action of a normal intubating dose lasting about 20–30 minutes. If the intubating dose is increased in order to secure a more rapid onset of block, then there is a proportionate increase in the duration of action.

Spontaneous recovery is reliable and vecuronium may not need reversal on occasions. The spontaneous recovery index is about 12–15 minutes. When accelerated recovery is required, this is readily achieved with the normal clinical dose of neostigmine.

Side-effects

Vecuronium is almost devoid of cardiovascular side-effects. There is no evidence of any activity at autonomic ganglia or vagal nerve endings. The clean cardiovascular profile has made vecuronium a popular relaxant for use in all patients, particularly those with compromised cardiovascular systems. Like atracurium, its lack of vagal action may unmask a bradycardia from surgical stimulation.

There is very little histamine release at normal clinical doses, but this may occasionally present a problem at higher doses.

Conclusion

It is thus apparent that a range of muscle relaxants exist at present. They differ principally with respect to their speed of onset, duration of action, routes of elimination and side-effects. There is probably no one ideal relaxant which would suit every anaesthetist for every operation. Few anaesthetists will choose a drug with adverse side-effects when one which is devoid of side-effects also exists. A rapid onset is clearly advantageous and the shorter-acting agents are more versatile than the longer-acting agents. Drugs which rely very little on the kidney or liver for their deactivation or elimination are at a clear advantage.

It is possible that we shall see a rationalization of the available relaxants over the next few years. Indeed, some anaesthetists are of the opinion that there are now too many relaxants available and it seems likely that some of the older, longer-acting ones will fall into disuse and disappear, their role being usurped by the newer, short- and intermediate-acting ones. Research still continues, however, and it is likely that we shall see yet more muscle relaxants arrive in the future. It will be interesting to see what novel molecules are developed for our consideration.

References

1. Griffith HR, Johnson GE. The use of curare in general anesthesia. *Anesthesiology* 1942; **3:** 418–420.
2. Walker JS, Shanks CA, Brown KF. Alcuronium kinetics and plasma concentration-effect relationship. *Clinical Pharmacology and Therapeutics* 1983; **33:** 510–516.
3. Stephens ID, Ho PC, Holloway AM, Bourne DWA, Triggs EJ. Pharmacokinetics of alcuronium in elderly patients undergoing total hip replacement or aortic reconstructive surgery. *British Journal of Anaesthesia* 1984; **56:** 465–471.
4. Parkin JE. Determination of alcuronium chloride in biological fluids by high performance liquid chromatography. *Journal of Chromatography* 1981; **225:** 240–242.
5. Kreig N, Crul JF, Booij LHDJ. Relative potency of Org NC45, pancuronium, alcuronium and tubocurarine in anaesthetized man. *British Journal of Anaesthesia* 1980; **52:** 783–787.
6. Hughes R, Chapple DJ. Effects of non-depolarizing neuromuscular blocking agents on peripheral autonomic mechanisms in cats. *British Journal of Anaesthesia* 1976; **48:** 59–68.
7. Kennedy BR, Kelman GR. Cardiovascular effects of alcuronium in man. *British Journal of Anaesthesia* 1970; **42:** 625–629.
8. Chan CS, Yeung ML. Anaphylactic reaction to alcuronium. *British Journal of Anaesthesia* 1972; **44:** 103–105.
9. Stenlake JB, Waigh RD, Uriwn J, Dewar GH, Coker GG. Atracurium: conception and inception. *British Journal of Anaesthesia* 1983; **55:** 3S–10S.
10. Lanier WL, Milde JH, Michenfelder JD. The cerebral effects of pancuronium and atracurium in halothane anesthetized dogs. *Anesthesiology* 1985; **63:** 589–597.

11. Nigrovic V, Klaunig JE, Smith SL, Schultz NE, Wajskol A. Comparative toxicity of atracurium and metocurine in isolated rat hepatocytes. *Anesthesia and Analgesia* 1983; **65**: 1107–1111.
12. Fisher DM, Canfell PC, Fahey MR, Rosen J, Rupp SM, Sheiner LB, et al. Elimination of aracurium in humans: Contribution of Hofmann elimination and ester hydrolysis versus organ-based elimination. *Anesthesiology* 1986; **65**: 6–12.
13. Ward S, Neill EAM. Pharmacokinetics of atracurium in acute hepatic failure (with acute renal failure). *British Journal of Anaesthesia* 1983; **55**: 1169–1172.
14. Healy TEJ, Pugh ND, Kay B, Sivalingam T, Petts HV. Atracurium and vecuronium: effect of dose on the time of onset. *British Journal of Anaesthesia* 1986; **58**: 620–624.
15. Katz RL, Stirt J, Murray AL, Lee C. Neuromuscular effects of atracurium in man. *Anesthesia and Analgesia* 1982; **61**: 730–734.
16. Healy TEJ, Palmer JP. *In vitro* comparison between the neuromuscular and ganglion blocking potency ratios of atracurium and tubocurarine. *British Journal of Anaesthesia* 1982; **54**: 1307–1311.
17. Hughes R, Chapple DJ. The pharmacology of atracurium: a competing neuromuscular blocking agent. *British Journal of Anaesthesia* 1981; **53**: 31–44.
18. Cook DR, Freeman JA, Lai AA, Robertson KA, Kang Y, Stiller RL, et al. Pharmacokinetics and pharmacodynamics of doxacurium in normal patients and in those with hepatic or renal failure. *Anesthesia and Analgesia* 1991; **72**: 145–150.
19. Lennon RL, Hosking MP, Houck PC, Rose SH, Wedel DJ, Gibson BE, et al. Doxacurium chloride for neuromuscular blockade before tracheal intubation and surgery during nitrous oxide-oxygen-narcotic-enflurane anesthesia. *Anesthesia and Analgesia* 1989; **68**: 255–260.
20. Murray DJ, Mehta MP, Choi WW, Forbes RB, Sokoll MD, Gergis SD, et al. The neuromuscular blocking and cardiovascular effects of doxacurium chloride in patients receiving nitrous oxide-narcotic anesthesia. *Anesthesiology* 1988; **69**: 472–477.
21. Emmott RS, Bracey BJ, Goldhill DR, Yate PM, Flynn PJ. Cardiovascular effects of doxacurium, pancuronium and vecuronium in anaesthetised patients presenting for coronary artery bypass surgery. *British Journal of Anaesthesia* 1990; **65**: 480–486.
22. Cullen DJ. The effect of pretreatment with nondepolarizing muscle relaxants on the neuromuscular blocking action of succinylcholine. *Anesthesiology* 1971; **35**: 572–578.
23. Ramzan MI, Snanks CA, Triggs EJ. Gallamine disposition in surgical patients with chronic renal failure. *British Journal of Clinical Pharmacology* 1981; **12**: 141–147.
24. Donlon JV, Ali HH, Savarese JJ. A new approach to the study of four nondepolarising relaxants in man. *Anesthesia and Analgesia* 1974; **53**: 934–939.
25. Stoelting RK. Hemodynamic effects of gallamine during halothane-nitrous oxide anesthesia. *Anesthesiology* 1973; **39**: 645–647.
26. Brotherton WP, Matteo RS. Pharmacokinetics and pharmacodynamics of metocurine in humans with and without renal failure. *Anesthesiology* 1981; **55**: 273–276.
27. Savarese JJ, Ali HH, Antonio RP. The clinical pharmacology of metocurine: dimethyltubocurarine revisited. *Anesthesiology* 1977; **47**: 277–285.
28. Ostergaard D, Jensen E, Jensen FS, Viby-Mogenson J. The duration of action of mivacurium-induced neuromuscular block in patients homozygous for the atypical plasma cholinesterase gene. *Anesthesiology* 1991; **75**: A774.
29. Phillips BJ, Hunter JM. Use of mivacurium chloride by constant infusion in the anephric patient. *British Journal of Anaesthesia* 1992; **68**: 492–498.
30. Basta SJ, Dresner DL, Shaff LP, Lai AA, Welch R. Neuromuscular effects and pharmacokinetics of mivacurium in elderly patients undergoing isoflurane anaesthesia. *Anesthesia and Analgesia* 1989; **68**: S18.

31. Caldwell JE, Heier T, Kitts JB, Lynam DP, Fahey MR, Miller RD. Comparison of the neuromuscular block induced by mivacurium, suxamethonium or atracurium during nitrous oxide-fentanyl anaesthesia. *British Journal of Anaesthesia* 1989; **63:** 393–399.
32. Kelman GR, Kennedy BR. Cardiovascular effects of pancuronium in man. *British Journal of Anaesthesia* 1971; **43:** 335–338.
33. Boros M, Szenorhadszky J, Marosi G, Toth I. Comparative clinical study of pipecuronium bromide and pancuronium bromide. *Arzneimittel Forschung* 1980; **30:** 389–393.
34. Caldwell JE, Canfell PC, Castagnoli KP, Lynam DP, Fahey MR, Fisher DM, et al. The influence of renal failure on the pharmacokinetics and duration of pipecuronium bromide in patients anesthetized with halothane and nitrous oxide. *Anesthesiology* 1989; **70:** 7–12.
35. Pittet JF, Tassonyi E, Schopfer C, Morel DR, Leemann P, Mentha G, et al. Dose requirements and plasma concentrations of pipecuronium during bilateral renal exclusion and orthoptic liver transplantation in pigs. *British Journal of Anaesthesia* 1990; **65:** 779–785.
36. Matteo RS, Schwartz AE, Ornstein E, Jamdar S, Diaz J, Rodrigues H, et al. Pharmacokinetics and pharmacodynamics of pipecuronium in elderly surgical patients. *Anesthesia and Analgesia* 1991; **72:** S172.
37. Stanley JC, Mirakhur RK, Bell PF, Sharpe TDE, Clark RSJ. Neuromuscular effects of pipecuronium bromide. *European Journal of Anaesthesiology* 1991; **8:** 151–156.
38. Tassonyi E, Neidhard E, Pittet JF, Morel DR, Gemperle M. Cardiovascular effects of pipecuronium and pancuronium in patients undergoing coronary artery bypass grafting. *Anesthesiology* 1988; **69:** 793–796.
39. Szenorhadszky J, Segredo V, Caldwell JE, Sharma M, Gruenke LD, Miller RD. Pharmacokinetics, onset and duration of action of Org 9426 in humans: normal vs absent renal function. *Anesthesia and Analgesia* 1991; **72:** S290.
40. Magorian T, Wood P, Caldwell JE, Szenorhadszky J, Segredo V, Sharma H, et al. Pharmacokinetics, onset, and duration of action of rocuronium in humans: normal vs hepatic dysfunction. *Anesthesiology* 1991; **75:** A1069.
41. Matteo RS, Ornstein E, Schwartz AE, Stone JG, Ostapkovich N, Spencer HK, et al. Pharmacokinetics and pharmacodynamics of Org 9426 in elderly surgical patients. *Anesthesiology* 1991; **75:** A1065.
42. Wierda JMKH, Kleef UW, Lambalk LM, Kloppenberg WD, Agoston S. The pharmacodynamics and pharmacokinetics of Org 9426, a new nondepolarizing neuromuscular blocking agent in patients anaesthetized with nitrous oxide, halothane. *Canadian Journal of Anaesthesia* 1991; **38:** 430–435.
43. Quill TJ, Begin M, Glass PSA, Ginsberg B, Gorback MS. Clinical responses to Org 9426 during isoflurane anesthesia. *Anesthesia and Analgesia* 1991; **72:** 203–206.
44. Booij LDHJ, Knape HTA. The neuromuscular blocking effect of Org 9426; a new intermediately-acting non-depolarising muscle relaxant in man. *Anaesthesia* 1991; **46:** 341–343.
45. Davis GK, Szlam F, Lowdon JD, Levy JH. Evaluation of histamine release following Org 9426 administration using a new radioimmunoassay. *Anesthesiology* 1991; **75:** A818.
46. Meijer DKF, Weitering JG, Vermeer GA, Scaf AKJ. Comparative pharmacokinetics of d-tubocurarine and metocurine in man. *Anesthesiology* 1979; **51:** 402–407.
47. Matteo RS, Backus WW, McDaniel DD, Brotherton WP, Abraham R, Diaz J. Pharmacokinetics and pharmacodynamics of d-tubocurarine and metocurine in the elderly. *Anesthesia and Analgesia* 1985; **64:** 23–29.
48. Bowman WC. Prejunctional and postjunctional cholinoceptors at the neuromuscular junction. *Anesthesia and Analgesia* 1980; **59:** 935–943.

49. Stoelting RK. The hemodynamic effects of pancuronium and d-tubocurarine in anesthetized patients. *Anesthesiology* 1972; **36:** 612–615.
50. Marshall IG, Gibb AJ, Durant NN. Neuromuscular and vagal blocking actions of pancuronium bromide, its metabolites, and vecuronium bromide (Org NC45) and its potential metabolites in the anaesthetized cat. *British Journal of Anaesthesia* 1983; **55:** 703–714.
51. Pollard BJ. Paralysis after long term administration of vecuronium. *Anesthesiology* 1990; **73:** 364.
52. Bevan DR, Donati F, Gyasi H, Williams A. Vecuronium in renal failure. *Canadian Anaesthetists Society Journal* 1984; **31:** 491–496.
53. Lebrault C, Berger JL, d'Hollander AA, Gomeni R, Henzel D, Duvaldestin P. Pharmacokinetics and pharmacodynamics of vecuronium (Org NC45) in patients with cirrhosis. *Anesthesiology* 1985; **62:** 601–605.

6

Monitoring Neuromuscular Blockade

NJN Harper

Introduction

The importance of neuromuscular monitoring during anaesthesia has been highlighted in recent years by the alarming observations from several major hospitals[1,2] that many patients who receive long-acting neuromuscular blocking drugs arrive in the recovery room still partially paralysed, to the extent that spontaneous ventilation[3] and the hypoxic drive[4] are impaired. In these studies, reversal of residual blockade by neostigmine had been judged to be adequate by clinical criteria, but neuromuscular monitoring with a nerve stimulator had not been used. The use of simple monitoring with a nerve stimulator was associated with a considerable improvement. These observations have been verified whenever the work has been repeated in other hospitals.

The introduction of new, shorter-acting neuromuscular blocking drugs into clinical practice has partially alleviated this particular problem,[5] but another difficulty has emerged – that of maintaining a consistent level of neuromuscular blockade during anaesthesia, avoiding undesirable peaks and troughs of effect. At the end of the surgical procedure, a short-acting neuromuscular blocking agent may provide the opportunity to avoid reversal with an anticholinesterase but the anaesthetist needs an objective measurement to provide certain knowledge that neuromuscular function has returned to normal.

Physiology and pharmacology

Fade and facilitation

Neuromuscular monitoring is dependent on measuring the response of a muscle to stimulation of its motor nerve. The response of the muscle, whether an index of the force of contraction or the evoked electrical activity, comprises the aggregated responses of all the individual muscle fibres, each of which is an all-or-none phenomenon. When a muscle relaxant is given, increasing numbers of individual muscle fibres 'drop out', starting with the most sensitive, as depolarization of successive fibres fails to reach the threshold for contraction. The response of the muscle progressively declines as neuromuscular blockade deepens. The extent of this process is often expressed as the 'percent block' by comparing the muscle response with a pre-relaxant value. Although neuromuscular blockade is the result of occupation of cholinoceptors by the muscle relaxant drug, the percent block is not proportional to the number of cholinoceptors that are occupied by the drug (Chapter 2). The contraction of a muscle is not demonstrably reduced until approximately 70% of its receptors (as distinct from 70% of its muscle fibres) are blocked,[6] so that the variation in neuromuscular blockade that is observed during anaesthesia represents only the tip of the pharmacological iceberg.

Fade

During normal muscle activity, the number of acetylcholine molecules liberated is considerably in excess of that needed to activate all the individual nerve fibres. Thus, if the voltage clamp technique is applied to an *in vitro* preparation, the muscle end-plate potential is maintained even at high stimulation frequencies.[7] During a competitive, non-depolarizing block, if neuromuscular transmission is stressed by applying stimuli to the nerve at a high frequency (above 1 Hz), the twitch responses diminish rapidly: this is known as 'fade'. The frequency of nerve action potentials is approximately 50 Hz during normal, voluntary movement. The artificial, repetitive stimuli used for neuromuscular monitoring are Train-of-Four, Double Burst, or tetanic. The non-depolarizing muscle relaxants are associated with significant fade, which is greater during the offset of action than during the onset. The degree to which fade becomes evident varies between the non-depolarizing drugs. Those agents which have a greater action than others at the prejunctional receptors are associated with more marked fade (Chapter 2). Fade is normally quantified by comparing the amplitude of the final response of the muscle to the train of stimuli with that of the first response, and expressing the proportion as a ratio or as a percentage. The depolarizing blockade produced by suxamethonium is associated with little or no fade unless phase ll block intervenes (Chapter 4).

Facilitation

If, during partial non-depolarizing neuromuscular blockade, the nerve is subjected to a burst of stimuli at a high frequency (e.g. 50 Hz) for a few seconds, the response of the muscle to subsequent low frequency stimulation (e.g. 1 Hz) is enhanced. This phenomenon is *post-tetanic facilitation*. During the tetanic stimulation, acetylcholine release by the motor nerve terminal is enhanced, probably as a result of increased calcium ion influx and the release of acetylcholine molecules that are immobilized within the nerve terminal. Post-tetanic facilitation is probably the consequence of persistence of this phenomenon after the tetanic stimulation has ceased. A subsequent, single stimulus will release a large quantity of acetylcholine which tends to overcome the barrier to neuromuscular transmission caused by the presence of a competitive drug within the neuromuscular junction. A slight increase in the force of contraction is observed following a tetanic stimulus even in the absence of a muscle relaxant drug. The compound muscle action potential (the surface electrical activity of the muscle representing all the events in the muscle with the exception of contraction) is not increased,[8] suggesting that a slight increase in contractility is responsible. This phenomenon is known as *post-tetanic potentiation* and is separate from the post-tetanic facilitation that is seen in the partially blocked muscle.

Monitoring techniques

Patterns of stimulation

No single pattern of nerve stimulation is appropriate to all levels of neuromuscular blockade: some patterns are suitable for monitoring 'light' blockade while others are more relevant to the assessment of profound blockade. A simple, schematic protocol for the most suitable patterns of stimulation at varying depths of neuromuscular blockade may be useful (Fig. 6.1).

Single stimuli

The response to a single stimulus in isolation is unhelpful unless a control response has been recorded before the administration of a muscle relaxant. In practice the response is difficult to evaluate visually or by manual assessment and this pattern of nerve stimulation is inappropriate unless the twitch

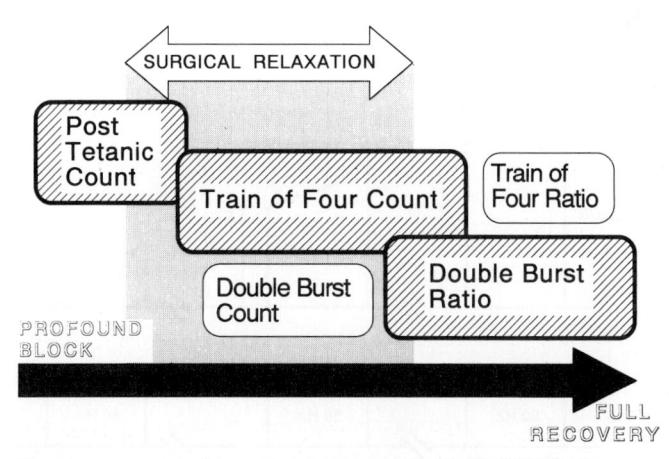

Fig. 6.1 The most appropriate stimulation techniques during the recovery of neuromuscular transmission from profound block to full neuromuscular function. The highlighted techniques are the most suitable when the muscle response is assessed by simple, clinical methods (visual or tactile).

response is being measured accurately. An interval of at least 10 seconds should be permitted between single stimuli if fade is to be avoided during partial non-depolarizing neuromuscular blockade.[9]

Tetanic stimulation

Tetanic fade is characteristic of a non-depolarizing block. A tetanus is a contraction that continues throughout repetitive stimulation. Theoretically, at low stimulation frequencies, the resulting force of contraction may oscillate above the baseline (unfused tetanus). If the frequency of stimulation is increased, the tetanus becomes fused, and the force of contraction increases to reach a sustained maximum of three to five times that of a single twitch response. In practice, it is impossible to detect unfused tetanus clinically. The most commonly used stimulus frequency for neuromuscular monitoring is 50 Hz. In the absence of neuromuscular disease or neuromuscular blocking drugs, an individual is able to sustain a contractile force for at least 5 seconds in response to 50 Hz stimulation.[10] This implies that sufficient acetylcholine is being manufactured, mobilized and released from the motor nerve terminal to fulfil the requirements for normal, sustainable neuromuscular transmission.

Unfortunately, tetanic stimulation has several, overwhelming disadvantages in clinical practice. A considerable interval between successive tetanic bursts must be permitted (5–6 minutes). Post-tetanic facilitation will influence the response to subsequent single, Train-of-Four or Double Burst stimulation unless an interval of at least 2 minutes is permitted between estimations when intermediate-duration relaxant drugs are used.[11] A greater time interval may be necessary during blockade with longer-acting drugs.[8] Using tactile assessment, tetanic fade (50 Hz) is detected reliably only when the tetanic fade ratio (Fig. 6.2) is 0.3 or less; it is no better than Train-of-Four stimulation in the clinical detection of residual blockade at the end of anaesthesia.[12] The limiting factor is the relatively poor performance of the anaesthetist's finger as a force transducer. Increasing the tetanic

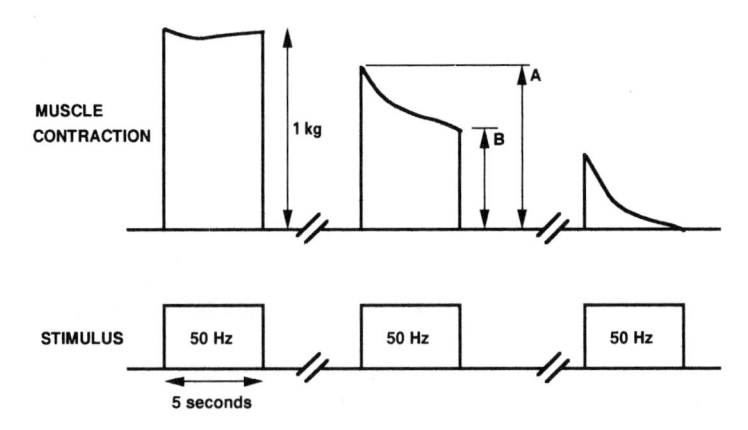

Fig. 6.2 The effect of non-depolarizing neuromuscular blockade on the muscle contraction produced by tetanic nerve stimulation. The tetanic fade ratio = B/A.

frequency to 100 Hz may induce fade even in the absence of a muscle relaxant.[13] A further disadvantage of tetanic stimulation is that it is extremely painful to the patient recovering from anaesthesia.

Train-of-Four ratio

Train-of-Four stimulation[14] was first described in 1970 and remains the standard method of monitoring non-depolarizing neuromuscular blockade. It represents a significant advance over single stimuli because it is not necessary to obtain a pre-relaxant control value for comparison. Each train comprises four stimuli at a frequency of 2 Hz. During non-depolarizing neuromuscular blockade there is progressive fade in the four responses (Fig. 6.3a). During depolarizing blockade there is very little fade in the Train-of-Four responses unless phase ll block develops (Fig. 6.3b). The amplitude of the fourth response of the train compared with the amplitude of the first response of the train is expressed as the Train-of-Four ratio. There is a linear, positive association between the Train-of-Four ratio and the ratio

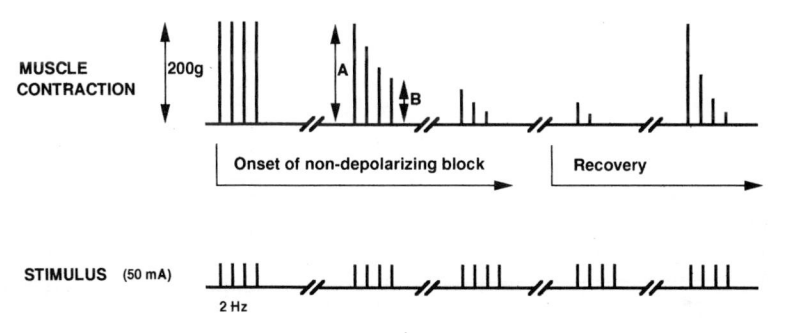

Fig. 6.3a Train-of-Four stimulation demonstrating the characteristic fade in the muscle responses during the onset and offset of non-depolarizing neuromuscular blockade. The Train-of-Four ratio = B/A. The Train-of-Four fade is less during the onset of blockade than during recovery.

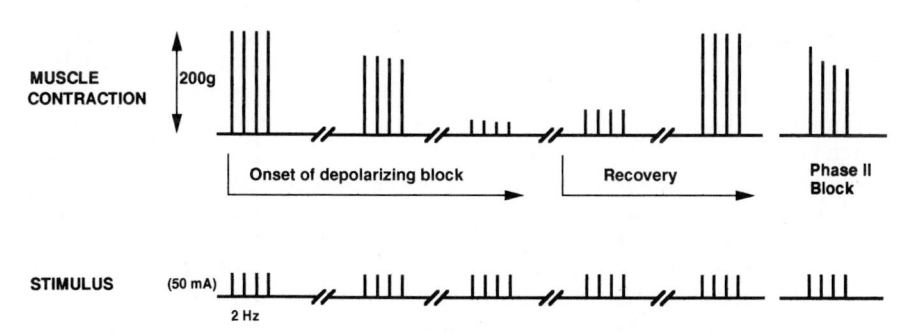

Fig. 6.3b Train-of-Four stimulation during depolarizing neuromuscular blockade. Train-of-Four fade is minimal unless phase ll block develops.

of the first response of the train to the pre-relaxant control response[15] and the first response of the Train-of-Four is often used in the place of single stimuli in the research setting. Nonetheless, the Train-of-Four ratio is a more sensitive and consistent index of returning muscle power than the first response ratio.[16] It is necessary to allow at least 10 seconds between Train-of-Four stimuli to ensure that each measurement has not been influenced by the previous train.[17] Train-of-Four fade is more evident during recovery from blockade than during onset. The extent to which the responses fade varies between muscle relaxant drugs. The magnitude of the Train-of-Four decrement has been shown to increase in the order: alcuronium< atracurium< tubocurarine< gallamine.[18]

Although Train-of-Four stimulation has become the standard method of assessing neuromuscular blockade, the ability of the anaesthetist to interpret the response of the muscle using simple, manual assessment of the force of contraction is severely limited. It is clearly important that any method of monitoring neuromuscular transmission should be capable of identifying the patient in whom residual blockade is of sufficient severity to impair spontaneous ventilation in the immediate postoperative period. It is generally accepted that, during recovery from non-depolarizing blockade, a Train-of-Four ratio of 0.7 corresponds with the ability to maintain adequate spontaneous ventilation.[19] At this level of blockade, the ability of the anaesthetist to detect fade in the Train-of-Four responses by clinical (non-instrumental) means is extremely poor: when the Train-of-Four ratio has recovered to 0.7, only approximately 25% of experienced observers are able to detect any fade in the four responses using tactile assessment of the force of contraction. It is possible to detect fade consistently only when the Train-of-Four ratio is as low as 0.3–0.4.[20] Lack of palpable Train-of-Four fade at the end of anaesthesia is no guarantee that neuromuscular transmission is adequate to support normal spontaneous ventilation.

Train-of-Four Count

After the administration of a non-depolarizing muscle relaxant drug, the fourth, third, second and first responses disappear in succession. The number of measurable responses is expressed as the Train-of-Four Count (0–4). The Train-of-Four Count is a useful, simple index for monitoring neuromuscular blockade at the level appropriate for abdominal surgery.[21] In general, during spontaneous recovery, the second, third and fourth twitches reappear at approximately 90%, 80% and 75% depression of the first twitch response: i.e. 10, 20 and 25% recovery.[21] Thus, at the time when the second twitch becomes palpable, the force of contraction of the first twitch will have recovered to approximately 10% of its pre-relaxant control value. When the third or fourth twitch first becomes apparent, neuromuscular transmission will have recovered to the limits of acceptability for abdominal surgery.[22] During continuing abdominal surgery, a Train-of-Four Count of greater than 2 or 3 signals the need for an incremental dose, or an increase in the infusion rate of the muscle relaxant. The exact Train-of-Four Count at which this action is necessary depends on the nature and the expected duration of the surgical procedure and the recovery profile of the muscle relaxant drug. For example, a Train-of-Four Count of 3 during mivacurium-induced blockade indicates a more urgent requirement to deepen the block than the same Train-of-Four Count during a pancuronium-induced neuromuscular blockade. During surgery which requires absolute confidence that the patient cannot hiccough; for example, neurosurgery, a Train-of-Four Count of zero may be appropriate and the Post-Tetanic Count becomes a more appropriate mode of monitoring (Fig. 6.1).

Double Burst stimulation

The ability of the anaesthetist to assess the Train-of-Four ratio accurately using simple tactile or visual assessment of the responses is very limited. For example, it is unusual to be able to distinguish clinically between a true Train-of-Four ratio of 0.8 and a true Train-of-Four ratio of 0.4:[20] the former indicating adequate reversal and the latter inadequate reversal. It appears that the assessment of a ratio between the last and the first of four twitch responses is relatively unreliable without the assistance of a force or acceleration transducer, or electromyography. Double Burst stimulation (DBS) was developed in response to the need for a mode of repetitive nerve stimulation that would elicit approximately the same amount of fade as the Train-of-Four ratio for a given degree of neuromuscular blockade, but the extent of the fade would be measurable more accurately by simple tactile or visual assessment.[23] The pattern of stimulation comprises two, very short tetanic bursts (only three stimuli at 50 Hz) separated by 0.75 seconds (Fig. 6.4). This is known as DBS $_{3,3}$. The responses of the muscle are felt as two, large discrete twitches (fused tetani). During partial non-depolarizing blockade, the second twitch is less forceful than the first, and a comparison of the two responses can be expressed as the Double Burst ratio. Although

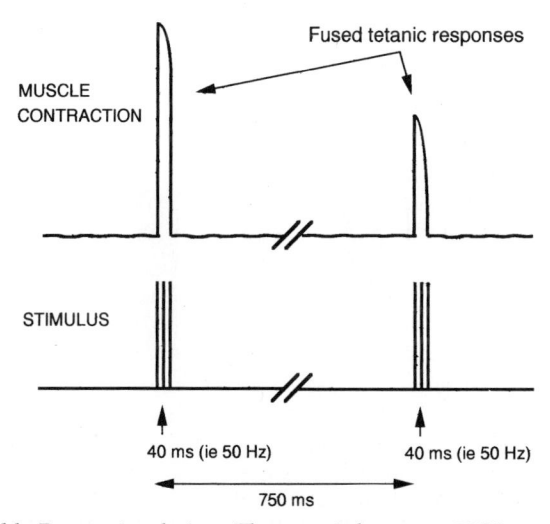

Fig. 6.4 Double Burst stimulation. The tetanic bursts at 50 Hz are separated by 0.75 seconds. The muscle response is detected as two, fused tetanic twitches. DBS$_{3,3}$ has three stimuli in each burst; DBS$_{3,2}$ has two stimuli in the second burst.

DBS was devised to improve the clinical detection of residual neuromuscular blockade, it has been suggested that DBS might be useful during the deeper blockade necessary for abdominal surgery. In this situation it is appropriate to maintain one DBS response and to deepen neuromuscular blockade when the second response reappears.[24]

Double Burst stimulation elicits a degree of fade that is almost identical to Train-of-Four stimulation; its advantage lies in the relative accuracy with which this fade is detectable by the palpating finger of the anaesthetist (Fig. 6.5). In comparative studies DBS $_{3,3}$ revealed tactile fade in almost all cases when the Train-of-Four ratio was below 0.5, whereas Train-of-Four fade was present in only three-quarters of the estimations even when the Train-of-Four ratio had fallen below 0.4.[25] DBS $_{3,3}$ represents a considerable improvement compared with the Train-of-Four ratio in routine, clinical use.

Changing the DBS pattern slightly so that there are only two stimuli in the second burst (DBS $_{3,2}$) increases the probability that tactile fade will be present when the Train-of-Four ratio is below 0.7. It has been suggested that fade in the twitch responses to DBS $_{3,2}$ may be present when the Train-of-Four ratio is above 0.7 (indicating satisfactory reversal).[23] No stimulation pattern is currently able to distinguish between satisfactory and unsatisfactory reversal using simple, tactile assessment of the force of muscular contraction with complete reliability.

Post-Tetanic Count

During profound neuromuscular blockade there is no twitch response to any of the patterns of nerve stimulation so far discussed. The Post-Tetanic

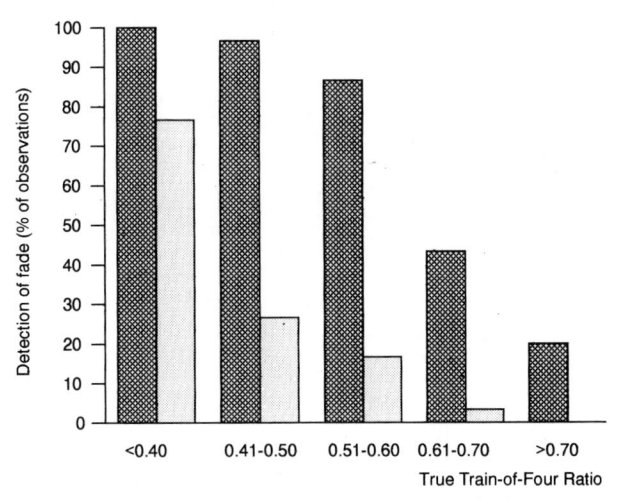

Fig. 6.5 A comparison of detectable fade in the Train-of-Four responses (lightly shaded bars) and the Double Burst stimulation (DBS) responses (heavily shaded bars) at varying depths of neuromuscular blockade. When the true Train-of-Four ratio is approximately 0.7, tactile Train-of-Four fade is detectable less than once in every ten attempts at detecting fade. The fade detection rate is improved to approximately five in every ten attempts using DBS. (From Drenk et al.[25])

Count (PTC) is the appropriate method of monitoring in this situation which might prevail shortly after the administration of a neuromuscular blocking drug or when the nature of the surgical procedure demands exceptionally deep non-depolarizing blockade.

When a tetanic stimulus has been applied to a motor nerve, the release of acetylcholine in response to subsequent low frequency stimulation is enhanced by post-tetanic facilitation. The phenomenon of post-tetanic facilitation may be employed in the measurement of profound blockade by applying a standardized tetanus followed by standardized stimuli at a low frequency. The responses to the low frequency stimuli will be enhanced by post-tetanic facilitation so that they become sufficiently forceful to be measured. Post-tetanic facilitation soon wanes so that the force of the post-tetanic twitches declines to zero. The stimulus pattern for the PTC is a 5-second tetanic burst (at 50 Hz) followed, after an interval of 3 seconds, by stimulation at 1 Hz until the responses have declined to zero.[26] The PTC is simply the number of measurable twitches following the tetanus (Fig. 6.6).

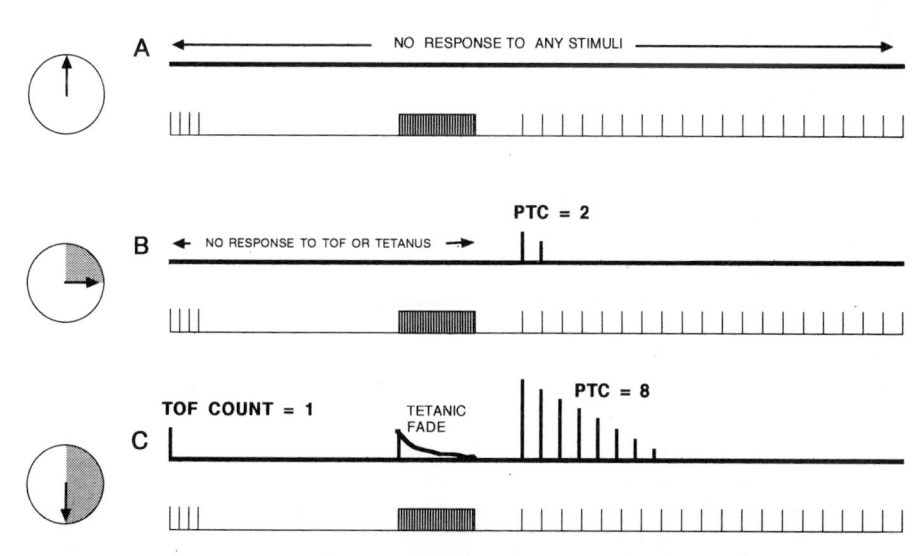

Fig. 6.6 Post-Tetanic Count (PTC) as a predictive measure of the rate of recovery from profound, atracurium-induced neuromuscular blockade. A. Extremely profound neuromuscular blockade: there is no twitch response to a Train-of-Four stimuli, tetanic stimuli for 5 seconds or subsequent single stimuli at 1 Hz. In this situation it would be useful to be able to predict the time at which the first twitch of the Train-of-Four can be expected to reappear. B. A period of 15 minutes has elapsed. There is still no response to a Train-of-Four stimuli or tetanic stimuli for 5 seconds. However, post-tetanic stimulation has enhanced the response to subsequent single stimuli and two, post-tetanic twitches are palpable. The PTC is 2. Previous experience indicates that the first twitch of the Train-of-Four can be expected to reappear in approximately 15 minutes when the PTC will be 8. C. A further 15 minutes has elapsed and the PTC has risen to 8. The first Train-of-Four twitch is, indeed, just palpable; i.e. the Train-of-Four Count is 1.

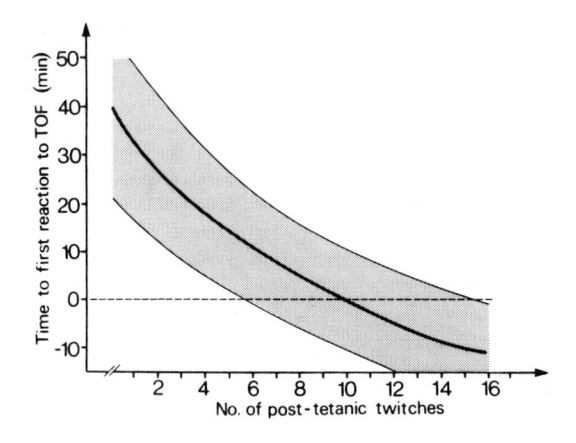

Fig. 6.7 A graph that can be used to predict the time at which the first twitch of the Train-of-Four can be expected to reappear by measuring the Post-Tetanic Count. This curve applies to profound, pancuronium-induced blockade. (From Viby-Mogensen et al.[26])

The PTC is able to quantify profound neuromuscular blockade and also to predict the duration of the period during which there is no response to Train-of-Four stimulation (Fig. 6.7). A low PTC is associated with a prolonged period before the first response of the Train-of-Four and, conversely, a greater PTC is associated with a shorter period. If tactile assessment is used for the PTC and the subsequent Train-of-Four responses, the predicted times (Table 6.1) are similar to those obtained using a force transducer.[27] There is a close linear relation between the time before the first response of the Train-of-Four becomes detectable and the square root of the PTC.[26] This relation can be expressed as $t = a + b\sqrt{PTC}$, where t is the time from a given PTC to the reappearance of the first response of the Train-of-Four, a is the intercept constant, and b is the slope of the graph: Time to first response vs PTC. These constants are characteristics of each individual muscle relaxant. Inhalational agents appear to increase the interval between the administration of the muscle relaxant drug and the appearance of the first post-tetanic

Table 6.1 The relation between the Post-Tetanic Count and the predicted time before the first response of the Train-of-Four can be expected to reappear

PTC	Pancuronium[26]	Atracurium[28]	Vecuronium[29]
1	35	9	16
2	25	7	14
4	20	4	10
6	12	2	6
8	5	0	4
10	0	—	2

twitch (i.e. PTC = 1). However, the subsequent progression through increasing numbers of post-tetanic twitches is only very slightly delayed.[28]

Post-Tetanic Count is the stimulation pattern of choice when there is a requirement for profound non-depolarizing neuromuscular blockade; for example, during microsurgery, craniotomy, or in situations of critically raised intracranial pressure in the Intensive Care Unit. During light halothane and vecuronium anaesthesia, the reappearance of the first Train-of-Four response is associated with a diaphragmatic response to carinal stimulation in approximately 50% of patients; the incidence is reduced to only 2% when the PTC = 2.[30] If it is necessary to reverse this profound degree of neuromuscular blockade, considerable spontaneous recovery of neuromuscular transmission must be permitted before neostigmine is given (Chapter 8). Adequate reversal with neostigmine from very intense blockade (PTC = 2) may take 30 minutes.[31] Tetanic stimulation enhances acetylcholine release for a considerable time and at least 5–6 minutes should be permitted to elapse between successive estimations of PTC.[26]

The optimum mode of stimulation

No single pattern of stimulation is suitable for monitoring all levels of neuromuscular blockade. Each of the methods described above is appropriate to part of the spectrum of neuromuscular blockade: from profound blockade to full recovery of neuromuscular transmission. The optimum pattern of stimulation is dependent on the method used to measure the twitch response. For example, a Train-of-Four ratio measured with a force transducer is appropriate at the same degree of blockade as a manually measured Double Burst ratio. This is better explained by referring to a diagram which indicates the usefulness of the various stimulation modes *when the response is assessed by the tactile or visual methods* (Fig. 6.1). Single twitch and tetanic stimulation offer no particular advantages and it can be argued that the Train-of-Four ratio can be replaced by the Double Burst ratio unless some form of instrument is employed to measure the response of the muscle. In the clinical situation, the most useful techniques, therefore, are Post-Tetanic Count, Train-of-Four Count and Double Burst Stimulation.

The technique of nerve stimulation

Stimulus current

When a brief current is applied to a motor nerve, the force of contraction of the muscle depends on the number of muscle fibres reaching the threshold for contraction; and this, in turn, depends on how many nerve fibres are depolarized sufficiently to generate an action potential. A small current, for example 10–20 mA, will activate only a small proportion of the motor units and will elicit a weak contraction. If the current is gradually increased, the force of contraction will also increase as more muscle fibres contract until the maximum force of contraction is reached. The current required to elicit the maximum force of contraction is the 'maximal current' and any current greater than this is 'supramaximal'. It is generally accepted that a supramaximal current should be used for monitoring neuromuscular transmission so that the entire range of motor fibres is monitored. A current of at least 65 mA is required to produce supramaximal stimulation in all patients.[32] Submaximal stimulation is less painful, and it has been suggested that a current of 20–30 mA may be used for Train-of-Four monitoring in the recovery room with reasonable accuracy.[11]

Stimulus duration

In addition to the current delivered, the quantity of electrical energy delivered to the nerve depends on the duration of the stimulating pulse. A pulse duration of 0.2 ms is commonly used both for research and for clinical monitoring. The pulse duration should be sufficiently great to deliver adequate energy to the nerve. However, under some circumstances a wide pulse, which exceeds the duration of the neuromuscular refractory period, may stimulate the nerve twice, leading to underestimation of the extent of neuromuscular blockade. If the second nerve action potential is generated 1–2 ms after the first, the muscle will not respond because the muscle is refractory. During partial, non-depolarizing neuromuscular blockade, a proportion of muscle fibres will not have contracted in response to the first stimulus due to the presence of the neuromuscular blocking drug. When the second stimulus arrives, the release of acetylcholine by the motor nerve terminal is enhanced (facilitation) and a greater number of muscle fibres will reach their depolarization threshold, resulting in a spuriously strong contraction.[33]

Electrode considerations

Surface, ECG-type electrodes are satisfactory and convenient for routine, clinical monitoring (Fig. 6.8). Paediatric or neonatal ECG electrodes may be preferable because the current density is greater over the nerve due to their smaller size. It is desirable, but not essential, to first degrease the skin with acetone or alcohol. The electrodes should be placed near the nerve, either parallel to, or straddling the nerve. If the electrodes are separated by more than a few centimetres, the positive electrode should be the proximal.[34] Some ECG electrodes are prone to evaporation of the electrolyte before use; a dry electrode is one cause of apparent complete failure of neuromuscular transmission. Needle electrodes may permit more exact nerve stimulation and are less likely to be subject to high impedance that will limit the output of some low power stimulators.[35] However, they are unnecessary if an adequate stimulator is used. Spherical 'ball-electrodes' are invariably associated with a variable, high impedance and cannot be recommended.

Nerve stimulators

Excessive complication should be avoided in a nerve stimulator designed for clinical use. It is desirable to equip every location where muscle relaxant

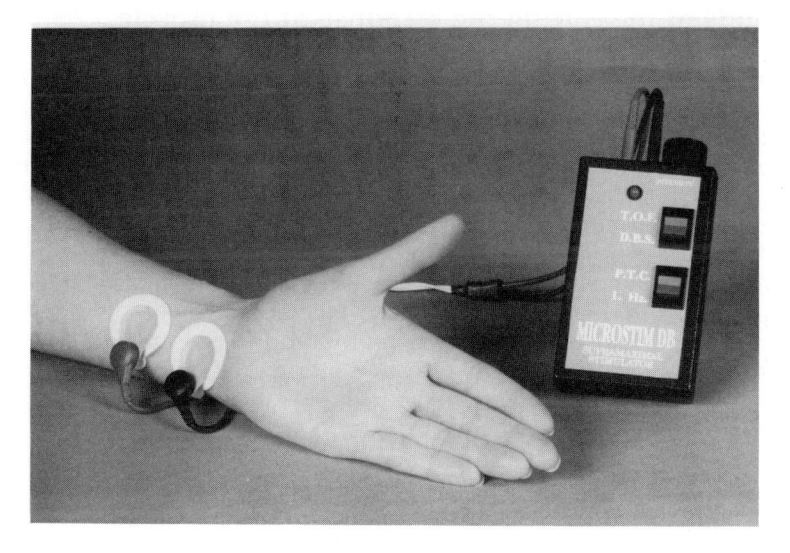

Fig. 6.8 A simple nerve stimulator providing Train-of-Four, Double Burst, tetanic and Post-Tetanic Count modes of stimulation. The ECG electrodes are placed over the ulnar nerve at the wrist.

drugs are given so that no patient is unmonitored. Cost is an important consideration, but it is desirable that a stimulator for clinical use should have certain features (Table 6.2). These should include Train-of-Four, Double Burst and Post-Tetanic Count. A stimulator designed for monitoring neuromuscular blockade should be capable of delivering at least 60 mA when connected to the patient. This is at least ten times the maximum current required when a needle is used to locate a nerve for a local anaesthetic nerve block, and if applied accidentally could damage the nerve. For this reason, 'dual-purpose' stimulators should be treated with circumspection and stimulators capable of a high output current should be used only with surface electrodes. A 'constant current' stimulator is less susceptible to changes in electrode impedance.

Table 6.2 Desirable features of a nerve stimulator for clinical use.

Simple controls
Variable output up to 60 mA (checked using a 1 kΩ test load)
Facility for Train-of-Four, Double Burst and Post-Tetanic Count
0.2 ms pulse duration
Long battery life and 'low-battery' indicator
Leads with ECG-type connectors, colour-coded for polarity

Methods of measuring the response of the muscle

The adductor pollicis muscle (the only muscle of the thenar eminence innervated by the ulnar nerve) is used commonly for routine, neuromuscular monitoring. Other muscles may be more appropriate in some situations (Table 6.3). The twitch response of the muscle can be measured in several ways. There is considerable variation between the various methods in the numerical values obtained and in their reproducibility. Three methods are in common use: measurement of the force of contraction of the muscle (mechanomyography), measurement of the associated acceleration (acceleromyography) and measurement of the electrical activity of the muscle (electromyography). Mechanomyography and electromyography have been used for many years in the research setting. The force of contraction of the muscle may be measured with a force transducer or by simple visual or tactile assessment of the twitch.

Visual and tactile assessment of the force of muscle contraction

Visual assessment of muscle contraction is possibly more related to the acceleration of the thumb than its force of contraction. Tactile assessment aims to emulate a force transducer. Visual or tactile assessment is inevitably less accurate than methods which use instrumentation. Using Train-of-Four stimulation, error is apparent (a) when assessing the return of the second, third and fourth twitches (Train-of-Four Count), (b) when assessing the Train-of-Four ratio, and (c) when assessing the force of contraction of the first twitch in the absence of a pre-relaxant control twitch. Visual or tactile assessment of the Train-of-Four ratio almost invariably underestimates the degree of fade and Double Burst stimulation appears to have considerable advantages in this context.

Direct muscle stimulation

During even profound neuromuscular blockade, electrical stimulation of a muscle will bypass the neuromuscular junction and produce a twitch response. This direct response may present problems of interpretation. When the ulnar nerve is stimulated at the wrist, direct stimulation of the superficial flexors of the forearm may occur, producing spurious movement of the hand. Under these circumstances a Train-of-Four stimuli will elicit four, small twitches of the hand even during profound neuromuscular blockade. Double Burst stimulation appears to cause greater direct muscle stimulation than Train-of-Four stimulation (personal observations). It is possible to minimize direct muscle stimulation by ensuring that the electrodes are accurately placed over the ulnar nerve, avoiding excessive stimulus current and positioning the positive electrode over the ulnar nerve at the elbow. Direct muscle stimulation is less likely to cause spurious underestimation of neuromuscular blockade if tactile, rather than visual assessment is used.

Visual assessment

Neuromuscular blockade measured by the Train-of-Four Count tends to be overestimated by visual assessment. When a force transducer is used to measure recovery from vecuronium blockade, the second twitch (T2) of the train reappears when there is 93% residual block, and the third twitch (T3) at 89% block.[36] If visual assessment is substituted, T2 reappears only when residual block has recovered to 84% (measured), and T3 reappears at 76% block. Thus, the reappearance of a second twitch may not be observed visually until the time when a force transducer detects two or three twitches. In addition, there is a greater variation between patients in the time taken for the twitches to reappear using visual assessment.[37]

Tactile assessment

Tactile assessment improves on the accuracy of visual assessment. In the absence of a documented pre-relaxant, control twitch, there is a tendency to overestimate the force of contraction of the first response of the Train-of-Four (T1) when non-instrumental techniques are used. Visual assessment makes this overestimation in approximately 30% of measurements and this can be reduced to approximately 10% if the tactile method is used.[38] Tetanic fade is detected reliably by the tactile method only when the tetanic fade ratio has fallen to less than 0.3 (measured with a force transducer).[12] If simple, clinical monitoring is used, tetanic stimulation appears to be no more sensitive to residual blockade than Train-of-Four stimulation.

Measurement with a force transducer

Measurement of the force of contraction of a muscle (mechanomyography) can be accomplished accurately only with a force transducer. The most commonly used technique is measurement of the force of adduction of the thumb (adductor pollicis) in response to ulnar nerve stimulation. The adductor pollicis is the only thenar muscle supplied by the ulnar nerve; because muscles vary in their sensitivity to relaxants, it is important to assess the response of a single muscle. For routine, clinical use a simple force transducer that uses the patient's vital-signs monitor to amplify and display the signal is adequate. Simple, visual or tactile assessment invariably underestimates the extent of Train-of-Four fade. For research applications, it is extremely important to stabilize the monitored hand and arm to minimize drift due to gradual displacement of the thumb in relation to the transducer. For this reason, mechanomyography is probably less appropriate than electromyography if movement of the patient is inevitable during the course of the monitoring period. Measurement should be isometric, i.e. very little movement of the thumb should be permitted. A tetanic contraction can generate a force of 2 kg and the apparatus should be accurate up to this level. The force of contraction varies in proportion to the preload applied to the thumb, within limits according to Starling's law. The optimum preload for the thumb appears to be in the range 200–300 g[39] and it is important to maintain this resting tension throughout the period of measurement. Drift may also occur when strain gauges in the transducer change their characteristics due to a heating effect; it is normal to arrange strain gauges in a Wheatstone-bridge configuration which minimizes this effect in addition to increasing sensitivity. The duration of a single muscle twitch is similar to the duration of the QRS complex of the electrocardiogram (50–100 ms) and the apparatus must have a similarly high frequency response, with sufficient damping to prevent overshoot. In common with all methods of measurement of neuromuscular transmission, it is important to insulate the limb against heat loss which can progressively reduce the force of muscle contraction.

Electromyography

The electrical activity of contracting muscles (EMG) has been used as an index of neuromuscular transmission for many years. In 1959, Churchill-Davidson and Christie used surface electrodes placed over the hypothenar eminence (abductor digiti minimi) in a wide-ranging study of the effects of different patterns of nerve stimulation on the EMG response during neuromuscular blockade, and the effects of anticholinesterases.[40] Since that time, EMG apparatus has developed such that electromyography is frequently more convenient that mechanomyography in the clinical setting. These two methods measure different aspects of neuromuscular function but the differences are relatively small and likely to be of significance only during neuromuscular research. Single-fibre electromyography is used for the diagnosis of neuromuscular disorders. For the purposes of measuring the extent of neuromuscular blockade during anaesthesia, the compound muscle action potential (CMAP) is recorded using a surface electrode positioned over the muscle.

The compound muscle action potential

Depolarization of a single muscle fibre membrane is an all-or-none phenomenon that results from a supra-threshold end-plate potential. The depolarization of an individual muscle fibre may be detectable by a fine, exploring needle which is insulated except at its tip as a single-fibre action potential. If a larger electrode is placed over the surface of the muscle, a CMAP is detected. The compound action potential comprises the vector of the summated electrical activity of the individual, single-fibre potentials in the direction of the surface recording electrode. A single-fibre action potential is a brief event in relation to the compound action potential but both exhibit the characteristic initial depolarization phase, followed by a repolarization phase (Fig. 6.9a). When a nerve is stimulated, the resulting electrical activity of the muscle, measured using a large electrode, is the evoked compound action potential (ECAP).

Although it is possible to measure the amplitude if the ECAP on a simple storage oscilloscope, newer electromyographs; for example, the Relaxograph®, designed for monitoring neuromuscular blockade convert the signal to a voltage by rectification and integration (Fig. 6.9a). The electrical artefact caused by the stimulus is excluded from this process and the voltage derived from the ECAP is displayed on a paper chart (Fig. 6.9b).

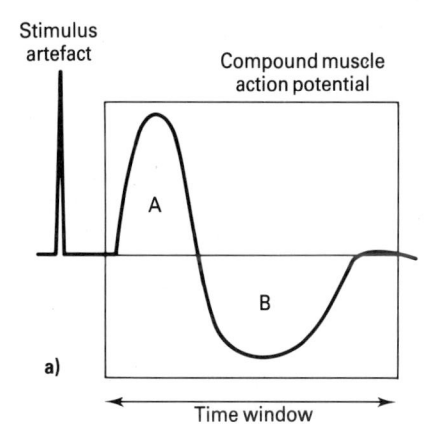

Fig. 6.9a The evoked compound muscle action potential (CMAP) showing the stimulus artefact, the depolarization phase (A) and the repolarization phase (B). The Relaxograph® integrates and rectifies the CMAP within a time window. The resulting voltage represents the total energy within the CMAP and is used to provide a display of the Train-of-Four ratio (as a percentage) and the percentage depression of the first response of the Train-of-Four compared with a pre-relaxant control value.

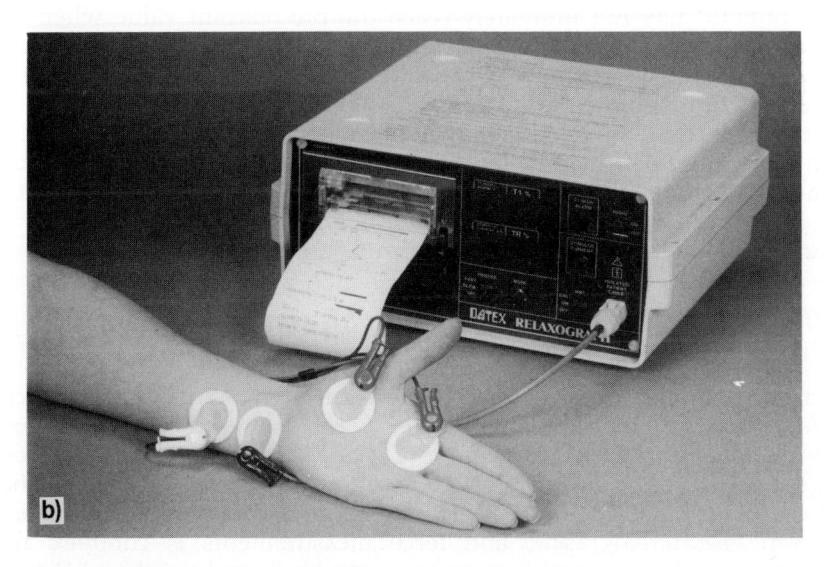

Fig. 6.9b The Datex Relaxograph® demonstrating satisfactory electrode positions for recording the electromyogram of the adductor pollicis muscle. The ground electrode is positioned on the dorsum of the hand.

In the absence of a neuromuscular blocking drug, and if the stimulation is supramaximal, all the muscle fibres reach their end-plate potential threshold. However, if neuromuscular transmission is impaired by the presence of a muscle relaxant drug, fewer fibres reach the threshold for generating a single-fibre action potential and the amplitude of the ECAP diminishes as the number of contributing fibres declines. The amplitude of both the depolarization phase and the repolarization phase is diminished, and the ECAP becomes slightly elongated[41] which may reflect increased variability in the synaptic delay at the neuromuscular junction[42] and is related to a change in the power spectrum of the ECAP to favour lower frequencies.[43]

The quality of the EMG signal is dependent on the care with which the recording electrodes are applied to the skin. It is important to degrease and slightly abrade the skin before applying the electrodes. It is often recommended that the ground electrode should be placed between the stimulating electrodes and the recording electrodes; this is less important with equipment that excludes the stimulus artefact electronically. The active recording electrode should be placed over the motor-point of the muscle where the electrical activity is maximal, and the indifferent electrode over a nearby bony point. A muscle should be chosen which gives a large EMG signal and this should be checked in the anaesthetized patient before calibrating the signal and administering a muscle relaxant. Some EMG equipment is able to operate in an uncalibrated mode in which the Train-of-Four ratio is reasonably accurate but it is not possible to compare the first response of the Train-of-Four with a pre-relaxant value.

It is relatively common for the EMG signal to become attenuated during a period of monitoring. The Train-of-Four ratio is unaffected but the absolute magnitude of the responses is diminished so that the first response of the Train-of-Four may not ultimately reach the pre-relaxant value when the muscle relaxant drug is no longer acting. The drift is probably multifactorial and appears to have reached its maximum approximately 10 minutes after the period of monitoring has begun.[44] A change in the impedance of the recording electrodes with time may not be responsible.[45] It is possible that both central and peripheral effects of general anaesthesia might contribute to the initial reduction in the integrated EMG signal. However, it is unlikely that this effect is mediated via the motor nerve.[46] Drift is also less likely to occur if the monitored hand is immobilized.[47]

Differences between electromyographic and mechanometric measurements

Electromyography measures only the electrical component of neuromuscular function. EMG measurements therefore reflect events occurring before and at the neuromuscular junction, but changes in the contractile process are not measured. However, electromyography has the advantage that muscles not amenable to force measurement can be monitored. The relation between electromyographic and force measurements is complex and depends upon the nature of the neuromuscular blocking drug, the inhalational anaesthetic agent and the nature of the reversal agent. In the absence of a neuromuscular blocking drug, post-tetanic facilitation of the force of the twitch response can be observed but there is no corresponding increase in the amplitude of the ECAP.[48] During voluntary contraction of a muscle, the integrated EMG increases with the force of contraction. If the initial fibre length is increased, the force of a maximal effort increases (Starling's law); however, this is not matched by a concomitant increase in the integrated EMG.[49]

The ECAP is depressed more rapidly by suxamethonium than the force of contraction and recovers more slowly. After the administration of suxamethonium, there is a characteristic brief increase in the twitch response which is not seen in the ECAP. The force of the twitch response frequently recovers to greater than the pre-suxamethonium value for a period of approximately 10 minutes after the ECAP has recovered to 100%.[8,50] The initial increase in contractile force is associated with a change in the shape of the ECAP suggesting repetitive firing of the muscle fibres.[51]

After the administration of a non-depolarizing neuromuscular blocking drug, the force of contraction is depressed more rapidly than the ECAP and takes longer to recover. When comparing methods of recording it is important to measure the responses of the same muscle. When the ECAP and the force of contraction of the adductor pollicis are measured simultaneously, the relation between the two methods appears to be drug specific, the discrepancy being considerably greater after alcuronium and tubocurarine than atracurium.[8,52] The difference between the methods is less pronounced during reversal with neostigmine.[52]

Accelerometry

This method uses Newton's second law (force = mass × acceleration) to measure the force of contraction of the thumb indirectly.[53] A small acceleration transducer containing a piezo-electric ceramic wafer is taped to the pad of the thumb. When the thumb moves, a voltage is generated that is proportional to the acceleration of the transducer. Providing that the hand is positioned so that the thumb is able to move freely, accelerometric measurements of neuromuscular blockade appear to correlate well with mechanometric measurements,[54] although accelerometry tends to underestimate the degree of blockade during onset and overestimate the degree of blockade during recovery.[55] A characteristic of accelerometry is that the first response of the Train-of-Four ratio, in the absence of neuromuscular blockade, is frequently greater than the subsequent three responses. The mechanism is obscure. Accelerometry appears to be unsuitable for Double Burst monitoring, which, in any case is more suited to visual or tactile assessment of the muscle response. In common with all methods of monitoring, accelerometry requires a period of stabilization before consistent recordings can be made. The device is available commercially as the TOF-Guard® (Fig. 6.10) which has the advantage of automated measurement of the Post-Tetanic Count, in addition to the Train-of-Four ratio.

Indications for monitoring specific muscles

Many different muscles are accessible for neuromuscular monitoring (Table 6.3) but there is considerable variation between muscles in their sensitivity

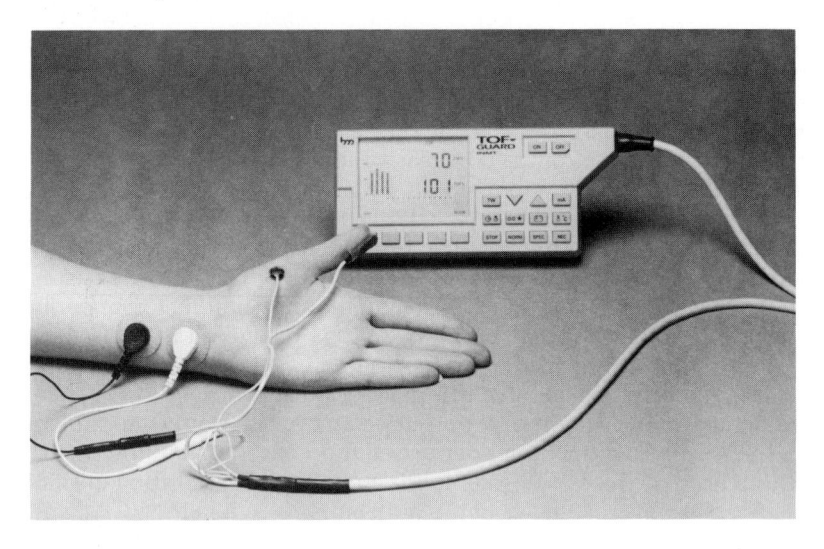

Fig. 6.10 Acceleromyography of the adductor pollicis. The TOF-Guard® employs an acceleration transducer fastened to the thumb (which must be free to move). Train-of-Four and Post-Tetanic Count modes are provided.

Table 6.3 Muscles (and their nerves) which may be used in monitoring neuromuscular blockade during anaesthesia

---

Muscles of the hand (ulnar nerve)
    Adductor pollicis
    First dorsal interosseous
    Abductor digiti minimi + other hypothenar muscles

    Abductor + flexor pollicis brevis + opponens (*median nerve*)

Muscles of the face (facial nerve)
    Orbicularis oculi
    Orbicularis oris
    Frontalis

Muscles of the leg and foot
    Gastrocnemius (post. tibial n.)
    Soleus (post. tibial n., med. tibial n.)
    Long flexors of the foot (post. tibial n.)
    Flexor hallucis brevis (med. plantar n.)

Abdominal muscles (intercostal nerves T7–T12)
    External oblique
    Rectus abdominis

Diaphragm (phrenic nerve)
    Crural fibres
    Costal fibres
Laryngeal adductors (recurrent laryngeal nerve)

---

to muscle relaxant drugs. It is well known that the diaphragm is relatively resistant to neuromuscular blockade, but it is not generally appreciated that the extent of the difference in sensitivity between two muscles is not constant and depends on the neuromuscular blocking agent that is administered to the patient. For example, the duration of effect of pancuronium at the adductor pollicis is approximately twice that at the diaphragm. If tubocurarine is substituted, the corresponding ratio is 3:1.[56] In clinical practice, it is not always possible to monitor the muscles which are most significant; for example, the masseter and laryngeal muscles for endotracheal intubation, the rectus abdominis during abdominal surgery, or the muscles of ventilation in the Intensive Care Unit. However, it is possible to establish by research the individual sensitivities of these muscles and how they relate to muscles which are easily monitored, e.g. the adductor pollicis and the facial muscles, and to use this information in clinical practice.

The muscles of the hand

The adductor pollicis is well suited to mechanomyography, electromyography and acceleromyography, and is usually accessible during anaesthesia (Fig. 6.11a). In the majority of individuals, the adductor pollicis is the only thenar muscle innervated by the ulnar nerve. In addition, it receives innervation from the median nerve in a proportion of individuals. Hand muscles other than the adductor pollicis may be used for electromyographic monitoring. The *first dorsal interosseous* muscle (Fig. 6.11b) gives a large surface EMG[57] which appears to be less vulnerable to movement artefact and otherwise appears to give identical results to the adductor pollicis.[58] The ECAP obtained over the *hypothenar eminence* (Fig. 6.11c) is also used for monitoring.[40] This site has the disadvantage of monitoring simultaneously several dissimilar muscles which may vary in their sensitivity to muscle relaxants. During recovery from non-depolarizing neuromuscular blockade, the hypothenar EMG recovers more rapidly than that of the adductor pollicis EMG.[8]

Abdominal muscles

The control of abdominal tone by manipulating neuromuscular blockade during abdominal surgery is not without difficulty. It is unfortunate that it is difficult to monitor the degree of paralysis of these muscles during abdominal surgery. During light general anaesthesia with spontaneous respiration, the activity of the abdominal muscles varies reciprocally with that of the diaphragm, increasing during expiration and decreasing during inspiration. This phasic activity can be demonstrated electromyographically and is superimposed on the normal tonic contraction of these muscles. Mechanical hyperventilation abolishes rhythmic EMG activity both in the diaphragm and in the abdominal wall muscles, but some tonic EMG activity persists.[59] Increasing the depth of anaesthesia also decreases the abdominal wall spontaneous EMG markedly,[60] which effectively precludes its use for monitoring neuromuscular transmission during anaesthesia.

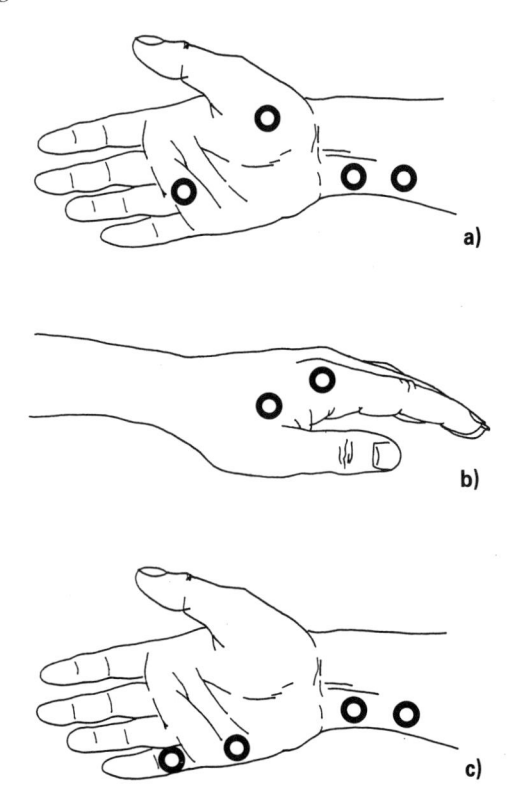

Fig. 6.11 Suitable electrode positions for electromyography of the (a) adductor pollicis, (b) first dorsal interosseous, and (c) hypothenar muscles. The active recording electrode must be positioned carefully, preferably over the motor-point of the muscle. The indifferent electrode is positioned over a bony (non-muscular) area and the ground electrode can be placed over the dorsum of the hand or between the stimulating and the recording electrodes.

Tonic contraction of the abdominal wall muscles (abdominal wall tension) has been studied using an abdominal retractor equipped with a sterile strain gauge force transducer. As neuromuscular blockade deepens, abdominal wall tension declines to a minimum before adductor pollicis blockade has exceeded approximately 80% depression of the control. This observation suggests that there is little value in maintaining profound neuromuscular blockade (greater than 80–90% twitch depression) for abdominal surgery.[61] The evoked EMG of the rectus abdominis has been studied after a single dose of atracurium during isoflurane anaesthesia, by stimulating the tenth intercostal nerve and recording the EMG from the skin overlying the long axis of the muscle. The onset of neuromuscular blockade is more rapid and recovery occurs more rapidly at the rectus than at the hypothenar muscles.[62] Further investigation is required to establish the relative sensitivity of these muscles during steady state neuromuscular blockade.

The diaphragm

It is an everyday observation that spontaneous ventilation recovers more rapidly than peripheral muscle function during the offset of neuromuscular blockade. The diaphragm is resistant to both depolarizing and non-depolarizing neuromuscular blockade. Every muscle has a considerable excess of cholinoceptors at the neuromuscular junction; this represents the 'margin of safety'. In a limb muscle, neuromuscular transmission is normal until approximately 75% of receptors are blocked.[6] The diaphragm has a greater margin of safety than peripheral muscles and, in laboratory studies, requires only half the number of unblocked receptors for neuromuscular transmission to proceed normally.[63] In man, when the adductor pollicis is 90% blocked by pancuronium, the EMG response of the diaphragm is reduced by only approximately 25%.[64] However, factors other than the number of receptors determine the pattern of the response of a muscle to a neuromuscular blocking drug. The relative blood supply to the muscle is very important in determining how rapidly the drug is conveyed to the neuromuscular junction, and the richness of diaphragmatic muscle blood flow might explain the rapidity with which the diaphragm becomes paralysed after the administration of a muscle relaxant drug. After pancuronium 0.08 mg kg^{-1}, maximum blockade of the diaphragm occurs in approximately 1 minute, 30% faster than the adductor pollicis. Spontaneous recovery is also more rapid in the diaphragm. The time taken to recover the first response of the Train-of-Four in the diaphragm is approximately half that of the adductor pollicis.[56]

Neuromuscular transmission in the diaphragm may be measured by stimulating one phrenic nerve using ECG electrodes at the neck, and measuring the EMG response using ECG electrodes in the eighth intercostal space (costal fibres), or using an oesophageal electrode (crural fibres). Alternatively, the force of contraction of the diaphragm may be measured indirectly using transdiaphragmatic pressure or inspiratory pressure. The diaphragm is not a homogeneous muscle. However, the costal fibres and the crural fibres appear to be affected equally by neuromuscular blockade in animal studies.[65]

The larynx

Although there is unlikely to be a requirement for monitoring these muscles in clinical practice, it is possible to measure neuromuscular blockade at the laryngeal adductor muscles for research purposes. Neuromuscular blockade of the laryngeal adductors may be measured by stimulating the recurrent laryngeal nerve using ECG electrodes placed over the thyroid notch and recording the resultant pressure changes in the partially inflated cuff of an endotracheal tube placed between the vocal cords.[66] Alternatively, electromyography of the vocal cords with needle electrodes has been used in the experimental situation.[67] The introduction of rocuronium has highlighted the prospect of a non-depolarizing neuromuscular blocking agent which permits early endotracheal intubation. Measurement of the onset time at the larynx is part of the evaluation of such drugs. The laryngeal muscles form only one of several muscle groups which must be relaxed before laryngoscopy and endotracheal intubation can be accomplished easily. Ideal conditions for endotracheal intubation can be obtained only if the masseter, muscles of the neck and diaphragm are blocked. In common with the diaphragm, depolarizing and non-depolarizing neuromuscular blockade of the laryngeal adductors is more rapid, both in onset and in recovery, than the adductor pollicis. However, the extent of the blockade is less intense. Suxamethonium 0.5 mg kg^{-1} produces maximum blockade of the laryngeal adductors in approximately 1 minute, and maximum blockade of the adductor pollicis in 1.7 minutes.[68] Although intubating conditions may be satisfactory before the adductor pollicis has become completely paralysed, ideal intubating conditions, suitable, for example, in a penetrating eye injury, may take a little longer to become established.[69] After rocuronium 0.5 mg kg^{-1}, maximum blockade occurs after 1.4 minutes at the larynx and 2.4 minutes at the adductor pollicis.[70] In some clinical situations, for example, where there is critical elevation of the intracranial pressure, any cough response to passing an endotracheal tube may be disastrous. In these circumstances, complete disappearance of both Double Burst responses at the adductor pollicis may be used as an indicator of optimum intubating conditions.[71]

Facial muscles

If the limbs are inaccessible to neuromuscular monitoring during anaesthesia, it may be convenient to monitor the response of the facial muscles to surface stimulation of the facial nerve. The stimulating electrodes are positioned just anterior to the tragus of the ear (Fig. 6.12). The facial muscles

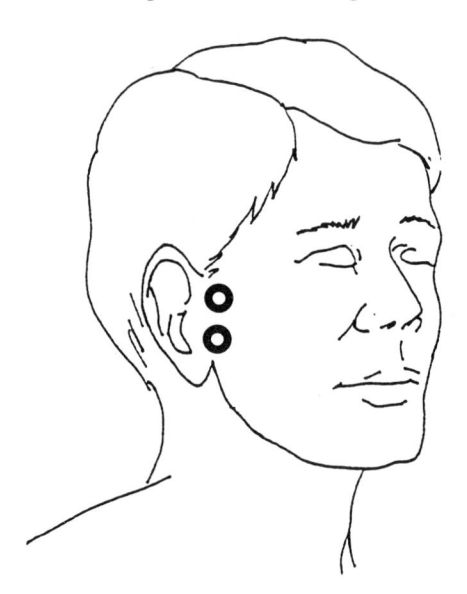

Fig. 6.12 Suitable electrode positions for monitoring the twitch response of the facial muscles by stimulating the facial nerve anterior to the ear.

appear to behave as 'central' muscles rather than 'peripheral muscles'; in comparison with the adductor pollicis, neuromuscular blockade is more rapid in onset[72] and more rapid in offset.[73] It is important to remember that the facial muscles will underestimate the extent of neuromuscular blockade in comparison with the adductor pollicis so that the unwary might be tempted to give an inadequate dose of relaxant. The extent of neuromuscular blockade at the facial muscles probably reflects the situation at the diaphragm with reasonable accuracy. However, full recovery of neuromuscular function at the diaphragm occurs earlier that in the limb muscles; and, if a patient is awakened as soon as the facial muscles demonstrate full Train-of-Four recovery, there may be significant residual blockade in other muscles and an unpleasant sensation of weakness.

Muscles of the leg

It may be more convenient to monitor the response of the muscles of the leg to nerve stimulation during some surgical procedures. The published work is limited but suggests that the sensitivity of the gastrocnemius muscle is similar to that of the adductor pollicis.[74] In a volunteer study, the volitional strength of the soleus muscle (predominantly slow fibres) was diminished to a lesser extent than the gastrocnemius muscle by a small dose of tubocurarine.[75] This observation supports the hypothesis that the predominant muscle fibre type is one determinant of the sensitivity of a muscle to muscle relaxants.

Stimulation posterior to the medial malleolus at the ankle depolarizes the posterior tibial nerve and its medial plantar division and elicits contraction of the short flexors of the foot and plantar-flexion of the great toe (flexor hallucis brevis) which is suitable for tactile assessment by the anaesthetist (Fig. 6.13). There are no clinically significant differences between the flexor hallucis brevis and the adductor pollicis in the pattern of neuromuscular blockade using Train-of-Four stimulation.[76] If the posterior tibial nerve is stimulated in the popliteal fossa, a satisfactory EMG can be measured over the gastrocnemius and soleus muscles in the calf.

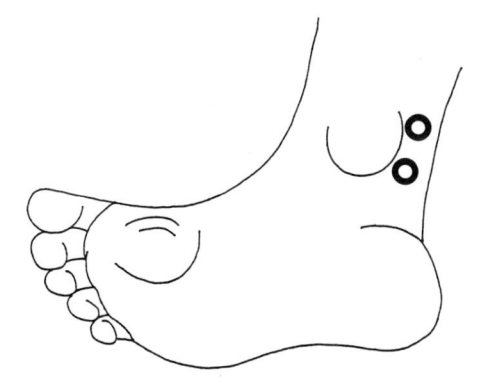

Fig. 6.13 Suitable electrode positions for monitoring the twitch response of the great toe by stimulating the posterior tibial nerve and its medial plantar branch.

Muscle fibre types

Muscle fibres can be classified in two ways: first, according to their ATPase activity and hence the maximum rate of cross-bridge formation (slow, type 1 or fast, type 2); and second, according to their method of synthesizing ATP (oxidative or glycolytic). Fast fibres may be oxidative (type 2a) or glycolytic (type 2b). During muscle contraction, slow fibres are recruited first, followed by fast-oxidative, then fast-glycolytic fibres as the force of contraction increases. Slow fibres fatigue slowly and comprise the greater proportion of the muscles of posture; for example, almost 90% of the soleus muscle fibres are type 1. In contrast, muscles with fast, phasic activity have a high type 2 content. The orbicularis oculi, responsible for blinking, is composed of 85% type 2 fibres in man.[77]

There is some evidence that the proportion of fast and slow muscle fibres in a muscle may influence its response to muscle relaxant drugs. The orbicularis oculi (15% type 1 fibres) is more resistant to atracurium than the adductor pollicis (80% type 1 fibres).[72] The gastrocnemius muscle (54% type 1 fibres) is more resistant than the soleus muscle (71% type 1 fibres) to tubocurarine,[75, 78] and pancuronium.[78] Clearly, many other factors are responsible for the vulnerability of an individual muscle to neuromuscular blockade; the gastrocnemius and adductor pollicis exhibit similar sensitivities to atracurium despite a considerable difference in their muscle fibre composition.[74] Recent work suggests that there is a relation between the diameter of type 1 muscle fibres and their sensitivity to both depolarizing and non-depolarizing neuromuscular blocking drugs in such a way that sensitivity increases with muscle fibre size.[79]

References

1. Viby Mogensen J, Jorgensen BC, Ording H. Residual curarisation in the recovery room. *Anesthesiology* 1979; **50:** 539–541.
2. Beemer GH, Rozental P. Postoperative neuromuscular function. *Anaesthesia and Intensive Care* 1986; **14:** 41–45.
3. Ali HH, Wilson RS, Savarese JJ, Kitz RJ. The effect of tubocurarine on indirectly elicited train-of-four muscle response and respiratory measurements in humans. *British Journal of Anaesthesia* 1975; **47:** 570–574.
4. Eriksson LI, Sato M, Severinghaus JW. Effect of vecuronium-induced partial neuromuscular block on hypoxic ventilatory response. *Anesthesiology* 1993; **78:** 693–699.
5. Bevan DR, Smith CE, Donati F. Postoperative neuromuscular blockade: a comparison between atracurium, vecuronium and pancuronium. *Anesthesiology* 1988; **69:** 272–276.
6. Paton WDM, Waud DR. The margin of safety of neuromuscular transmission. *Journal of Physiology* 1967; **191:** 59–90.
7. Glavinovic MI. Presynaptic action of curare. *Journal of Physiology* (Lond) 1979; **290:** 499–506.
8. Katz RL. Electromyographic and mechanical effects of suxamethonium and tubocurarine on twitch, tetanic and post-tetanic responses. *British Journal of Anaesthesia* 1973; **45:** 849–859.

9. Ali HH, Savarese JJ. Stimulus frequency and dose-response curve to d-tubocurarine in man. *Anesthesiology* 1980; **52:** 36–39.
10. Stanec A, Heyduk J, Stanec G, et al. Tetanic fade and post-tetanic tension in the absence of neuromuscular blocking agents in anesthetized man. *Anesthesia and Analgesia* (Cleve) 1978; **57:** 102–107.
11. Brull SJ, Ehrenwerth J, Silverman DG. Stimulation with submaximal current for train-of-four monitoring. *Anesthesiology* 1990; **72:** 629–632.
12. Dupuis JY, Martin R, Tessonnier JM, Tetrault JP. Clinical assessment of the muscular response to tetanic nerve stimulation. *Canadian Journal of Anaesthesia* 1990; **37:** 397–400.
13. Kopman AF, Epstein RH, Flashburg MH. Use of 100-Hertz tetanus as an index of recovery from pancuronium-induced nondepolarizing neuromuscular blockade. *Anesthesia and Analgesia* 1982; **61:** 439–441.
14. Ali HH, Utting JE, Gray C. Stimulus frequency in the detection of neuromuscular block in humans. *British Journal of Anaesthesia* 1970; **42:** 967–977.
15. Ali HH, Utting JE, Gray TC. Quantitative assessment of residual antidepolarizing block (Part I). *British Journal of Anaesthesia* 1971; **43:** 473–476.
16. Ali HH, Utting JE, Gray TC. Quantitative assessment of residual antidepolarizing block (Part II). *British Journal of Anaesthesia* 1971; **43:** 478–485.
17. Lee CM. Train-of-4 quantitation of competitive neuromuscular block. *Anesthesia and Analgesia* 1975; **54:** 649–653.
18. Williams NE, Webb SN, Calvey TN. Differential effects of myoneural blocking drugs on neuromuscular transmission. *British Journal of Anaesthesia* 1980; **52:** 1111–1114.
19. Brand JB, Cullen DJ, Wilson NE, Ali HH. Spontaneous recovery from nondepolarizing neuromuscular blockade: correlation between clinical and evoked responses. *Anesthesia and Analgesia* 1977; **56:** 55–58.
20. Viby-Mogensen J, Jensen NH, Engbaek J, Ording H, Skovgaard LT, Stat C, et al. Tactile and visual evaluation of the response to train-of-four nerve stimulation. *Anesthesiology* 1985; **63:** 440–443.
21. DeJong RH. Controlled relaxation. 1. Quantitation of electromyogram with abdominal relaxation. *Journal of the American Medical Association* 1966; **197:** 113–115.
22. Ham J, Redpath JH. Comparative monitoring of non-depolarizing neuromuscular blockade in humans. *Anesthesia and Analgesia* 1985; **64:** 225.
23. Engbaek J, Ostergaard D, Viby-Mogensen J. Double burst stimulation (DBS): a new pattern of nerve stimulation to identify residual neuromuscular block. *British Journal of Anaesthesia* 1989; **62:** 274–278.
24. Braude N, Vyvyan HAL, Jordan MJ. Intraoperative assessment of atracurium-induced neuromuscular block using double-burst stimulation. *British Journal of Anaesthesia* 1991; **67:** 574–578.
25. Drenk NE, Ueda N, Olsen NV, et al. Manual evaluation of residual curarization using double burst stimulation: a comparison with train-of-four. *Anesthesiology* 1989; **70:** 578–581.
26. Viby-Mogensen J, Howardy-Hansen P, Chraemmer-Jorgensen B, et al. Post-tetanic count (PTC): a new method of evaluating an intense nondepolarizing neuromuscular blockade. *Anesthesiology* 1981; **55:** 458–461.
27. Howardy-Hansen P, Viby-Mogensen J, Gottschau A, Skovgaard LT, Chraemmer-Jorgensen B, Engbaek J. Tactile evaluation of posttetanic count (PTC). *Anesthesiology* 1984; **60:** 372–374.
28. Bonsu AK, Viby-Mogensen J, Fernando PUE, Muchhal K, Tamilarasan A, Lambourne A. Relationship of post-tetanic count and train-of-four response during intense neuromuscular blockade caused by atracurium. *British Journal of Anaesthesia* 1987; **59:** 1089–1092.

29. Eriksson LI, Lennmarken C, Staun P, Viby-Mogensen J. Use of post-tetanic count in assessment of a repetitive vecuronium-induced neuromuscular block. *British Journal of Anaesthesia* 1990; **65:** 487–493.
30. Fernando PUE, Viby-Mogensen J, Bonsu AK, Tamilarasan A, Muchhal KK, Lambourne A. Relationship between posttetanic count and response to carinal stimulation during vecuronium-induced neuromuscular blockade. *Acta Anaesthesiologica Scandinavica* 1987; **31:** 593–596.
31. Engbaek J, Østergaard D, Skovgaard LT, Viby-Mogensen J. Reversal of intense neuromuscular blockade following infusion of atracurium. *Anesthesiology* 1990; **72:** 803–806.
32. Kopman AF, Lawson D. Milliamperage requirements for supramaximal stimulation of the ulnar nerve with surface electrodes. *Anesthesiology* 1984; **61:** 83–85.
33. Berry FR. Detection of neuromuscular block in man. *British Journal of Anaesthesia* 1966; **38:** 929–935.
34. Berger JJ, Gravenstein JS, Munson ES. Electrode polarity and peripheral nerve stimulation. *Anesthesiology* 1982; **56:** 402–404.
35. Caplan LM, Satyanarayana T, Patel KP, Turndorf H, Ramanathan S. Assessment of neuromuscular blockade with surface electrodes. *Anaesthesia and Analgesia* 1981; **60:** 244–245.
36. O'Hara DA, Fragen RJ, Shanks CA. Reappearance of the train-of-four after neuromuscular blockade induced with tubocurarine, vecuronium or atracurium. *British Journal of Anaesthesia* 1986; **58:** 1296–1299.
37. O'Hara DA, Fragen RJ, Shanks CA. Comparison of visual and measured train-of-four recovery after vecuronium-induced neuromuscular blockade using two anaesthetic techniques. *British Journal of Anaesthesia* 1986; **58:** 1300–1302.
38. Tammisto T, Wirtavuori K, Linko K. Assessment of neuromuscular block: comparison of three clinical methods and evoked electromyography. *European Journal of Anaesthesia* 1988; **5:** 1–8.
39. Donlon JV, Savarese JJ, Ali HH. Cumulative dose-response curves for gallamine: effect of altered resting thumb tension and mode of stimulation. *Anesthesia and Analgesia* 1979; **58:** 377–391.
40. Churchill-Davidson HC, Christie TH. The diagnosis of neuromuscular blockade in man. *British Journal of Anaesthesia* 1959; **31:** 290–301.
41. Pugh ND, Harper NJN, Healy TEJ, Petts HV. Effects of atracurium on the latency and the duration of the negative deflection of the evoked compound action potential of the adductor pollicis. *British Journal of Anaesthesia* 1987; **59:** 195–199.
42. Ekstedt J, Stålberg E. The effect of non-paralytic doses of d-tubocurarine on individual motor end-plates in man, studied with a new electrophysiological method. *Electroencephalography and Clinical Neurophysiology* 1969; **27:** 557–562.
43. Harper NJN, Pugh ND, Healy TEJ, Petts HV. Changes in the power spectrum of the evoked compound potential of the adductor pollicis with the onset of neuromuscular blockade. *British Journal of Anaesthesia* 1987; **59:** 200–205.
44. Meretoja OA, Brown TCK. Drift of the thenar EMG signal. *Anesthesiology* 1989; **71:** A824.
45. Smith DC. Recording electrode impedance has no effect on evoked EMG responses. *British Journal of Anaesthesia* 1992; **69:** 535P.
46. Smith DC. Central enhancement of evoked electromyographic monitoring of neuromuscular function. *British Journal of Anaesthesia* 1991; **66:** 562–565.
47. Kosek PS, Sears DH, Ruberstein EH. Minimizing movement-induced changes in twitch response during integrated electromyography. *Anesthesiology* 1988; **69:** 142–143.
48. Botelho SY. Comparison of simultaneously recorded electrical and mechanical activity in myasthenia gravis patients and in partially curarized humans. *American Journal of Medicine* 1955; **19:** 693–696.

49. Inman VT, Ralston HJ, Saunders JB de CM, Feinstein B, Wright EW. Relation of human electromyogram to muscular tension. *Electroencephalography and Clinical Neurophysiology* 1952; **4:** 187–194.
50. Shanks CA, Jarvis JE. Electromyographic and mechanical twitch responses following suxamethonium administration. *Anaesthesia and Intensive Care* 1980; **8:** 341–344.
51. Donati F, Bevan DR. Muscle electromechanical correlations during succinylcholine infusions. *Anesthesia and Analgesia* 1984; **63:** 206.
52. Harper NJN, Bradshaw EG, Healy TEJ. Evoked electromyographic and mechanical responses of the adductor pollicis compared during the onset of neuromuscular blockade by atracurium or alcuronium, and during antagonism by neostigmine. *British Journal of Anaesthesia* 1986; **58:** 1278–1284.
53. Viby-Mogensen J, Jensen E, Werner M, Kirkegaard Nielsen H. Measurement of acceleration: a new method of monitoring neuromuscular function. *Acta Anaesthesiologica Scandinavica* 1988; **32:** 45–48.
54. Jensen E, Viby-Mogensen J, Bang U. The Accelograph®: a new neuromuscular transmission monitor. *Acta Anaesthesiologica Scandinavica* 1988; **32:** 49–52.
55. Harper NJN, Martlew R, Strang T, Wallace M. Monitoring neuromuscular blockade by acceleromyography: a comparison of the Mini-Accelograph with the Myograph 2000. *British Journal of Anaesthesia* 1994; **72:** 411–414.
56. Derrington MC, Hindocha N. Comparison of neuromuscular block in the diaphragm and hand after administration of tubocurarine, pancuronium and alcuronium. *British Journal of Anaesthesia* 1990; **64:** 294–299.
57. Kalli I. Effect of surface electrode position on the compound action potential evoked by ulnar nerve stimulation during isoflurane anaesthesia. *British Journal of Anaesthesia* 1990; **65:** 494–499.
58. Harper NJN. Comparison of the adductor pollicis and the first dorsal interosseous muscles during atracurium and vecuronium blockade: an electromyographic study. *British Journal of Anaesthesia* 1988; **61:** 477–478.
59. Fink BR. Electromyography in general anaesthesia. *British Journal of Anaesthesia* 1961; **33:** 555–559.
60. Katz RL. Comparison of electrical and mechanical recording of spontaneous and evoked muscle activity. *Anesthesiology* 1965; **26:** 204–211.
61. Weber S, Muravchick S, DeFeo SP, Rosato EF. Correlation of evoked twitch response to abdominal wall tension during surgery. *Anesthesiology* 1985; **63:** A325.
62. Saddler JM, Marks LF, Norman J. Comparison of atracurium-induced neuromuscular block in rectus abdominis and hand muscles of man. *British Journal of Anaesthesia* 1992; **69:** 26–28.
63. Waud BE, Waud DR. The margin of safety of neuromuscular transmission in the muscle of the diaphragm. *Anesthesiology* 1972; **37:** 417–422.
64. Donati F, Antzaka C, Bevan DR. Potency of pancuronium at the diaphragm and the adductor pollicis muscle in humans. *Anesthesiology* 1986; **65:** 1–5.
65. Goldman E, Road J, Grassino A. Recovery of costal and crural diaphragmatic contractility from partial paralysis. *Anesthesiology* 1991; **75:** 123–129.
66. Donati F, Plaud B, Meistelman C. A method to measure elicited contraction of laryngeal adductor muscles during anaesthesia. *Anesthesiology* 1991; **74:** 827–832.
67. Gilly H, Redi G, Werba A, *et al.* Pharmacodynamics of vecuronium in two muscle groups: vocal cords versus thenar neuromuscular blockade in man. *Anesthesiology* 1987; **67:** A614.
68. Meistelman C, Plaud B, Donati F. Neuromuscular effects of succinylcholine on the vocal cords and adductor pollicis muscles. *Anesthesia and Analgesia* 1991; **73:** 278–282.

69. Bencini A, Newton DEF. Rate of onset of good intubating conditions, respiratory depression and hand muscle paralysis after vecuronium. *British Journal of Anaesthesia* 1984; **56:** 959–965.
70. Meistelman C, Plaud B, Donati F. Rocuronium (ORG 9426) neuromuscular blockade at the adductor muscles of the larynx and adductor pollicis in humans. *Canadian Journal of Anaesthesia* 1992; **39:** 665–669.
71. Ueda N, Muteki T, Tsuda H, *et al.* Determining the optimal time for endotracheal intubation during the onset of neuromuscular blockade. *European Journal of Anaesthesiology* 1993; **10:** 3–8.
72. Harper NJN, Wilson A. Onset of atracurium-induced blockade in the muscles of the face: orbicularis oculi and adductor pollicis compared. *British Journal of Anaesthesia* 1987; **59:** 932P–933P.
73. Caffrey R, Warren ML, Becker KE. Neuromuscular blockade monitoring comparing the orbicularis oculi and adductor pollicis muscles. *Anesthesiology* 1986; **65:** 95–97.
74. Harper NJN, Wilson A. Characteristics of atracurium block in the gastrocnemius muscle. *British Journal of Anaesthesia* 1989; **63:** 240P–241P.
75. Secher NH, Rube N, Secher O. Effect of tubocurarine on human soleus and gastrocnemius muscles. *Acta Anaesthesiologica Scandinavica* 1982; **26:** 231–234.
76. Sopher MJ, Sears D, Walts L. Neuromuscular function monitoring comparing the flexor hallucis brevis and adductor pollicis muscles. *Anesthesiology* 1988; **69:** 129–131.
77. Johnson MA, Polgar J, Weightman D, Appleton D. Data on the distribution of fibre types in thirty-six human muscles: an autopsy study. *Journal of the Neurological Sciences* 1973; **18:** 111–129.
78. Day NS, Blake GJ, Standaert FG, Dretchen KL. Characterization of the Train-of-Four response in fast and slow muscles: effect of d-tubocurarine, pancuronium, and vecuronium. *Anesthesiology* 1983; **58:** 414–417.
79. Ibebunjo C, Hall LW. Muscle fibre diameter and sensitivity to neuromuscular blocking drugs. *British Journal of Anaesthesia* 1993; **71:** 732–733.

7

Techniques of Administering Muscle Relaxants

François Donati

Muscle relaxants are used to (1) facilitate endotracheal intubation, (2) provide muscle relaxation for surgical procedures and (3) enable ventilatory support. For all these applications, the basic requirements are the same: (1) a rapid onset of paralysis, (2) duration of blockade which accurately matches the requirements and (3) a rapid recovery. There are many practical means which meet these requirements, and the method chosen will depend on the clinical circumstances, the skill and experience of the physician, and the pharmacological properties of the available drugs.

Clinically, the duration of required paralysis varies widely. Some procedures require muscle relaxation for only a few minutes, whereas this duration can be as long as several hours or sometimes several days. In certain circumstances, it is difficult to foresee the exact duration of a procedure. Thus, a certain element of flexibility is needed. It is possible to antagonize neuromuscular blockade and restore neuromuscular function with anticholinesterase drugs. However, these reversal drugs are effective only if some degree of spontaneous recovery is manifest.[1] In other words, reversal of neuromuscular blockade is possible only if some degree of spontaneous recovery is present.

Thus, it is essential to base administration of muscle relaxants such that a certain degree of spontaneous recovery will be present at the end of the procedure for which relaxants are indicated. Since recovery occurs when plasma concentration of the muscle relaxant drug decreases below a certain level, it is important to know the pharmacokinetics of the relaxants.

Pharmacokinetics

Muscle relaxants can be classified according to their duration of action into short-, intermediate- and long-acting drugs (Table 7.1). For most drugs, the duration of action correlates well with the elimination half-life, i.e. the time it takes for plasma concentration to decrease by one half during the elimination

Table 7.1 Pharmacokinetics of various agents

	Drug	Elimination half-life (min)	Duration depends on
Short-acting	Suxamethonium	2–4	Elimination
	Mivacurium	3–6	Elimination
Intermediate-acting	Atracurium	20	Elimination
	Vecuronium	60–120	Redistribution
	Rocuronium	60–120	Redistribution
Long-acting	d-Tubocurarine	90–180	Elimination
	Gallamine	120	Elimination
	Alcuronium	60	Elimination
	Pancuronium	90–120	Elimination
	Metocurine	120	Elimination
	Doxacurium	60–90	Elimination
	Pipecuronium	60–120	Elimination

phase. However, there are a few exceptions. For example vecuronium and rocuronium have a very important redistribution phase, and plasma concentrations decrease because the drug is redistributed.[2,3] Thus, termination of effect of vecuronium or rocuronium normally depends upon redistribution rather than elimination, unless redistribution sites are saturated as might be the case with high doses and/or prolonged administration.

Factors affecting response of the patient

The response to relaxants can be modified by many factors. Some of these factors pertain only to one drug and are treated elsewhere. For example, suxamethonium and mivacurium metabolism is slowed in patients with abnormal plasma cholinesterase.[4] Drugs can affect the response to relaxants, and the most clinically relevant interactions are produced by other relaxants and inhalational anaesthetic agents (Chapter 10).

The response to relaxants also depends on the patient. Neuromuscular relaxants are polar, ionized compounds which have a volume of distribution approximately equal to that of extracellular fluid.[5] Thus, the dose is normally adjusted according to the weight of the patient, assuming that the extracellular fluid volume changes in proportion to body weight. This assumption is not always true. For example, obese individuals have less extracellular fluid volume, expressed as a percentage of total body weight. Thus, dosage of relaxants should be calculated on an ideal body weight basis.

Children have a greater extracellular fluid volume, kg^{-1} body weight, than adults, and require slightly greater doses, expressed as $mg\ kg^{-1}$.[6] Infants have a still greater proportion of extracellular fluid, but in this case their neuromuscular junction is particularly sensitive to relaxants, so a smaller concentration is required for the same effect, when compared with adults. As a result, infants require approximately the same dose, in mg kg^{-1} as adults (Chapter 11).[7]

Elderly patients require the same dose as younger adults to establish neuromuscular blockade, but the duration of action is usually prolonged because the function of organs of elimination (liver and kidney) is reduced in the elderly (Chapter 9).[8] Similarly, the duration of neuromuscular effect is expected to be prolonged in patients with liver and/or kidney disease. These changes are especially important for drugs which depend on hepatic or renal function for their elimination.

Requirements for muscle paralysis

The degree of muscle relaxation required depends on many factors, which are related to the patient, and the type of procedure performed. Moreover, the requirements for muscle relaxation vary with time within a same procedure.

For example, large, muscular individuals usually require more relaxant than frail, old or malnourished patients. Abdominal surgery and orthopaedic procedures are usually performed with a greater degree of muscle relaxation than superficial operations. Also, requirements may change with time. Tracheal intubation, normally performed before surgery starts, requires very intense neuromuscular blockade. The requirements for relaxation during abdominal surgery are usually fewer. The degree of relaxation may increase again for peritoneal closure, and no muscle paralysis may be needed for skin closure.

Management of muscle relaxation

To meet the requirements for muscle relaxation a variety of drugs with different durations of action can be administered by either bolus or infusion. It is also possible to use a fast-onset, short-acting drug at the beginning to provide adequate conditions for tracheal intubation, and then to switch to a longer-acting drug. At present, the only drug with fast-onset and short-duration characteristics is suxamethonium which produces onset of paralysis within 1.0–1.5 minutes, and a duration of action of 5–10 minutes.[9] At present, no non-depolarizing agent has a comparable onset and all these drugs have a longer duration of action. Thus, if a non-depolarizing drug is given as a bolus at induction of anaesthesia, adequate conditions for intubation might not be present for 2–4 minutes after injection, a longer interval than suxamethonium (1–1.5 minutes).

The other problem associated with the administration of a non-depolarizing drug is its duration of action. One must be careful not to produce intense paralysis in excess of the expected duration of the surgery. On the other hand, the administration of too small a dose might be associated with inadequate paralysis to perform tracheal intubation. The other major problem associated with the administration of large doses of non-depolarizing drugs arises when the patient cannot be intubated or ventilated. Although this event is rare, it is catastrophic. Thus, suxamethonium is recommended for emergency surgery and in cases where a difficult

Table 7.2 Methods of administration

	Duration of procedure			
	Short (< 30 min)	Intermediate (30–120 min)	Long (2–4 h)	Very long (> 4 h)
Loading dose				
Suxamethonium	X	X	X	X
Mivacurium	X	X		
Intermediate-acting drugs		X	X	X
Long-acting drugs				X
Maintenance				
Suxamethonium infusion	X			
Mivacurium infusion	X	X		
Intermediate-acting drug bolus or infusion		X	X	
Long-acting drug bolus			X	X

An X indicates that the use of the drug may be indicated or appropriate.

intubation is a possibility. If tracheal intubation is unsuccessful, the patient can be allowed to wake up and an alternate plan of management considered. Provided that the patient has no contraindications to suxamethonium, this depolarizing agent may be given at the beginning of the case, irrespective of its duration (Table 7.2).

Loading dose

The initial dose of relaxant is normally given immediately after induction of anaesthesia, and should be large enough to provide adequate relaxation to facilitate tracheal intubation. Laryngoscopy and intubation usually require relaxation of the vocal cords and respiratory muscles, which are known to be resistant to the effect of muscle relaxants.[10] Thus, a relatively large dose must be given (Table 7.3). Usually, a dose equivalent to at least twice the ED95 is required, the ED95 being the dose which corresponds to 95% neuromuscular blockade at the adductor pollicis muscle. This standard was chosen because blockade at the adductor pollicis can be measured easily. The drug is chosen according to its duration of action. Cardiovascular side-effects must also be taken into consideration. As a rule, variability in the duration of action is greater with long-acting drugs.

Maintenance dose

If suxamethonium was used to facilitate tracheal intubation, maintenance of relaxation may be achieved by switching to a non-depolarizing drug. In this case, it is necessary to give a loading dose of the non-depolarizing agent, because the effect of suxamethonium wears off rapidly. However,

Table 7.3 Drugs for intubation

Drug	ED95 (mg kg^{-1})	Typical intubating doses (mg kg^{-1})	Onset time of intubating doses (min)	Duration of intubating doses (min)
Suxamethonium	0.3–0.5	1.0–1.5	1.0–1.5	8–12
Mivacurium	0.08	0.15–0.2	3.0–4.0	20–25
Atracurium	0.20–0.25	0.5–0.6	3.0–4.0	40–50
Vecuronium	0.05	0.1–0.15	3.0–4.0	40–50
Rocuronium	0.3–0.4	0.6	2.0–3.0	40–50
Pancuronium	0.07	0.1–0.15	3.0–5.0	60–180
Doxacurium	0.025	0.04–0.05	7.0–15.0	60–180

this loading dose should be considerably less than the 2 × ED95 which would have been required if the non-depolarizer had been chosen for intubation. This is because surgical relaxation is normally adequate when 95% block is reached, i.e. with an ED95 dose, and also because suxamethonium increases the effect of non-depolarizing drugs given subsequently.[11]

Maintenance of relaxation can be approached by either injecting repeated bolus doses or by giving an infusion. The frequency and size of bolus doses depends on the pharmacokinetics of the drug and its potency. However, an important principle must be kept in mind: repeat doses have a much greater effect, and duration of action is considerably greater than loading doses of the same size. This is because the 'therapeutic window' for muscle relaxants is very narrow. Animal experiments have shown that neuromuscular blockade is not detectable until approximately 75% of receptors are occupied. When 92% of receptors are bound to a relaxant molecule, blockade is complete.[12] Surgical relaxation is adequate when 75–95% of twitch height at the adductor pollicis is depressed. This probably corresponds to 85–90% receptor occupancy range. Thus, the loading dose should be large enough to increase receptor occupancy from zero to more than 90%. However, when additional doses are given, a large proportion of receptors remain occupied, and the patient might require only enough relaxant drug to increase receptor occupancy, say, from 85 to 90%. As a result, only small doses should be given, especially for intermediate- and long-acting drugs. Typical top-up doses do not exceed 0.25 × ED95 (Table 7.4). Because of their short duration of action, suxamethonium and mivacurium are not often given by repeat injections. The duration of action of additional doses of intermediate-acting agents is shorter (10–20 minutes) than that of long-acting agents (20–45 minutes).

For drugs which are not eliminated in the plasma, i.e. all drugs except suxamethonium, mivacurium and atracurium, there is a tendency for the duration of action of successive doses to increase. This phenomenon, called cumulation, is more important in the case of long-acting drugs.[13]

Infusions

An increasingly popular method of drug administration for maintenance of muscle relaxation is the infusion technique. The relaxant is given via a pump

Table 7.4 Drugs for maintenance

Drug	ED95	Typical incremental doses (mg kg^{-1})	Function of incremental dose (min)	Typical infusion dose (mg kg^{-1} min^{-1})
Suxamethonium	0.3–0.5	—	—	50–100
Mivacurium	0.08	—	—	5–10
Atracurium	0.2–0.25	0.05–0.10	10–20	3–7
Vecuronium	0.05	0.01–0.02	10–20	0.5–1.0
Rocuronium	0.3–0.4	0.1	10–20	5–10
Pancuronium	0.07	0.01–0.02	20–45	1
d-Tubocurarine	0.4	0.05	20–45	—
Doxacurium	0.025	0.005	20–45	—

and the rate of infusion is adjusted to achieve the desired level of paralysis. Neuromuscular blockade should be monitored frequently and the infusion rate should be modified accordingly.[5, 14] Only drugs which are cleared rapidly from plasma are suitable for administration by infusion. Long-acting agents do not belong to this category. Once a stable level of relaxation has been reached, infusion rates of mivacurium or atracurium require little adjustment, unless high concentrations of potent vapours (enflurane or isoflurane) are given.[14] Vecuronium requirement tends to decrease because of accumulation of the drug in the body.[14] Suxamethonium infusion rates need to be increased to produce the same effect. This phenomenon, called tachyphylaxis, which occurs 30–60 minutes after the start of suxamethonium administration, makes frequent adjustments in the rate of infusion necessary.[15] For this reason, suxamethonium infusions are usually limited to procedures less than 30 minutes in duration (Chapter 4).

Monitoring is essential in the management of relaxation (Chapter 6). Surgical relaxation is normally sufficient when the response to Train-of-Four stimulation yields one to two discernible twitches, which corresponds to 80–95% blockade. Recovery after continuous infusions might, in certain instances, be longer than after bolus doses. The recovery from suxamethonium is prolonged as time increases. Prolonged infusions of vecuronium have also been associated with an increase in recovery time. Recovery from mivacurium and atracurium are usually time independent.[14, 16]

Future trends

Recently, there has been a tendency to use shorter and shorter-acting drugs, even for long procedures. The reason for this preference is a greater flexibility in the intensity of neuromuscular blockade which can be obtained at any time. The only short-acting drug whose popularity has decreased is suxamethonium, which is avoided because of its side-effects, and because infusion rate and type of blockade change with time. As the duration of action of available drugs decreases, it becomes more convenient to give them by infusion instead of bolus doses.

Such rapid changes in the level of relaxation make the use of monitoring mandatory. Devices which record neuromuscular blockade and feed back the information to an infusion device to adjust drug administration have been proposed. It is possible that some of these computer-controlled devices might be available commercially in the near future. With or without the use of sophisticated pumps and computer-controlled devices, the clinician will be the person responsible for a safe and smooth management of neuromuscular blockade.

References

1. Rupp SM, McChristian JW, Miller RD, Taboada JA, Cronnelly R. Neostigmine and edrophonium antagonism of varying intensity neuromuscular blockade induced by atracurium, pancuronium, or vecuronium. *Anesthesiology* 1986; **64:** 711–717.
2. Miller RD, Rupp SM, Fisher DM, Cronnelly R, Fahey MR, Sohn YJ. Clinical pharmacology of vecuronium and atracurium. *Anesthesiology* 1984; **61:** 444–453.
3. Mark J. Wierda KH, Kleef UW, Lambalk LM, Kloppenburg WD, Agoston S. The pharmacodynamics and pharmacokinetics of Org 9426, a new non-depolarizing neuromuscular blocking agent, in patients anaesthetized with nitrous oxide, halothane and fentanyl. *Canadian Journal of Anaesthesia* 1991; **38:** 430–435.
4. Ostergaard D, Jensen FS, Jensen E, Skovgaard LT, Viby-Mogensen J. Influence of plasma cholinesterase activity on recovery from mivacurium-induced neuromuscular blockade in phenotypically normal patients. *Acta Anaesthesiologica Scandinavica* 1992; **36:** 702–706.
5. Shanks CA. Pharmacokinetics of the non-depolarizing neuromuscular relaxants applied to calculation of bolus and infusion dosage regimens. *Anesthesiology* 1986; **64:** 72–86.
6. Meretoja OA, Wirtavuori K, Neuvonen PJ. Age-dependence of the dose-response curve of vecuronium in pediatric patients during balanced anesthesia. *Anesthesia and Analgesia* 1988; **67:** 721–26.
7. Meretoja OA. Neuromuscular blocking agents in paediatric patients: influence of age on the response. *Anaesthesia and Intensive Care* 1990; **18:** 440–448.
8. Koscielniak-Nielsen ZJ, Law-Min JC, Donati F, Bevan DR, Clement P, Wise R. Dose-response relations of doxacurium and its reversal with neostigmine in young adults and healthy elderly patients. *Anesthesia and Analgesia* 1992; **74:** 845–850.
9. Vanlinthout LEH, Egmond JV, De Boo T, Lerou JGC, Wevers RA, Booij LHD. Factors affecting magnitude and time course of neuromuscular block produced by suxamethonium. *British Journal of Anaesthesia* 1992; **69:** 29–35.
10. Donati F, Meistelmans C, Plaud B. Vecuronium neuromuscular blockade at the adductor muscles of the larynx and adductor pollicis. *Anesthesiology* 1991; **74:** 833–837.
11. Donati F, Gill SS, Bevan DR, Ducharme J, Theoret Y, Varin F. Pharmacokinetics and pharmacodynamics of atracurium with and without previous suxamethonium administration. *British Journal of Anaesthesia* 1991; **66:** 557–561.
12. Paton WDM, Waud DR. The margin of safety of neuromuscular transmission. *Journal of Physiology* 1967; **191:** 59–90.
13. Fahey MR, Morris RB, Miller RD, Sohn YJ, Cronnelly R, Gencarelli P. Clinical pharmacology of ORG NC45 (Norcuron™): a new non-depolarizing muscle relaxant. *Anesthesiology* 1981; **55:** 6–11.

14. Martineau RJ, St-Jean B, Kitts JB, Curran MC, Lindsay P, Hull KA, et al. Cumulation and reversal with prolonged infusions of atracurium and vecuronium. *Canadian Journal of Anaesthesia* 1992; **39:** 670–676.
15. Donati F, Bevan DR. Long-term succinylcholine infusion during isoflurane anesthesia. *Anesthesiology* 1983; **58:** 6–10.
16. Diefenbach C, Mellinghoff H, Lynch J, Buzello W. Mivacurium: dose-response relationship and administration by repeated injection or infusion. *Anesthesia and Analgesia* 1992; **74:** 420–423.

8

Reversal of Neuromuscular Blockade

NJN Harper

Introduction

This chapter describes the physiology of acetylcholinesterase at the neuromuscular junction, the pharmacology of the antagonism of neuromuscular blockade in the normal individual and in disease and, finally, sets out some suggested guidelines for clinical practice.

The role of acetylcholine and acetylcholinesterase in normal neuromuscular transmission

The action of acetylcholinesterase at the neuromuscular junction

Before considering the effects of anticholinesterases it is useful to review the function of the enzyme acetylcholinesterase during normal neuromuscular transmission. When each of the two binding sites on the cholinoceptor (one on each alpha subunit) is occupied by a molecule of acetylcholine, the ion channel opens. Sodium and calcium ions enter the post-synaptic muscle membrane and potassium ions leave. The current generated by this torrential (but unequal) movement of ions depolarizes the adjacent membrane. The current continues to flow until the channel closes. The simultaneous opening of many thousands of ion channels at each neuromuscular junction depolarizes a large area of the post-synaptic membrane to the extent that an end-plate potential is generated. Normally the end-plate potential reaches threshold and an action potential is triggered which spreads over the surface of the muscle fibre.

Mechanism of inactivation of acetylcholine

The enzyme acetylcholinesterase is distributed on the post-synaptic membrane and, to a lesser extent, on the pre-synaptic membrane where it may reduce the impact of released acetylcholine on the prejunctional receptors. It is a large molecule with two active sites. The negatively charged *anionic site* binds the positively charged nitrogen atom in acetylcholine. The *esteratic site* forms a bond with the acetate component of the acetylcholine molecule which initiates cleavage at the ester linkage, releasing choline. The acetylated enzyme then combines with water to release acetic acid, regenerating itself in the process. This process is extremely rapid and efficient: only approximately 50% of the released acetylcholine molecules successfully penetrate the acetylcholinesterase barrier to reach the cholinoceptors and, of these, very few are able to combine with the alpha sites on more than two, brief occasions. When sufficient acetylcholine has been hydrolysed to favour closure of a relatively small proportion of the ion channels (there is a considerable 'margin of safety'), repolarization of the muscle membrane begins to restore the resting transmembrane potential. Diffusion of acetylcholine from the synaptic cleft also occurs but is of minor importance. Acetylcholine bound to the cholinoceptors is in equilibrium with free acetylcholine in the synaptic cleft. However, restoration of the normal transmembrane potential is probably limited by the rate of dissociation of acetylcholine from the alpha subunits rather than the rate of hydrolysis of free acetylcholine by acetylcholinesterase.

Anticholinesterase drugs

Action at the neuromuscular junction

There are at least four possible mechanisms whereby anticholinesterases can influence neuromuscular transmission. Inactivation of acetylcholinesterase is the most important mechanism by a large margin.

Inactivation of acetylcholinesterase

Anticholinesterase drugs are able to reverse neuromuscular blockade by temporarily inactivating acetylcholinesterase: this is the most important of the possible mechanisms. A greater proportion of the released acetylcholine is then available to interact with cholinoceptors and, at any moment in time, it becomes more probable that the alpha subunits are being occupied by acetylcholine rather than being occupied by the non-depolarizing muscle relaxant drug. It is important to appreciate that the increase in ambient acetylcholine concentration affects not only the post-synaptic receptors, but also the pre-synaptic receptors that are concerned with the positive-feedback modulation of acetylcholine release from the nerve terminal, so that each nerve impulse releases a greater quantity of acetylcholine into the synaptic cleft (Chapter 1). There are two mechanisms whereby anticholinesterase drugs inhibit the action of acetylcholinesterase. Neostigmine and pyridostigmine combine with both the anionic and esteratic sites in the acetylcholinesterase molecule. Their quaternary ammonium group binds with the anionic site whilst their carbamate group binds with the esteratic site and inactivates the enzyme by carbamylation. Reactivation by decarbamylation is a relatively slow process during which acetylcholine is free to exist unmolested in the synaptic cleft. Edrophonium combines only with the esteratic site on the acetylcholinesterase molecule, forming a loose, electrostatic bond. There is competition between acetylcholine and edrophonium for this esteratic site and the action of edrophonium is terminated as the drug simply diffuses away from the synaptic cleft into the extracellular fluid.

Modification of the depolarization of the motor nerve terminals

In vitro studies have demonstrated that neostigmine and edrophonium may change the response of the motor nerve to a single electrical stimulus so that repetitive nerve action potentials are briefly observed and the twitch response of the muscle is more forceful because the single stimulus has been converted into a tetanic burst. The likely mechanisms are as follows:

1. The nerve terminal, unprotected from the effects of acetylcholine because local acetylcholinesterase has been inactivated, is exposed to a relative excess of acetylcholine each time a nerve action potential reaches it.
2. The high, local concentration of acetylcholine augments the negative afterpotential that normally occurs when the arrival of a nerve action potential depolarizes the nerve terminal.
3. The duration of the large afterpotential exceeds the refractory period of the adjacent nerve membrane in a particularly susceptible nerve terminal, triggering impulses which travel antidromically (i.e. cephalad) and become amplified by generating action potentials in many more nerve terminals. This effect occurs partially at spinal cord level and partially by axon reflex to the remaining nerve terminals in the motor unit.
4. The short burst of repetitive nerve action potentials elicits repetitive muscle action potentials which are manifest as a short, fused, tetanic contraction which is more forceful than a single twitch.

Increased release of acetylcholine

In low concentrations, neostigmine or edrophonium appear to have a direct effect on the motor nerve terminal, which may enhance the release of acetylcholine. (NB Suxamethonium in low concentration is also able to enhance muscle contraction by causing repetitive nerve action potentials.)

Direct effect at post-synaptic cholinoceptors

Neostigmine exhibits complex direct effects at the post-synaptic muscle membrane and it is unclear whether they have any clinical importance. *In vitro*, the close arterial injection of high concentrations of neostigmine in a preparation in which acetylcholinesterase is absent produces muscle contraction, possibly by depolarizing the end-plate.

Action at other sites

Cardiac side-effects

Anticholinesterases increase the concentration of acetylcholine at both nicotinic and muscarinic receptors. Parasympathetic fibres are distributed to the sinoatrial and atrioventricular nodes and Purkinje tissue. Stimulation of cardiac muscarinic receptors classically produces bradycardia and slowing of conduction. In addition, the refractory period is increased in the atrioventricular node and Purkinje fibres. A negative inotropic effect of neostigmine with atropine might be expected, but this has not been demonstrated.[1] Numerous arrhythmias have been described in association with the administration of neostigmine with atropine; various techniques have been suggested to minimize the problem and administration over 3 minutes attenuates the changes in cardiac rate.[2] If the administration of atropine precedes neostigmine, the incidence of cardiac dysrhythmia is increased and this practice cannot be supported.[3]

Neostigmine and pyridostigmine require the administration of atropine in a similar dose ratio but a smaller dose ratio may be used safely with edrophonium (atropine 7 $\mu g\,kg^{-1}$ + edrophonium 0.5 $mg\,kg^{-1}$).[4] An initial, transient tachycardia (2–4 minutes) is almost invariable when sufficient atropine is administered with neostigmine to avoid bradycardia throughout the reversal period. This undesirable effect is minimized if neostigmine is administered with glycopyrrolate (neostigmine 50 $\mu g\,kg^{-1}$ + glycopyrrolate 10 $\mu g\,kg^{-1}$)[5] or edrophonium is administered with atropine (Table 8.1).[6] Glycopyrrolate is less rapidly acting than edrophonium and their simultaneous administration is associated with an initial bradycardia unless an excess of glycopyrrolate is given so that a persistent tachycardia develops after 4–5 minutes.[6] Pyridostigmine is more slowly acting than neostigmine. Administration with glycopyrrolate causes less initial tachycardia than atropine and protection against subsequent bradycardia appears to be superior.[7]

Alimentary side-effects

Muscarinic effects on the bowel include increase in tone and peristaltic activity. The increase in motility is more pronounced in the proximal than

Table 8.1 The cardiac effects of various combinations of anticholinesterases and anticholinergic agents when given to reverse neuromuscular blockade

Anticholinesterase	Anticholinergic	Effect
Neostigmine (0.05 mg kg^{-1})	Atropine (20 $\mu g\,kg^{-1}$)	Initial tachycardia
Neostigmine (0.05 mg kg^{-1})	Glycopyrrolate (10 $\mu g\,kg^{-1}$)	Cardiostable
Edrophonium (0.5 mg kg^{-1})	Atropine (10 $\mu g\,kg^{-1}$)	Cardiostable
Edrophonium (0.5 mg kg^{-1})	Glycopyrrolate (5 $\mu g\,kg^{-1}$)	Initial bradycardia
Edrophonium (0.5 mg kg^{-1})	Glycopyrrolate (10 $\mu g\,kg^{-1}$)	Delayed tachycardia
Pyridostigmine (0.25 mg kg^{-1})	Atropine (20 $\mu g\,kg^{-1}$)	Tachy then brady
Pyridostigmine (0.25 mg kg^{-1})	Glycopyrrolate (10 $\mu g\,kg^{-1}$)	Slight initial tachycardia

the distal gut. It has been suggested that the incidence of anastomotic breakdown may be increased in those patients who have received neostigmine to reverse neuromuscular blockade after colonic surgery.[8, 9] However, there is little convincing evidence and further investigation is needed before the use of anticholinesterases should be discouraged in these circumstances. Furthermore, it is possible that adequate and rapid recovery of neuromuscular function might be jeopardized if reliance were to be placed on spontaneous recovery of muscle function unless an infusion of a short-acting relaxant, e.g. mivacurium, were being used. The antisialagogue activity of glycopyrrolate is four to five times as potent as atropine and this action is apparent at a dose that has little effect on heart rate.

Pulmonary side-effects

Stimulation of muscarinic receptors in bronchial smooth muscle might be expected to result in bronchoconstriction and the avoidance of neostigmine in asthmatic patients has been suggested. However, in non-asthmatic patients, measurement of specific airways conductance has so far demonstrated only small changes. Nonetheless, the time course and direction of these changes corresponded with the changes observed in heart rate. Atropine was associated with an initial rise in airways conductance which was not observed after glycopyrrolate. There is no published evidence that indicates that anticholinesterases should be avoided in asthmatic patients.[10]

Factors which influence the effectiveness of reversal

The following factors are important in determining the quality of the reversal of neuromuscular blockade.

> Method of measurement of the return of neuromuscular function.
> Choice of neuromuscular blocking drug.
> Choice of antagonist.
> Dose of antagonist.
> Intensity of neuromuscular blockade.
> Plasma concentration of relaxant.
> Anaesthetic agent.
> Organ function.
> Age.
> Drug interactions.
> Electrolyte and acid-base balance.
> Plasma cholinesterase concentration (mivacurium).

Method of measurement of the return of neuromuscular function

It is generally accepted that the optimum index of returning neuromuscular function is the Train-of-Four ratio. However, in the absence of an accurate method of measuring the response of the muscle, the insensitivity of the visual and tactile methods of assessing the twitch response unless the Train-of-Four ratio is less than 0.4–0.5, necessitate using Double Burst stimulation to be certain that there is no significant residual neuromuscular blockade before discontinuing anaesthesia. The patient can be said to be fully reversed when there is no Double Burst fade. The single twitch response may not improve at the same rate as the Train-of-Four ratio. In comparison with the single twitch ratio, the Train-of-Four ratio initially accelerates more rapidly after edrophonium[11, 12] but lags behind after neostigmine and to a greater extent after pyridostigmine.[12] These differences, however, are important only during 3–4 minutes after the reversal agent has been given and there is no risk of the Train-of-Four ratio underestimating the extent of residual neuromuscular blockade when its value approaches 0.7.

Choice of neuromuscular blocking drug

With the recent introduction of several new neuromuscular blocking drugs, it has become clearly apparent that the most important determinant of the rapidity of reversal is the choice of relaxant, not the choice of reversal agent. The observed rate of reversal with anticholinesterases must be seen in the context of the rate of the spontaneous recovery that continues in the background. If rapid reversal of adequate, surgical neuromuscular blockade is a clinical requirement, it can be obtained more easily if the drug occupying the synaptic cleft at the end of the procedure is of the short-acting type. It is extremely difficult to make a direct comparison of different neuromuscular drugs from published data (Table 8.2) because different methods and criteria have been used to investigate their reversal characteristics. In general, the short-acting and intermediate-acting relaxant drugs appear to form one group in relation to their speed of reversal, and the long-acting drugs a separate group. Although mivacurium displays a rapid spontaneous recovery, the rate of antagonism with neostigmine from a twitch height of 10% of the control value does not appear to be more rapid than that of rocuronium. Increasing the dose of neostigmine does not necessarily hasten reversal because a 'ceiling effect' may be encountered.

Choice of reversal agent, dose and extent of existing neuromuscular blockade

Several studies have sought to standardize conditions at the neuromuscular junction by testing the effectiveness of a reversal agent against a background of a continuous infusion of neuromuscular blocking drug which is not discontinued when the reversal agent is given. In these circumstances, neuromuscular function may be partially or completely restored but the effect of the reversal agent may subsequently wane so that spurious re-curarization may observed. In this type of experiment, the term 'onset of effect' has been used to describe the time from the administration of the reversal agent to the peak effect, and the term 'duration' has been used to describe the time taken for the twitch response to wane to a predetermined proportion of the peak effect.[13] It is desirable to investigate the effects of reversal agents when neuromuscular blockade has been maintained at a stable level for some time.

Table 8.2 A comparison of the rate of reversal of neuromuscular blockade after various doses of neostigmine

Drug	Ref. Source	Recovery index	Time to T1 = 0.9	Time to ToF = 0.7	Reversal drug	Dose (mg kg^{-1})	Anaesthesia
Mivacurium (at 10% recovery)	91	9.7	15.7	15.4	None		Isoflurane
		5.5	9.4	8.1	Neo.	0.02	
		3.5	5.4	7.5	Neo.	0.02	
		2.8	5.2	6.9	Edro.	0.4	
		4.8	5.6	5.9	Edro.	1.0	
Rocuronium (at 10% recovery)	61		31.4	36.1	None		Halothane
			7.9	7.5	Neo.	0.05	
			4.1	9.3	Edro.	1.0	
Atracurium (at 5–10% recovery)	26		4.2[a]	11.3	Neo.	0.02	Isoflurane
		2.3		8.3	Neo.	0.04	
		1.0		5.2	Neo.	0.08	
Vecuronium (at 10% recovery)	39	2.2		8.9	Neo.	0.07	Halothane
Pancuronium (at 5% recovery[b])	92		22.8		Neo.	0.03	Intravenous
			15.8[c]		Neo.	0.06	
			14.8[c]		Neo.	0.08	
Alcuronium (at 0–10% recovery)	43			31.4	Neo.	0.04	Mainly volatile
				25.4	Neo.	0.08	
				53.4[d]	Edro.	0.5	
				39.4[d]	Edro.	1.0	
Pipecuronium (at 20% recovery[e])	41			8.3	Neo.	0.06	Halothane
				12.8	Edro.	1.0	
Doxacurium (at 25% recovery)	93		1.7	11.6	Neo.	0.06	Isoflurane

ToF = Train-of-Four.
[a] Significantly slower than 0.04 mg kg^{-1}.
[b] Time to TI = 0.95, ToF = 0.75.
[c] Significantly faster than neostigmine 0.03 mg kg^{-1}.
[d] Significantly different from neostigmine 0.08 mg kg^{-1}.
[e] Time to ToF = 0.75.

However, unless the infusion of muscle relaxant is discontinued simultaneously, the measured extent and persistence of reversal will be meaningless in relation to clinical practice. The discrepancy between the two investigational techniques will clearly be greater when the muscle relaxant normally exhibits rapid spontaneous recovery and caution must be applied when comparing the results of different studies.

Neostigmine

Neostigmine remains the most commonly used reversal agent. In common with pyridostigmine and edrophonium, its chemical structure is related to physostigmine (eserine), an alkaloid derived from the calabar bean. Physostigmine is a tertiary amine and, consequently, penetrates the blood–brain barrier to a significant extent. It has been used as an analeptic and to control the central nervous system manifestations of poisoning by anticholinergic drugs (central anticholinergic syndrome) but exhibits slow onset and inadequate potency as an anti-curare agent.[14] The mechanism of the anticholinesterase activity of neostigmine is described above. Unlike edrophonium, the molecule does not survive its interaction with acetylcholinesterase, and its metabolite (3-hydroxy-phenyl-trimethylammonium) has no anti-curare activity[15] although it has sufficient anticholinesterase activity to increase twitch tension in the absence of a neuromuscular blocking drug in experimental animals.[16] Neostigmine, in common with pyridostigmine and edrophonium occupies a volume of distribution that is larger than might be expected of these polar, lipophobic non-protein bound drugs: this phenomenon appears to be due to tissue sequestration, particularly in the liver (Table 8.3).[17] Clearance from the plasma is rapid in comparison with the neuromuscular blocking drugs and the elimination half-time ($T_{1/2}\beta$ 5 mg dose) is approximately 77–80 minutes[18, 19] which exceeds that of the short- and intermediate-duration relaxants but is less than that of pancuronium, pipecuronium and doxacurium (Chapter 3).

The relation between the plasma concentration of neostigmine and its activity in reversing neuromuscular blockade has not yet been determined in man. However, in the rat, there is a significant linear relationship between the logarithm of the plasma concentration of neostigmine, pyridostigmine and edrophonium, and the increase in tibialis anterior twitch tension in the absence of a neuromuscular blocking drug.[16]

Neostigmine blockade

The administration of neostigmine in the absence of a neuromuscular blocking drug will reduce neuromuscular transmission and produce muscle weakness. When used to reverse non-depolarizing neuromuscular blockade, an excess of neostigmine in relation to the extent of the pre-existing neuromuscular blockade may transiently hinder the process of reversal or even increase the depth of blockade.[21] The mechanism is likely to

Table 8.3 Pharmacokinetic variables of neostigmine, pyridostigmine and edrophonium in normal, adult patients (mean and standard deviation)[19, 20]

	$t_{1/2}\beta$	Vd_{ss} (ml kg^{-1})	Cl (ml kg^{-1} min^{-1})
Neostigmine (0.07 mg kg^{-1})	77 (47)	0.7 (0.2)	9.2 (2.6)
Pyridostigmine (0.35 mg kg^{-1})	112 (12)	1.1 (12)	8.6 (1.7)
Edrophonium (0.5 mg kg^{-1})	110 (34)	1.1 (0.2)	9.6 (2.7)

be indirect via the accumulation of a large concentration of acetylcholine in the synaptic cleft which itself produces a well described, agonist-type of neuromuscular blockade. This is supported by the observation that suxamethonium enhances the block in man.[21] Hughes and Payne[22] have demonstrated that a second dose of neostigmine may enhance a partially reversed block. Subsequently, the circumstances in which this potential hazard might appear have been more closely defined, but it is becoming clear that each muscle relaxant drug has slightly different characteristics in this regard. The most important characteristics of 'neostigmine blockade' are as follows:

1. Neostigmine block is more likely to occur if two, divided doses of 2.5 mg are given, separated by 2-5 minutes, rather than a single dose of 5 mg.[21, 23]
2. The blockade due to neostigmine is more readily demonstrated using tetanic nerve stimulation (tetanic fade) than single stimuli at 1 Hz.[21] This implies that there are prejunctional as well as postjunctional components.
3. Halothane appears to enhance neostigmine blockade.[21]
4. Paradoxically, the block can be partially reversed by tubocurarine or gallamine, presumably because molecules of the non-depolarizing drug reduce the probability of acetylcholine combining with cholinoceptors to produce a depolarizing neuromuscular blockade.
5. Neostigmine blockade has been demonstrated using the Train-of-Four technique but only when two doses of neostigmine have been separated by approximately 2 minutes.[23] Even when spontaneous recovery from vecuronium or atracurium is exceptionally well advanced, a single dose of neostigmine up to 80 μg kg^{-1} (approximately 5.0 mg) has been demonstrated not to cause 'neostigmine blockade'.[23, 24, 25, 26]

Edrophonium

The ideal reversal agent should be able to restore sustained neuromuscular function rapidly even when neuromuscular blockade is profound at the time of administration. Edrophonium has enjoyed a limited renaissance in recent years following several encouraging reports suggesting that some of its characteristics are superior to neostigmine. However, it has become apparent that, although edrophonium reverses light or moderate neuromuscular blockade with greater rapidity than neostigmine, reversal of more intense block may be inadequate.

A clinical study of the use of edrophonium to reverse neuromuscular blockade was published by Hunter in Manchester in 1952.[27] Using maximum doses of only 20 mg in an attempt to reverse surgical levels of neuromuscular blockade due to tubocurarine, metocurine or gallamine, Hunter observed that the persistence of reversal was jeopardized if a large total dose of muscle relaxant was given during the course of anaesthesia and he concluded that, 'There were too many cases in which its effect was incomplete.' Limited and transient reversal of tubocurarine was also observed by Katz[28] using up to 100 mg of edrophonium. These and other workers consistently concluded that edrophonium could be expected to reverse tubocurarine satisfactorily only if a small dose of muscle relaxant had been given during the course of the anaesthetic.

Interest in edrophonium was rekindled when Bevan[29] and Kopman[30] published studies of the reversal of pancuronium-induced neuromuscular blockade with large doses of edrophonium using the Train-of-Four technique. Bevan delayed administering edrophonium (up to 50 mg) until the first response to Train-of-Four stimulation had reappeared, and commented that recovery of single twitch was similar to that observed after 2.5 mg of neostigmine at a similar degree of spontaneous recovery from pancuronium, with the advantage that the initial rate of reversal was more rapid and less atropine was needed. Restoration of the Train-of-Four ratio was more rapid if the dose of pancuronium had been restricted. Kopman[30] also attempted to ensure that a measurable Train-of-Four ratio existed before administering edrophonium – on this occasion in an incremental fashion up to a dose of 0.5 mg kg^{-1}. Reversal of pancuronium-induced neuromuscular blockade was adequate provided that the Train-of-Four ratio exceeded 0.1 at the time of administration of the drug: reversal was incomplete in those patients in whom four twitches were not discernible before edrophonium was given.

Several, subsequent publications concerning the rapidity with which edrophonium antagonized pancuronium,[31] tubocurarine,[19] alcuronium,[32] atracurium[33] and vecuronium[34] were optimistic that there may be significant advantages in comparison with neostigmine. In addition, it was established that the optimum dose of atropine required in association with edrophonium was only 7 µg kg^{-1}.[35]

Despite the initial enthusiasm, it became obvious that reversal with edrophonium was superior to neostigmine only if significant spontaneous recovery had already occurred: otherwise antagonism was unacceptably slow. Foldes[36] concluded that reversal of vecuronium neuromuscular blockade with edrophonium 0.5 mg kg^{-1} may be unsatisfactory if the pre-existing twitch height was less than 40% of the control value. Hughes[37] demonstrated that, despite extremely rapid reversal at 40–50% spontaneous recovery, the time taken for edrophonium to abolish tetanic fade during profound block due to atracurium was approximately twice that for neostigmine. Several subsequent publications concluded that edrophonium was inferior to neostigmine if reversal was attempted at 90% blockade or greater, or if less than four responses of the Train-of-Four were discernible. This stricture appears to apply to the intermediate-duration relaxants[38,39] in addition to the longer-acting drugs.[40,41] To summarize the place for edrophonium in routine practice: provided that (a) all four responses of the Train-of-Four can be easily detected using a simple, nerve stimulator and (b) the total dose of relaxant has been relatively small, edrophonium 0.5 mg kg^{-1} appears to offer advantages by increasing the rapidity of reversal and decreasing the muscarinic challenge. For example, during atracurium blockade, if there is 25% spontaneous recovery of the twitch response, a Train-of-Four ratio of 0.7 can be achieved in approximately 5.5 minutes after edrophonium 0.5 mg kg^{-1},[42] compared with 7.0 minutes after neostigmine 0.04 mg/kg^{-1},[43] and 30 minutes if no antagonist is administered.[39]

Priming with anticholinesterases

The administration of a small proportion of an 'intubating dose' of a neuromuscular blocking drug a few minutes before inducing anaesthesia and giving the remaining larger portion hastens the onset of neuromuscular blockade (Chapter 10). This principle of 'priming' has been extended to the reversal of neuromuscular blockade with anticholinesterases, although it is clear that the mechanism whereby the rate of onset of neuromuscular blocking drugs is increased, i.e., a deliberate reduction in the 'margin of safety', is not applicable to their reversal. Opinion is divided concerning the value of this technique. Divided doses of neostigmine[44] and edrophonium[45] have been demonstrated by Abdulatif et al. to accelerate the reversal of atracurium. Other workers have concluded that priming with neostigmine[46] or edrophonium[47] offers no advantages compared with a single dose technique. It is probable that the priming method is appropriate only when considerable pre-existing recovery has occurred and when potentiating inhalational agents are avoided.[48] It should be borne in mind that the exact proportions of the divided doses are critical and a second dose of neostigmine 2.5 mg given 2 minutes after a similar dose may actually increase the depth of neuromuscular blockade.[23] Routine clinical monitoring of neuromuscular blockade using visual or tactile assessment of the twitch response is probably too inaccurate to facilitate the use of priming techniques and, in these circumstances, their use is not recommended.

Aminopyridines

It is possible to antagonize non-depolarizing neuromuscular blockade with drugs other than anticholinesterases. The aminopyridines have several pharmacological actions which tend to restore neuromuscular function. The predominant effect is at the motor nerve terminal where aminopyridines decrease potassium conductance. The repolarization phase of the nerve action potential is lengthened, resulting in a greater influx of calcium ions into the nerve terminal. As a consequence, the release of acetylcholine is enhanced and a non-depolarizing block is antagonized. A second, minor effect is to make available more calcium for the process of muscle contraction by a similar effect on potassium channel conductance at the sarcoplasmic reticulum. 4-Aminopyridine and, to a lesser extent, 3-4 aminopyridine cross the blood–brain barrier resulting in symptoms which vary from a mild analeptic effect to major cerebral excitation. It is this latter phenomenon which has prevented the introduction of this class of compounds into clinical practice.

Cholinesterase

Established phase ll block due to suxamethonium may be reversed with plasma cholinesterase derived from plasma donations (Chapter 4). However, it is generally considered that the expense and the possible risks of the transmission of viral disease associated with this preparation effectively precludes its use. The reversal of mivacurium-induced blockade with biologically engineered pure cholinesterase is an intriguing possibility (Chapter 5).

Plasma concentration of relaxant

Drug-induced reversal of neuromuscular blockade is normally superimposed on the process of spontaneous recovery; the rate of reversal will necessarily be influenced by the rate at which recovery would have proceeded if no reversal agent had been given. This is self-evident: short-acting relaxants reverse more rapidly than long-acting relaxants. Two factors (at least) determine the rate of *spontaneous* recovery from neuromuscular blockade: drug/receptor affinity and the extent of the concentration gradient from the receptor to the plasma. The relative influence of the plasma concentration of neuromuscular blocking drug at the time when reversal is attempted is the subject of much debate. It is clear that the concentration gradient of relaxant from neuromuscular junction to plasma does influence the rapidity of spontaneous recovery of neuromuscular function to some extent. Under conditions of equilibrium, the degree of neuromuscular blockade and the plasma concentration are closely correlated. However, a situation of equilibrium rarely exists in clinical practice, unless the relaxant is infused continuously. If a repeated bolus technique is used, the direction of the gradient is continually changing. It has been argued that, after a prolonged continuous infusion or after numerous boluses, the plasma concentration associated with a given level of neuromuscular blockade might be expected to be greater than after a single bolus during the offset of blockade. It is not surprising that the recovery index (25% recovery to 75% recovery) was considerably more rapid after a single bolus of pancuronium compared with a prolonged continuous infusion.[49] However, this observation does not apply to all relaxants: for example, the recovery index after a prolonged infusion of vecuronium is little different from that following a single bolus,[50] suggesting that the concentration gradient from receptor to plasma is less important in determining the rate of spontaneous recovery from vecuronium in comparison with pancuronium. In addition, doubt has been cast on the basic assumption that different dosage regimens can affect the relation between plasma concentration and depth of neuromuscular blockade, notably after tubocurarine[51] and atracurium.[52]

Attempts have been made to investigate this phenomenon by standardizing the prevailing concentration gradient and measuring the (artificial) rate of recovery. This can be accomplished by administering a small dose of relaxant into a vein of the forearm of a volunteer after the circulation has been isolated by inflating a tourniquet above arterial pressure. When the tourniquet is released, the plasma concentration in the forearm rapidly diminishes to almost zero. This type of experiment has been used to support the contention that the most important determinant of the rate of recovery from neuromuscular blockade is the magnitude of the prevailing concentration gradient as opposed to differences between relaxants in their affinity for the receptors. Pancuronium is apparently transformed by this technique into a rapid-offset drug and the recovery index is reduced to 10 minutes compared with 46 minutes after a continuous infusion (when the plasma concentration might be expected to remain high due to its long elimination half-life) suggesting that the concentration gradient is an important factor.[53] However, it is clear that the concentration gradient is of very little importance with regard to vecuronium: the isolated forearm recovery index is similar to that obtained after either a single bolus or an infusion in anaesthetized patients.[54]

The process of evoked reversal appears also to be broadly governed by the same interplay of receptor affinity and concentration gradient. The importance of a significant concentration gradient for some relaxants was again emphasized by the classical isolated forearm experiments. Neostigmine had little effect in accelerating the normal rate of spontaneous recovery of thumb twitch when given simultaneously with a small dose of tubocurarine into the isolated forearm. However, in volunteers pretreated with i.v. neostigmine, reversal was rapid when the tourniquet was released if tubocurarine alone had been given into the isolated forearm; suggesting that reversal of tubocurarine is significantly concentration gradient dependent.[55] This hypothesis is supported by Baraka[56] who demonstrated that reversal of tubocurarine was impossible until the plasma concentration had decreased to a low level. In addition, a small total dose of pancuronium was more easily reversed than a large total dose *at a given degree of spontaneous recovery*.[29] The situation regarding tubocurarine is less clear. Katz[28] found that reversal was difficult if a large, total dose had been given, but more recent work by Ham[51] demonstrated that the dose regime had no effect on the reversal characteristics.

The intermediate-acting and short-acting drugs appear to be different, at least from pancuronium: the speed of reversal with neostigmine appears to be similar regardless of whether the patient has received a single small bolus, several boluses or a continuous infusion. This is probably explained by the short elimination half-times of these drugs which are rapidly cleared from the plasma. In these circumstances, it is clear that the ability of these drugs to establish rapidly and maintain a large concentration gradient away from the neuromuscular junction is an important contributory factor to their rapid reversal with neostigmine. For example, Gencarelli and colleagues demonstrated that, when the effects of pharmacokinetics and duration of action were minimized by administering neostigmine during a continuous infusion, the reversal characteristics of vecuronium became indistinguishable from those of pancuronium,[57] in contrast to the usual situation where vecuronium requires less neostigmine to produce equal reversal.[58] Reversal of profound vecuronium- and atracurium-induced blockade (no Train-of-Four response) is relatively slow. In one study of both drugs, a Train-of-Four ratio of 0.7 was not achieved until almost 45 minutes after neostigmine 0.07 mg kg^{-1}. There was no evidence that neostigmine increased the recovery time compared with a group who were permitted to recover spontaneously.[39]

Rocuronium exhibits a slower rate of spontaneous recovery than mivacurium. The recovery index is approximately 10 minutes following a single bolus during fentanyl-nitrous oxide anaesthesia[59] and approximately 14 minutes following a single bolus during halothane anaesthesia: a pattern of recovery that is similar to vecuronium.[60] Neostigmine significantly accelerates the reversal of rocuronium so that a Train-of-Four Ratio of 0.7 is achieved after approximately 8 minutes; a rate comparable with vecuronium.[61]

Short-acting relaxants are, by definition, cleared extremely rapidly from the plasma and, consequently, spontaneous recovery is very rapid. It has been suggested that the use of neostigmine is unnecessary with these agents.[62] However, during short surgical procedures when rapid patient turnround is required, it may be desirable to accelerate the process of recovery of neuromuscular function. Mivacurium has the shortest recovery index of the agents currently in use (approximately 7 minutes).[62] In contrast to the longer-acting agents, it is possible to promptly reverse profound blockade (> 95% twitch depression) using neostigmine 0.05 mg kg^{-1},[63] suggesting that inactivation of plasma cholinesterase by neostigmine is less important than its effect at the neuromuscular junction in determining the rate of recovery from mivacurium-induced blockade. During enflurane anaesthesia, spontaneous recovery from an infusion of mivacurium at approximately 90% blockade to a Train-of-Four ratio of 0.7 takes approximately 17 minutes. This is reduced to 11 minutes after neostigmine 0.04 mg kg^{-1} and 8 minutes after edrophonium 0.75 mg kg^{-1}.[64] During propofol anaesthesia, spontaneous recovery from approximately 90% blockade (continuous infusion) requires 12 minutes and the administration of edrophonium at approximately 80% block rapidly reverses the Train-of-Four ratio to 0.7 in 5.4 minutes.[65] There are limited data concerning the reversal of mivacurium with anticholinesterases under conditions of low plasma cholinesterase activity, either inherited or acquired, but it would seem prudent to defer their administration until all four responses of the Train-of-Four are present.

Anaesthetic agent

There is a paucity of published work on the influence of different anaesthetic agents on the process of reversal of neuromuscular blockade. If the inhalational agent is continued throughout the process of reversal, enflurane, unlike halothane, reduces the effectiveness of neostigmine[66] and edrophonium.[67] Anaesthetic agents which potentiate neuromuscular blocking drugs, notably enflurane, may paradoxically *increase* the rapidity of reversal because it is likely that the tension of inhalational agents in blood will be reduced by the anaesthetist at the completion of the surgical procedure concurrently with the process of reversal of residual neuromuscular blockade.[68]

Renal failure

The process of reversal is largely governed by a balance between the plasma concentration of neuromuscular blocking drug and the plasma concentration of anticholinesterase. If muscle relaxant drugs that are predominantly excreted by the kidney (especially gallamine, see Chapter 9) are avoided, problems with reversal should be rare. Anticholinesterases are actively secreted into the renal tubules and their plasma clearance is reduced in renal failure. The consequent increase in plasma elimination half-times (two-fold for neostigmine[19] and three-fold for edrophonium[69]) represents an advantage when reversing residual neuromuscular blockade in patients who have renal failure. The rate of reversal of vecuronium, the elimination of which is delayed in renal failure, appears to be unaffected by impaired renal function.[70] Reports in the 1970s of inadequate reversal or residual neuromuscular blockade in patients in renal failure[71, 72] were generally characterized by monitoring techniques that would be considered to be inadequate today. Hence, reversal may have been attempted at very profound blockade. The apparent phenomenon of 're-curarization' is unlikely to occur unless muscle function is diminished by factors other than neuromuscular blockade;[73] for example, respiratory muscle fatigue, critical illness polyneuropathy which is manifest by muscle weakness without Train-of-Four fade,[74] or conditions which downgrade acetylcholine receptors, e.g. myasthenia gravis. It is of interest that a subparalysing plasma concentration of muscle relaxant does not appear to enhance fatigue of the diaphragm.[75] Nonetheless, it should be emphasized that neuromuscular blockade should be monitored routinely, especially when there is organ dysfunction.

Age

The plasma clearance of anticholinesterases is reduced in the elderly[76] in common with the plasma clearance of the non-depolarizing muscle relaxants.[77–80] It is worthy of note that even atracurium exhibits a reduced clearance and prolonged elimination half-time in the elderly.[81] It has been suggested that it is unnecessary to increase the dose of neostigmine in the elderly[76] or, indeed, that it is desirable to reduce the dose because this group of patients exhibit a left-shift in the dose–response relationship.[82] However, these observations were generally made when neostigmine was administered experimentally against the background of an infusion of muscle relaxant drug which was not discontinued at the time when the anticholinesterase was given. Recent evidence suggests that when neostigmine is administered against the background of a reducing plasma concentration of neuromuscular blocking drug in the usual fashion, the clearance of the muscle relaxant becomes important in determining the relative rate of antagonism of blockade in the young and the elderly. For example, the relatively prolonged clearance, and hence spontaneous recovery profile, of vecuronium appears to necessitate a larger dose of neostigmine in the elderly than in younger patients.[83] The plasma clearance of rocuronium is also delayed in the elderly, probably the result of impaired renal and hepatic perfusion[84] and its reversal is delayed by approximately 40% in elderly patients.[85]

Drug interactions

The most important drug interactions relevant to the administration of anticholinesterases are likely to involve agents that are administered intravenously during anaesthesia. Although there is a paucity of published data, any drug which potentiates neuromuscular blocking drugs (Chapter 10) will impede the reversal of neuromuscular blockade. Doxapram is a respiratory stimulant that is occasionally administered at the end of anaesthesia to counteract opioid respiratory depression. Recent evidence suggests that the recovery index (the time taken for the twitch response to recover from 75% depression to 25% depression) is prolonged by approximately 45% when neostigmine 0.05 mg kg^{-1}, given at 20% recovery from vecuronium-induced blockade, is immediately followed by doxapram 0.5 mg kg^{-1}.[86] The rate of spontaneous recovery from vecuronium (but not atracurium) also appears to be delayed by the administration of doxapram.[87] The mechanisms involved are unclear.

Electrolyte and acid-base imbalance

Normal neuromuscular function is dependent on pH and the intracellular and extracellular concentrations of the ions that contribute to the processes of neuromuscular transmission and muscle contraction. Respiratory acidosis potentiates neuromuscular blockade and may impede the reversal of residual paralysis. The antagonism of pancuronium is retarded in the presence of respiratory acidosis and metabolic alkalosis.[88,89] Metabolic alkalosis is often the result of hypokalaemia, which exacerbates the reduction in neuromuscular function.[90] The degradation of atracurium is partially pH dependent (Hofmann elimination, Chapter 5), but there is no evidence that acidosis impedes its antagonism by neostigmine in the clinical situation.

Clinical guidelines for reversal of non-depolarizing neuromuscular blockade

1. A peripheral nerve stimulator should be used whenever a muscle relaxant drug is given.
2. It is inadvisable to reverse residual blockade until at least one twitch of the Train-of-Four has returned. This constraint applies even to the short-acting muscle relaxants.
3. Always give a reversal agent unless neuromuscular function has recovered spontaneously to the extent that there is absolutely no fade in the response to Double Burst stimulation.
4. Choose a dose of the antagonist drug which is commensurate with the extent of the neuromuscular blockade and give this dose as a single bolus. A divided dose, separated by a few minutes, may result in a less satisfactory reversal than a single dose of the same total amount.
5. A single dose of neostigmine 0.08 mg kg^{-1} (5 mg) will not result in 'neostigmine blockade' in normal, clinical conditions.
6. Long-acting muscle relaxants reverse more slowly than intermediate- or short-acting drugs; therefore, allow more time for antagonism to take place before discontinuing general anaesthesia.
7. Antagonism of neuromuscular blockade is a less rapid process than is generally believed.

References

1. d'Hollander AA, Dewachter B, Deville A, Vaisiere D. Haemodynamic changes associated with atropine/neostigmine administration. *Acta Anaesthiologica Scandinavica* 1981; **25:** 187–192.
2. Harper KW, Bali IM, Gibson FM, Carlisle R, Black IHC, Grainger DJ, *et al.* Reversal of neuromuscular block: heart rate changes with slow injection of neostigmine and atropine mixtures. *Anaesthesia* 1984; **39:** 772–775.
3. Ovassapian A. Effects of administration of atropine and neostigmine in man. *Anesthesia and Analgesia* 1969; **48:** 219–223.
4. Cronnelly R, Morris RB, Miller RD. Edrophonium: duration of action and atropine requirements in humans during halothane anaesthesia. *Anesthesiology* 1982; **57:** 261–266.
5. Mirakhur RK, Dundee JW, Clarke RSJ. Glycopyrrolate–neostigmine mixture for antagonism of neuromuscular blockade: comparison with atropine–neostigmine mixture. *British Journal of Anaesthesia* 1977; **49:** 825–829.
6. Mirakhur RK. Antagonism of the muscarinic effects of edrophonium with atropine or glycopyrrolate. *British Journal of Anaesthesia* 1985; **57:** 1213–1216.
7. Mirakhur RK, Briggs LP, Clarke RSJ, Dundee JW, Jonston HML. Comparison of atropine and glycopyrrolate in a mixture with pyridostigmine for the antagonism of neuromuscular block. *British Journal of Anaesthesia* 1981; **53:** 1315–1320.
8. Bell CMA, Lewis CB. Effect of neostigmine on integrity of ileorectal anastomoses. *British Medical Journal* 1968; **2:** 587–588.
9. Aitkenhead AR, Wishart HY, Brown DA. High spinal block for large bowel anastomosis: a retrospective study. *British Journal of Anaesthesia* 1978; **50:** 177–183.
10. Hammond J, Wright D, Sale J. Pattern of bronchomotor tone following reversal of neuromuscular blockade. *British Journal of Anaesthesia* 1983; **55:** 955–959.
11. Harper NJN, Bradshaw EG, Healy TEJ. Antagonism of alcuronium with edrophonium or neostigmine. *British Journal of Anaesthesia* 1984; **56:** 1089–1094.
12. Donati F, Ferguson A, Bevan DR. Twitch depression and train-of-four ratio after antagonism of pancuronium with edrophonium, neostigmine or pyridostigmine. *Anesthesia and Analgesia* 1983; **62:** 314–316.
13. Cronnelly R, Morris RB, Miller RD. Edrophonium: duration of action and atropine requirement in humans during halothane anesthesia. *Anesthesiology* 1982; **57:** 261–266.
14. Baraka A. Antagonism of neuromuscular blockade by physostigmine in man. *British Journal of Anaesthesia* 1978; **50:** 1075–1077.
15. Hennis PJ, Cronnelly R, Sharma M, *et al.* Metabolites of neostigmine and pyridostigmine do not contribute to antagonism of neuromuscular blockade in the dog. *Anesthesiology* 1984; **61:** 534–539.
16. Barber HE, Calvey TN, Muir KT. The relationship between the pharmacokinetics, cholinesterase inhibition and facilitation of twitch tension of the quaternary ammonium anticholinesterase drugs, neostigmine, pyridostigmine, edrophonium and 3-hydroxyphenyltrimethylammonium. *British Journal of Pharmacology* 1979; **66:** 525–530.
17. Cronnelly R, Morris RB. Antagonism of neuromuscular blockade. *British Journal of Anaesthesia* 1982; **54:** 183–194.
18. Cronnelly R, Stanski DR, Miller RD, Sheiner LB, Sohn YJ. Renal function and the pharmacokinetics of neostigmine in anesthetized man. *Anesthesiology* 1979; **51:** 222–226.
19. Morris RB, Cronnelly R, Miller RD, Stanski DR, Fahey MR. Pharmacokinetics of edrophonium and neostigmine when antagonizing d-tubocurarine neuromuscular blockade in man. *Anesthesiology* 1981; **54:** 399–402.

20. Cronnelly R, Stanski DR, Miller RD, et al. Pyridostigmine kinetics with and without renal function. *Clinical Pharmacology and Therapeutics* 1980; **28**: 78–81.
21. Payne JP, Hughes R, Al Azawi S. Neuromuscular blockade by neostigmine in anaesthetized man. *British Journal of Anaesthesia* 1980; **52**: 69–76.
22. Hughes R, Payne JP. Interaction of halothane with non-depolarizing neuromuscular blocking drugs in man. *British Journal of Clinical Pharmacology* 1979; **7**: 485.
23. Fox MA, Keens SJ, Utting JE. Neostigmine in the antagonism of action of atracurium. *British Journal of Anaesthesia* 1987; **59**: 468–472.
24. Jones JE, Hunter JM, Utting JE. Use of neostigmine in the antagonism of residual neuromuscular blockade produced by vecuronium. *British Journal of Anaesthesia* 1987; **59**: 1454–1458.
25. Beemer GH, Bjorksten AR, Dawson PJ, et al. Determinants of the reversal time of competitive neuromuscular block by anticholinesterases. *British Journal of Anaesthesia* 1991; **66**: 469–475.
26. Harper NJN, Wallace M, Hall IA. Optimum dose of neostigmine at two levels of atracurium-induced neuromuscular block. *British Journal of Anaesthesia* 1994; **72**: 82–85.
27. Hunter AR. Tensilon: a new anti-curare agent. *British Journal of Anaesthesia* 1952; **24**: 175–186.
28. Katz RL. Neuromuscular effects of d-tubocurarine, edrophonium and neostigmine in man. *Anesthesiology* 1967; **28**: 327–336.
29. Bevan DR. Reversal of pancuronium with edrophonium. *Anaesthesia* 1979; **34**: 614–619.
30. Kopman AF. Edrophonium antagonism of pancuronium-induced neuromuscular blockade in man: a reappraisal. *Anesthesiology* 1979; **51**: 139–142.
31. Ferguson A, Egerszegi P, Bevan RD. Neostigmine, pyridostigmine, and edrophonium as antagonists of pancuronium. *Anesthesiology* 1980; **53**: 390–394.
32. Harper NJN, Bradshaw EG, Healy TEJ. Antagonism of alcuronium with edrophonium or neostigmine. *British Journal of Anaesthesia* 1984; **56**: 1089–1094.
33. Jones RM, Pearce AC, Williams JP. Recovery characteristics following antagonism of atracurium with neostigmine or edrophonium. *British Journal of Anaesthesia* 1984; **56**: 453–457.
34. Baird WLM, Bowman WC, Kerr WJ. Some actions of ORG NC45 and of edrophonium in the anaesthetized cat and in man. *British Journal of Anaesthesia* 1982; **54**: 375–384.
35. Morris RB, Cronnelly R, Miller RD. Atropine requirement for edrophonium. *Anesthesiology* 1981; **55**: A206.
36. Foldes FF, Yun H, Radnay PA, Badola RP, Kaplan R, Nagashima H. Antagonism of the NM effect of ORG-NC45 by edrophonium. *Anesthesiology* 1981; **55**: A201.
37. Hughes R, Astley BA, Payne JP. Reversal of atracurium by neostigmine and edrophonium. *British Journal of Anaesthesia* 1984; **56**: 796P.
38. Lavery GG, Mirakhur RK, Gibson FM. A comparison of edrophonium and neostigmine for the antagonism of atracurium-induced neuromuscular block. *Anesthesia and Analgesia* 1985; **64**: 867–870.
39. Caldwell JE, Robertson EN, Baird WLM. Antagonism of profound neuromuscular blockade induced by vecuronium or atracurium. *British Journal of Anaesthesia* 1986; **58**: 1285–1289.
40. Kopman AF. Recovery times following edrophonium and neostigmine reversal of pancuronium, atracurium, and vecuronium steady-state infusions. *Anesthesiology* 1986; **65**: 572–578.
41. Abdulatif M, Naguib M. Neostigmine and edrophonium for reversal of the neuromuscular blocking effect of pipecuronium. *Canadian Journal of Anaesthesia* 1991; **38**: 159–163.

42. Cashman JN, Jones RM, Adams AP. Atracurium recovery: prediction of safe reversal times with edrophonium. *Anaesthesia* 1989; **44**: 805–807.
43. Beemer GH, Bjorksten AR, Dawson PJ, Dawson RJ, Heenan PJ, Robertson BA. Determinants of the reversal time of competitive neuromuscular block by anti-cholinesterases. *British Journal of Anaesthesia* 1991; **66**: 469–475.
44. Abdulatif M, Naguib M. Accelerated reversal of atracurium blockade with divided doses of neostigmine. *Canadian Anaesthetists' Society Journal* 1986; **33**: 723–728.
45. Naguib M, Abdulatif M. Accelerated reversal of atracurium blockade with priming doses of edrophonium. *Anesthesiology* 1987; **66**: 397–400.
46. Donati F, Smith CE, Wiesel S, Bevan DR. 'Priming' with neostigmine: failure to accelerate reversal of single twitch and train-of-four responses. *Canadian Journal of Anaesthesia* 1989; **36**: 30–34.
47. Szalados JE, Donati F, Bevan DR. Edrophonium priming for antagonism of atracurium neuromuscular blockade. *Canadian Journal of Anaesthesia* 1990; **37**: 197–201.
48. Naguib M, Abdulatif M. Edrophonium priming alters the course of neuromuscular recovery from a pipecuronium blockade. *Canadian Journal of Anaesthesia* 1991; **38**: 722–777.
49. Agoston S, Feldman SA, Miller RD. Plasma concentrations of pancuronium and neuromuscular blockade after injection into the isolated arm, bolus injection and continuous infusion. *Anesthesiology* 1979; **51**: 119–122.
50. Agoston S, Salt P, Newton D, Bencini A, Boomsma W, Erdmann W. The neuromuscular blocking action of ORG 45, a new pancuronium derivative, in anaesthetic practice. *British Journal of Anaesthesia* 1980; **52**: 53S–59S.
51. Ham J, Miller RD, Sheiner LB, Matteo RS. Dose-schedule independence of d-tubocurarine pharmacokinetics and pharmacodynamics. and recovery of neuromuscular function. *Anesthesiology* 1979; **50**: 528–533.
52. Weatherley BC, Williams SG, Neill EAM. Pharmacokinetics, pharmacodynamics and dose-response relationships of atracurium administered i.v. *British Journal of Anaesthesia* 1983; **55**: 39S–45S.
53. Agoston S, Feldman SA, Miller RD. Plasma concentrations of pancuronium and neuromuscular blockade after injection into the isolated arm, bolus injection and continuous infusion. *Anesthesiology* 1979; **51**: 119–122.
54. Bencini A, Agoston S, Ket J. Use of human 'isolated arm' preparation to indicate qualitative aspects of a new neuromuscular blocking agent, Org NC45. *British Journal of Anaesthesia* 1980; **52**: 43S–47S.
55. Feldman SA, Agoston S. Failure of neostigmine to prevent tubocurarine neuromuscular blockade in the isolated arm. *British Journal of Anaesthesia* 1980; **52**: 119–1203.
56. Baraka A. Irreversible tubocurarine neuromuscular block in the human. *British Journal of Anaesthesia* 1967; **39**: 891–894.
57. Gencarelli PJ, Miller RD. Antagonism of ORG NC45 vecuronium and pancuronium neuromuscular blockade by neostigmine. *British Journal of Anaesthesia* 1982; **54**: 53–61.
58. Fahey MR, Morris RB, Miler RD, Sohn YJ, Cronnelly R. Clinical pharmacology of Org NC45 Norcuron. *Anesthesiology* 1981; **55**: 6–11.
59. Booij LHD, Knape HTA. The neuromuscular blocking effect of Org 9426. *Anaesthesia* 1991; **46**: 341–343.
60. Booth MG, Marsh B, Bryden FMM, Robertson EN, Baird WLM. A comparison of the pharmacodynamics of rocuronium and vecuronium during halothane anaesthesia. *Anaesthesia* 1992; **47**: 832–834.
61. McCoy E, Maddineni VR, Mirakhur RK. Reversal of rocuronium block with edrophonium or neostigmine. *British Journal of Anaesthesia* 1993; **71**: 765P.

62. Savarese JJ, Ali HA, Basta SJ, Embree PB, Scott RPF, Sunder N, et al. The clinical neuromuscular pharmacology of mivacurium chloride BW B1090U. *Anesthesiology* 1988; **68:** 723–732.
63. Curran MJ, Shaff L, Saverese JJ, Ali HH, Risner M. Comparison of spontaneous recovery and neostigmine-accelerated recovery from mivacurium neuromuscular blockade. *Anesthesiology* 1988; **69:** A528.
64. Goldhill DR, Whitehead JP, Emmott RS, Griffith AP, Bracey BJ, Flynn PJ. Neuromuscular and clinical effects of mivacurium chloride in healthy adult patients during nitrous oxide-enflurane anaesthesia. *British Journal of Anaesthesia* 1991; **67:** 289–295.
65. Wrigley SR, Jones RM, Harrop-Griffiths AW, Platt MW. Mivacurium chloride: a study to evaluate its use during propofol-nitrous oxide anaesthesia. *Anaesthesia* 1992; **47:** 653–657.
66. Delisle S, Bevan DR. Impaired neostigmine antagonism of pancuronium during enflurane anaesthesia in man. *British Journal of Anaesthesia* 1982; **54:** 441–445.
67. Dernovoi B, Agoston S, Baurain M, et al. Neostigmine antagonism of vecuronium paralysis during fentanyl, isoflurane and enflurane anaesthesia. *Anesthesiology* 1987; **66:** 698–701.
68. Gencarelli PJ, Miller RD, Eger II EI, Newfield P. Decreasing enflurane concentrations and d-tubocurarine neuromuscular blockade. *Anesthesiology* 1982; **56:** 192–194.
69. Morris RB, Cronnelly R, Miller RD, Stanski DR, Fahey MR. Pharmacokinetics of edrophonium in anephric and renal transplant patients. *British Journal of Anaesthesia* 1981; **53:** 1311–1314.
70. Bevan DR, Donati F, Gyasi H, Williams A. Vecuronium in renal failure. *Canadian Anaesthetists Society Journal* 1984; **31:** 491–496.
71. Riordan DD, Gilbertson AA. Prolonged curarization in a patient with renal failure. *British Journal of Anaesthesia* 1971; **43:** 506–508.
72. Miller RD, Cullen DJ. Renal failure and postoperative respiratory failure: recurarization? *British Journal of Anaesthesia* 1976; **48:** 253–256.
73. Lee C, Mok MS, Barnes A, Katz RL. Absence of 'recurarization' in patients with demonstrated prolonged neuromuscular block. *British Journal of Anaesthesia* 1977; **49:** 485–489.
74. Witt NJ, Zochodne DW, Bolton CE, Grand Maison F, Wells G, et al. Peripheral nerve function in sepsis and multiple organ failure. *Chest* 1991; **99:** 176–184.
75. Dureuil B, Lebrault C, Boczkowski J, Aubier M, Duvaldestin P. Does a subparalysing dose of vecuronium enhance diaphragm fatigue? *British Journal of Anaesthesia* 1992; **68:** 352–355.
76. Young WL, Backus W, Matteo RS, Ornstein E, Diaz J. Pharmacokinetics and pharmacodynamics of neostigmine in the elderly. *Anesthesiology* 1984; **51:** A300.
77. McCleod K, Hull CJ, Watson MJ. Effects of ageing on the pharmacokinetics of pancuronium. *British Journal of Anaesthesia* 1979; **51:** 435–438.
78. Duvaldestin P, Saada J, Berger JL, d'Hollander A, Desmonts JM. Pharmacokinetics, pharmacodynamics and dose–response relationships of pancuronium in control and elderly subjects. *Anesthesiology* 1982; **56:** 36–40.
79. Rupp SM, Castagnoli KP, Fisher DM, Miller RD. Pancuronium and vecuronium pharmacokinetics and pharmacodynamics in younger and elderly patients. *Anesthesiology* 1987; **647:** 45–49.
80. Kent AP, Parker CJR, Hunter JM. Pharmacokinetics of atracurium and laudanosine in the elderly. *British Journal of Anaesthesia* 1989; **63:** 661–666.
81. Parker CJR, Hunter JM, Snowdon SL. Effect of age, sex and anaesthetic technique on the pharmacokinetics of atracurium. *British Journal of Anaesthesia* 1992; **69:** 439–443.

82. Miller RD. Pharmacokinetics of reversal agents and clinical considerations in their use. In Agoston S, Bowman WC eds *Muscle Relaxants: Monographs in Anaesthesiology* 1990, Amsterdam: Elsevier, pp. 503–514.
83. McCarthy GJ, Cooper R, Stanley JC, Mirakhur RK. Dose–response relationships for neostigmine antagonism of vecuronium-induced neuromuscular block in adults and the elderly. *British Journal of Anaesthesia* 1992; **69:** 281–283.
84. Matteo RS, Ornstein E, Schwartz AE, *et al.* Pharmacokinetics and pharmacodynamics of rocuronium ORG 9426 in elderly surgical patients. *Anesthesiology* 1993; **77:** 1193–1197.
85. Bevan DR, Fiset P, Balendran P, *et al.* Pharmacodynamic behaviour of rocuronium in the elderly. *Canadian Journal of Anaesthesia* 1993; **40:** 127–132.
86. Orlowski M, Pollard BJ. Effect of doxapram on neostigmine evoked antagonism of vecuronium neuromuscular block. *British Journal of Anaesthesia* 1992; **68:** 418–419.
87. Cooper R, McCarthy G, Mirakhur RK, Maddineni VR. Effect of doxapram on the rate of recovery from atracurium and vecuronium neuromuscular block. *British Journal of Anaesthesia* 1992; **68:** 527–528.
88. Miller RD, Roderick LL. Acid-base balance and neostigmine antagonism of pancuronium neuromuscular blockade. *British Journal of Anaesthesia* 1978; **50:** 317–324.
89. Wirtavuori K, Salmenpera M, Tammisto T. Effect of hypocarbia and hypercarbia on the antagonism of pancuronium-induced neuromuscular blockade in man. *British Journal of Anaesthesia* 1982; **54:** 57–61.
90. Miller RD, Roderick LL. Diuretic-induced hypokalaemia, pancuronium neuromuscular blockade and its antagonism by neostigmine. *British Journal of Anaesthesia* 1978; **50:** 541–544.
91. Naguib M, Abdulatif A, Al-Ghamdi A, *et al.* Dose–response relationships for edrophonium and neostigmine antagonism of mivacurium-induced neuromuscular block. *British Journal of Anaesthesia* 1993; **71:** 704–714.
92. Goldhill DR, Embree PB, Ali HH, Savarese JJ. Reversal of pancuronium: neuromuscular and cardiovascular effects of a mixture of neostigmine and glycopyrronium. *Anaesthesia* 1988; **43:** 443–475.
93. Martlew RA, Harper NJN. Pharmacodynamics of doxacurium in the elderly. *Anaesthesia* (in press).

9

Relaxants in Specific Clinical Situations

NJN Harper

The elderly

The normal ageing process imposes many physiological changes which directly influence the effect of neuromuscular blocking drugs (Table 9.1). Almost without exception, these changes might be expected to delay the onset and to increase the duration of neuromuscular blockade (Table 9.2). The increase in duration of neuromuscular blockade commonly found in the elderly is frequently accompanied by an increase in variability between patients and neuromuscular monitoring is desirable. The choice of neuromuscular blocking drug has also to take into account cardiovascular stability and the ease of reversal. Residual blockade persisting into the recovery room is likely to depress the efficiency of respiration to a greater extent in the elderly patient in whom respiratory reserve is already compromised. The effects of ageing on the reversal of neuromuscular blockade are discussed in Chapter 8.

Table 9.1 The physiological effects of ageing which affect the pharmacokinetics and pharmacodynamics of neuromuscular blocking drugs. In general, the dose–response relation is little changed but the onset of action is delayed and the duration is prolonged

Increased	Decreased
Body fat	Lean body mass
	Total body water
	Protein binding
	Plasma cholinesterase
	Hepatic blood flow
	Renal blood flow and GFR
	Muscle blood flow
	Number of cholinoceptors

Table 9.2 The effect of ageing on the onset, duration and recovery of non-depolarizing neuromuscular blockade and the associated changes in elimination half-time and plasma clearance

	Ref. Source	Age	Onset time (min)	Clinical duration to T1 = 25% (min)	Recovery index (min)	Elimination half-time (min)	Plasma clearance (ml min^{-1} kg^{-1})
Altracurium	4	Y		15.			
		E		14.5			
	2	Y				20	5.9
		E				23*	5.4
Vecuronium	7	Y			32	15	78
		E			74*	49*	125
Mivacurium	8	Y	3.3	18			
		E	5.0	20			
Rocuronium	9	Y	181				
		E	256*				
	10	Y		26	13	56	5.8
		E		43*	21*	137*	3.4*
Pancuronium	12	Y		55	39	107	1.8
		E		73*	62*	201*	1.2*
Pipecuronium	14	Y			49	114	2.6
		E			62	122	2.7
Doxacurium	17	Y	8.2	80			
		E	10.5*	66			
	16	Y		68		86	2.2
		E		97		96	2.5
Alcuronium	19	Y				176	1.4
	18	E				439	0.53
Tubocurarine	20	Y		64	94	173	1.7
		E		43	48*	268*	0.8*

Y young adult patients
E elderly patients
* $p < 0.05$ compared with young adult patients

Suxamethonium

The synthesis of plasma cholinesterase is impaired in the elderly. However, there is normally a considerable excess of this enzyme in relation to the concentration of the drug. The factors that might influence the extent of suxamethonium blockade, the onset time and the duration have been investigated in a large study.[1] There was no significant correlation between age and any of these variables. It is probable that the variation in plasma cholinesterase in the population as a whole is greater than any difference due to age.

Atracurium

The effect of age on the characteristics of atracurium blockade has been studied extensively. In clinical practice the potency, duration, infusion requirements, spontaneous recovery rate and the reversal of atracurium are the same in the elderly as in a younger population. The results of early pharmacokinetic studies were conflicting. Subsequent studies have established that the pharmacokinetic indices are unaffected by old age[2] with the exception of the elimination half-time which is slightly prolonged (Table 9.2). Hofmann degradation, which is independent of age, accounts for the majority of the elimination of atracurium and the slightly prolonged elimination half-time in the elderly is probably the result of impaired renal excretion of the unchanged drug. The same workers demonstrated that the elimination half-time of laudanosine is also increased in the elderly suggesting that accumulation might occur during very prolonged infusion of atracurium in the Intensive Care Unit. The relation between the plasma concentration of atracurium and its neuromuscular blocking action is unaffected by old age.[3] The rate of infusion of atracurium required to produce stable neuromuscular blockade is the same in young adults and elderly patients.[4] Atracurium is the only non-depolarizing neuromuscular blocking drug to display these favourable characteristics in the elderly.

Vecuronium

Early work demonstrated that a significantly lower infusion rate of vecuronium is required to maintain constant neuromuscular blockade in the elderly and, subsequently, recovery was significantly prolonged.[5] After a short infusion the recovery index and elimination half-time are unchanged in the elderly but the plasma clearance and volume of distribution are reduced.[6] Subsequent work using a single bolus technique that more closely imitates the clinical situation[7] demonstrated a significantly greater clinical duration (74 vs 32 minutes to 25% recovery of T1) and a considerably prolonged recovery index (49 vs 15 minutes) (Table 9.2). The volume of distribution was unchanged in the elderly but the clearance and elimination half-time were adversely affected. In summary, spontaneous recovery and subsequent reversal (Chapter 8) of vecuronium may be prolonged in the elderly and, in common with the muscle relaxant drugs discussed below, it should be used with caution in this group of patients.

Mivacurium

Because it is eliminated largely as a result of hydrolysis by plasma cholinesterase, the characteristics of mivacurium would be expected to be relatively independent of increasing age (see the section on suxamethonium above).

Indeed, the clinical indices and the pharmacokinetics of mivacurium blockade appear to be unchanged in the elderly, with the possible exception of a slightly greater onset time.[8] Larger studies are required to investigate fully the influence of old age on the effects of repeated doses and prolonged infusions of mivacurium.

Rocuronium

The onset of neuromuscular blockade after rocuronium appears to be considerably delayed in the elderly. However, one study used a cumulative dose technique which may have overestimated the effect of age.[9] In a second study the clinical duration (to 25% recovery of T1) was prolonged by 65% and the recovery index was increased by a similar amount.[10]

Pancuronium

Delayed elimination of pancuronium in the elderly accounts for the considerable prolongation of its neuromuscular blocking effects in this group of patients. Approximately 60% of a dose of pancuronium is excreted unchanged by the kidney and, as renal function progressively declines, the elimination of the unchanged drug is reduced by approximately one-half to two-thirds.[11, 12] The clearance falls by approximately 500 ml h^{-1} for each decade. The clinical duration (to 25% recovery of T1) is increased by two-thirds in a group of elderly patients (Table 9.2). In common with the majority of non-depolarizing muscle relaxant drugs, the potency of pancuronium is not increased in the elderly but the onset time may be delayed.[12]

Pipecuronium

Pipecuronium, like its analogue pancuronium, is largely excreted unchanged by the kidney. The volume of distribution and the clearance are greater than pancuronium but the elimination half-time is similar to pancuronium. The potency of pipecuronium is unaltered in the elderly but the duration of neuromuscular blockade after repeated doses may be increased. The kinetic indices after a modest dose do not appear to be changed significantly.[13, 14]

Doxacurium

The potency of doxacurium is unaltered in the elderly,[15] but the normally slow onset time is prolonged further[16] so that conditions for endotracheal intubation after 4 minutes are considerably worse in the elderly.[17] Approximately 50% of a dose of doxacurium is excreted via the renal route and the normal deterioration of renal function that occurs with age would be expected to prolong its neuromuscular blocking action. Initial pharmacokinetic studies suggested that it may be unnecessary to modify the dose regimen of doxacurium in the elderly but subsequent pharmacodynamic data indicate that the neuromuscular blocking effects of doxacurium may be prolonged.

The rate of spontaneous recovery to 25% of the control twitch height after a single dose may be slower in the elderly[15] but some studies suggest that the difference does not reach significance.[16, 17] If reversal with neostigmine is attempted at 25% recovery of T1, attainment of a Train-of-Four ratio of 0.7 is very variable and may take a considerable period of time (mean 17 minutes) and a second dose of neostigmine may be required after 10 minutes.[17]

Older muscle relaxants

There are many anecdotal reports suggesting that the neuromuscular blockade produced by alcuronium is prolonged in the elderly. The pharmacokinetic parameters have been investigated in a single study[18] which demonstrated that the elimination half-time was more than doubled compared with a younger group.[19] Recovery from tubocurarine blockade is also prolonged: 25% recovery of T1 is reached more slowly and the recovery index is almost doubled from 48 to 94 minutes as a result of reduced plasma clearance.[20]

Organ failure

Muscle relaxant drugs are highly charged, hydrophilic compounds which are completely ionized at physiological pH. There is free filtration at the glomeruli and very little reabsorption from the renal tubules. Renal excretion of the parent drug is reduced in renal failure. Several additional mechanisms contribute to the overall changes in the clinical response to muscle relaxant drugs in patients with renal or hepatic dysfunction: for example, changes in the volume of distribution, accumulation of active drug metabolites, changes in cardiac output and hepatic blood flow, inhibition of hepatic enzymes such as cytochrome P450, and derangement of hydrogen ion and electrolyte homeostasis. The concentration of plasma cholinesterase is more variable in hepatic and, to a lesser extent, renal failure and the duration of suxamethonium and mivacurium can be expected to be less predictable than in patients with normal organ function.

Approximately 30–50% of a dose of neuromuscular blocking drug is sequestrated by binding to plasma proteins and is therefore unavailable for interaction with cholinoceptors or glomerular filtration. However, protein binding is modest in comparison with some other classes of drugs used in anaesthesia and it is unlikely that changes in plasma protein concentrations significantly affect the response to neuromuscular blocking drugs.[21] Early work suggested that muscle relaxant drugs are bound to albumin or globulin. More recent investigations have demonstrated that the principal binding proteins are within the alpha 1 globulin fraction: alpha 1 acid glycoprotein appears to be the most important determinant of the extent of binding of basic drugs in disease. It has been suggested that an increase in alpha 1 acid glycoprotein in some malignancies is responsible for the apparent resistance to atracurium.[22] Plasma concentrations of alpha 1 acid glycoprotein are commonly elevated in chronic renal failure but it is unclear whether this factor contributes to the initial resistance to some muscle relaxant drugs which may be observed in renal failure.

Renal impairment is a common accompaniment of hepatic failure[23] and each organ system has an effect on the other. The volume of distribution of muscle relaxant drugs tends to be increased in cirrhotic liver disease and, to a lesser extent, in chronic renal failure. In these circumstances, the effect of a single dose of relaxant drug would be expected to be decreased. Initial resistance to non-depolarizing neuromuscular blocking drugs is well described in cirrhotic liver disease but there is no clinical evidence of this phenomenon in chronic renal failure in isolation. Any tendency for resistance to muscle relaxant drugs is more than offset by impaired plasma clearance, with the exception of atracurium (see below).

A general characteristic of organ failure is that the normal variability in the response of patients to neuromuscular blocking drugs in respect of onset, duration and offset is exaggerated to the point where the unexpected might occur. Neuromuscular monitoring is mandatory in these circumstances.

Renal failure

Suxamethonium

The increased variability of plasma cholinesterase concentration seen in chronic renal failure is unlikely to affect the duration of action of suxamethonium in clinical practice. The suxamethonium-induced increase in serum potassium is no greater in patients with chronic renal failure than in healthy patients: the mean increase is approximately 0.5 mmol l^{-1}.[24] Suxamethonium is not contraindicated in these patients unless the serum potassium concentration is above 5.0 mmol l^{-1}.

Atracurium

Atracurium has become well established as a non-depolarizing muscle relaxant of choice in renal failure. Hofmann elimination permits atracurium to be cleared satisfactorily from the plasma in the absence of renal function: less than 10% is excreted in the urine. In chronic renal failure the duration of action of atracurium is not increased even after many increments or a continuous infusion.[25] One of the major metabolites of atracurium, laudanosine, has no significant neuromuscular blocking properties but has been associated with cerebral excitability in animals (Chapter 5). The plasma concentration of laudanosine is significantly increased 90 minutes after a single dose of atracurium in patients undergoing renal transplantation.[26] There is no evidence to suggest that this is of clinical significance even after prolonged administration of atracurium.

Vecuronium

Vecuronium is excreted mainly via the hepatobiliary route; considerable metabolism to the desacetyl derivative occurs in the liver. Only approximately 20–30% of the parent drug is excreted unchanged by the kidney. The extent and the duration of the neuromuscular blockade induced by a single dose of vecuronium is unchanged in anephric patients and the duration with small, repeated doses (40 μg kg^{-1}) is not increased, indicating that accumulation does not occur. In addition, reversal with neostigmine is not jeopardized in this group of patients.[25] However, problems may arise when vecuronium is given in large, repeated doses or by continuous infusion and, in these circumstances, the duration may be considerably increased as a result of accumulation of both the parent drug and the desacetyl derivative (see below) which is renally excreted and has neuromuscular blocking properties.[27] Prolonged administration of vecuronium in patients with impaired renal function cannot be recommended for this reason.[28]

Mivacurium

Mivacurium is a new, intermediate-onset, short-acting non-depolarizing muscle relaxant which depends on hydrolysis by plasma cholinesterase for its inactivation. It is hydrolysed at approximately 90% of the rate of suxamethonium.[29] Renal excretion of the unchanged drug is minimal.[30, 31] The duration of action of mivacurium is critically dependent on the concentration of cholinesterase in the plasma (Chapter 5). Plasma cholinesterase may be reduced in renal failure but not to the extent that is observed in hepatic failure,[32] although the normal variation between patients is exaggerated in chronic renal impairment.[33] There is some disagreement concerning the effect of chronic renal failure on the duration of action of a single dose of mivacurium but it appears that it is unlikely to be prolonged to a clinically relevant extent.[31, 33] The infusion rate of mivacurium required to maintain a steady 95% twitch depression is reduced by approximately 40% in renal failure.[33] Mivacurium appears to offer few advantages over atracurium in patients with chronic renal failure.

Rocuronium

Rocuronium is a new rapid-onset, intermediate-duration non-depolarizing, steroidal muscle relaxant drug. It appears that the hepatobiliary route of excretion predominates, although there are few published data. Renal excretion accounts for only 10–20% of the total dose administered.[34] The mean neuromuscular blocking effect and duration of action of initial and repeated doses are unchanged in renal failure but there is wide variation between patients.[35] The plasma clearance is reduced by approximately one-third in chronic renal failure but the volume of distribution is unchanged.[36] It might be predicted that accumulation would result if a prolonged continuous infusion were given to a patient with significant renal impairment.

Pancuronium

Approximately 60% of a dose of pancuronium is excreted unchanged by the kidney. A proportion of the remainder is eliminated unchanged in the bile but there is significant biotransformation to an active (desacetyl) metabolite which may accumulate if renal function is impaired. The plasma clearance of pancuronium is reduced by two-thirds in renal failure.[37] It is not surprising

Organ failure 163

that the duration of pancuronium may be extremely prolonged in chronic renal failure if large or repeated doses are given.[38]

Pipecuronium

Pipecuronium in a single dose exhibits a near-normal duration of action in renal failure.[39] However, this characteristic is derived from the opposing influences of an increased volume of distribution and a reduced plasma clearance. Accumulation might be expected with repeated doses.

Doxacurium

Doxacurium is largely eliminated unchanged in the bile and in the urine. The renal route probably accounts for 50% of the total dose and in anephric patients the mean residence time (Chapter 3) is increased by approximately 70%. The duration of a single dose tends to be variable and prolonged in chronic renal failure.[40]

Older muscle relaxants

Approximately 80% of the elimination of alcuronium is via the kidney as the unchanged drug. The plasma elimination half-time is prolonged four-fold in patients with impaired renal function in whom alcuronium should be avoided.

Over 90% of gallamine is excreted unchanged in the urine and the effect of a single dose may last for several days in an anephric patient.[41] Currently, the use of gallamine is confined to pre-curarization to reduce the incidence of suxamethonium myalgia (Chapter 4). If a small dose such as this were to be administered inadvertently to a patient with impaired renal function it is likely that redistribution would limit its duration.[42]

The clinical use of fazadinium was terminated largely because of the high incidence of severe tachycardia. It is predominantly excreted by the kidney and its elimination is prolonged in renal failure.[43]

Approximately two-thirds of the excretion of tubocurarine is via the kidney. It has been used successfully in patients with renal dysfunction but there are several reports of apparent re-curarization as a result of prolonged elimination.[44] The effect of a single, small dose may be terminated by redistribution but the elimination route cannot be relied upon in renal failure and tubocurarine is considered by most to be unsuitable in this situation. Metocurine is even more dependent on renal elimination than tubocurarine.

Antagonism of neuromuscular blockade

Renal dysfunction and cardiac impairment coexist in many disease states, e.g. diabetes mellitus, severe hypertension, peripheral vascular disease, pre-eclampsia and severe sepsis. Stability of the cardiovascular system is therefore a second, significant factor in choosing a muscle relaxant drug in these clinical situations. It is also important to avoid the excessive tachycardia and hypertension sometimes associated with the administration of neostigmine and atropine. Glycopyrrolate is to be preferred to atropine (Chapter 8). The plasma clearance of neostigmine is reduced in anephric patients to a greater extent than that of atracurium or vecuronium. This phenomenon represents a safety factor provided that two Train-of-Four responses are present before reversal is attempted.

Hepatic failure

Suxamethonium

The concentration of plasma cholinesterase may be reduced by 70% in hepatic failure and the duration of action of suxamethonium is increased. However, the plasma cholinesterase concentration must be considerably depleted before clinically significant prolongation of neuromuscular blockade occurs and hepatic failure is not a contraindication to the use of suxamethonium provided that neuromuscular monitoring is used.

Non-depolarizing neuromuscular blocking drugs

Although the liver plays only a subsidiary part in the elimination of muscle relaxant drugs (mivacurium is a special case, see below), severe hepatic dysfunction may have clinically significant effects. The steroidal muscle relaxants undergo biotransformation in the liver in addition to excretion of the unchanged drug in the bile. Approximately 10–20% of the total dose of these drugs is metabolized in this way by hydrolysis of the acetyl group at the 3 and 17 positions. The 3-OH desacetyl derivatives have neuromuscular blocking properties and may accumulate in renal failure (see above). It is useful to make the distinction between obstructive liver disease (cholestatic) and cirrhotic liver disease.

Cholestatic liver disease

In obstructive liver disease biliary excretion is reduced without severe impairment of drug metabolism. Those relaxants which are more dependent on biliary excretion, notably the aminosteroid drugs, exhibit a prolonged duration of action. The duration of pancuronium-induced blockade may be increased by 60%[45] and that of vecuronium by 45%.[46] Opinion is divided concerning the extent of the increase in volume of distribution of pancuronium in patients with cholestasis. It has been suggested that high concentrations of bile salts may inhibit biliary uptake of steroid muscle relaxant drugs.[47] The clinical characteristics of atracurium are unchanged in cholestatic liver disease.

Cirrhotic liver disease

Hepatic blood flow and hepatocyte function are reduced in cirrhosis. In general, cirrhosis is associated with an initial resistance to non-depolarizing muscle relaxant drugs as a result of the increased volume of distribution.

Subsequent doses may have a prolonged duration of effect as a consequence of reduced plasma clearance. It is now considered that the reduction in the concentrations of plasma proteins in cirrhotic patients plays only a small part in the resistance to muscle relaxants and this factor cannot be responsible for the observed increase in their volume of distribution.

Atracurium

There is no evidence of initial resistance to atracurium and the duration of blockade is unaffected in cirrhosis.[48] The development of resistance to atracurium during a prolonged infusion in the Intensive Care Unit has been described (see Chapter 12) but this phenomenon does not appear to be important in patients with fulminant hepatic failure receiving a prolonged infusion.[49] It has been demonstrated that some of the breakdown products of atracurium, particularly the acrylates, may cause damage to isolated hepatocytes *in vitro* in high concentrations.[50] However, this effect cannot be reproduced in the perfused rat liver model.[51]

Vecuronium

Both the extent and duration of effect of a single dose of vecuronium (0.1–0.15 mg kg^{-1}) are slightly reduced in cirrhotic patients. Larger doses should be contemplated with caution because neuromuscular blockade may be greatly prolonged and the duration to 25% recovery of a dose of 0.2 mg kg^{-1} may be increased from 60 to 90 minutes.[48, 52]

Mivacurium

The rate of metabolism of mivacurium is critically dependent on the concentration of plasma cholinesterase which may be markedly reduced in cirrhotic liver disease. The clinical duration of a single dose of mivacurium (0.15 mg kg^{-1}) is prolonged by 70–15% depending upon the severity of the disease (Table 9.3). In this situation mivacurium loses its short-acting status and its duration is similar to that of atracurium or a single dose of vecuronium.[53]

Rocuronium

Rocuronium is mainly eliminated in the bile.[54] There is some initial resistance to rocuronium as a result of a larger initial volume of distribution. There is some evidence that accumulation of rocuronium may be less extensive than that observed after repeated doses of the other steroidal neuromuscular blocking drugs.[55]

Pancuronium

The volume of distribution of pancuronium is increased by approximately 50% in cirrhosis and it would be necessary to increase the initial dose by approximately 10–20% to produce the expected degree of neuromuscular

Table 9.3 The effect of renal failure and hepatic failure on the plasma cholinesterase concentration and the pharmacodynamics and pharmacokinetics of mivacurium which is dependent on plasma cholinesterase for its metabolism (mean and SD)

	Healthy (n = 9)	Hepatic failure (n = 9)	Renal failure (n = 9)
Plasma cholinesterase (iu ml^{-1})	4.9 (1.3)	1.4* (0.9)	4.9** (1.5)
Dibucaine number (% inhibition)	82.9 (1.2)	47.4* (12.4)	83.0** (3.7)
Maximum block (%)	99.8 (0.7)	100 (0)	100 (0)
Duration to 25% T1 (min)	18.7 (6.2)	57.2* (18.6)	30.1 (12.8)
Recovery index (min)	5.5 (2.5)	15.7* (9.3)	11.5 (3.6)
Plasma clearance (ml kg min^{-1})	70.4 (28.1)	33.3* (13.8)	76.6** (43.6)

* $p < 0.05$ compared with healthy patients.
** $p < 0.05$ compared with hepatic failure patients.
After Cook DR et al. (1992).[31]

blockade. Only approximately 10% of a dose of pancuronium is excreted in the bile. However, measurement of biliary excretion alone invariably underestimates the role of the liver in sequestering the drug by hepatic uptake; the terminal elimination half-time is increased by approximately 80% and blockade may be extremely prolonged.[56]

Pipecuronium

Only approximately 2% of a dose of pipecuronium is excreted in the bile. In patients with cirrhosis, the onset of neuromuscular blockade is prolonged but the duration is unchanged.[57]

Doxacurium

Doxacurium is largely excreted by the kidney and the pharmacokinetics are not significantly changed in patients presenting for liver transplantation.[40]

Older muscle relaxants

There is very little information available about the pharmacokinetics of alcuronium and tubocurarine in patients with hepatic failure. If renal function is preserved, it might be expected that the clinical course of neuromuscular blockade after alcuronium or gallamine would be unchanged, with the exception of some initial resistance and this is supported by the limited data available. Resistance to tubocurarine in cirrhotic patients is a well-known phenomenon but the mechanism is still unclear. Early work demonstrated a positive relation between the requirements for tubocurarine and the plasma globulin concentration[58] and it was supposed that large quantities of the drug were sequestered by protein binding and effectively removed from the biophase. However, subsequent work cast doubt on the importance of plasma protein binding as a determinant of the extent of the observed neuromuscular blockade.[59] Biliary excretion accounts for only approximately 10% of the excretion of tubocurarine although this may be increased three-fold in the absence of renal function.

Denervation

The risk of precipitating severe hyperkalaemia associated with the administration of suxamethonium has been discussed in Chapter 4. Non-depolarizing muscle relaxants may be used safely in this situation, but resistance to this group of drugs is evident in the affected limb irrespective of whether the denervation is functional or anatomical, e.g. stroke[60] or immobility due to a plaster cast[61] or prolonged mechanical ventilation in the Intensive Care Unit. Neuromuscular monitoring should be performed on the normally functioning side in hemiplegic patients. If a muscle in the paretic limb is monitored, the twitch response will be greater than that of normally innervated muscles (including the muscles of respiration) and the unwary anaesthetist may be encouraged to give excessive doses of muscle relaxant.

Burns

The use of suxamethonium in burned patients is discussed in Chapter 4. Patients with severe burns (greater than 20% of the body surface area) are resistant to non-depolarizing muscle relaxants. Two factors contribute: changes in plasma protein binding and up-regulation of the acetylcholine receptors. The latter mechanism is the more important. Resistance appears after the first week and the dose requirement is related to the extent of the burn injury. The ED95 of atracurium may be increased ten-fold in a 60% burn.[62] Lesser resistance has been demonstrated with pancuronium.[63]

The changes observed at receptor level are shared by patients with burns and those who have denervated muscle (see above). There is an increase in the number of cholinoceptors which spread over the surface of the muscle and are not confined to the neuromuscular junction. The cholinoceptors in normal, adult muscle comprise alpha (two), beta, epsilon and delta units. Cholinergic agonists bind at the two alpha subunits but the nature of the binding process is influenced by the surrounding subunits such that the two alpha subunits have different binding affinities (Chapter 2). The cell nuclei of normal, adult muscle cells retain the potential to synthesize the gamma subunits that are normally found only in infancy. When muscle is deprived of normal, frequent neuromuscular transmission, e.g. after burn injury or denervation, gamma subunits are synthesized and become substituted for the adult-type epsilon subunits. The normal adult cholinoceptor has a half-time of approximately 2 weeks which may explain why the clinical changes in the sensitivity of the cholinoceptor to muscle relaxants are not observed for some days after the injury. The abnormal receptors are not confined to the neuromuscular junction and proliferate progressively throughout the muscle membrane as extrajunctional receptors. It is important to note that the number of intrajunctional receptors is also increased in these clinical conditions (see below). A very small dose of agonist, e.g. suxamethonium, will cause depolarization in the muscle membrane surrounding the extrajunctional receptors. These gamma-substituted receptors have different binding characteristics and a greater concentration of non-depolarizing neuromuscular blocking drug is required to inhibit the binding of acetylcholine.

Four distinct mechanisms may contribute to the relative resistance of patients with burns (or denervation) to non-depolarizing muscle relaxants.

1. The extrajunctional receptors (which do not take part in neuromuscular transmission) act as a sump for these agents by sequestering sufficient molecules of the drug to reduce its concentration available to the intrajunctional receptors.
2. Those gamma-substituted cholinoceptors which lie within the neuromuscular junction and take part in neuromuscular transmission bind less firmly with non-depolarizing muscle relaxant drugs.
3. The 'margin of safety' is increased in those muscles with large numbers of intrajunctional receptors. Because agonist and antagonist are in dynamic equilibrium at the neuromuscular junction, the *proportion* of receptors occupied will be constant for a given ratio of agonist and antagonist molecules. If the number of functional, intrajunctional receptors is now increased by burn injury or denervation (see above), the number of receptors remaining unblocked is also increased. The process of neuromuscular transmission depends on the *number* of unblocked cholinoceptors. If the release of acetylcholine from the motor nerve terminal is unaltered, a given dose of non-depolarizing blocking drug will diminish neuromuscular transmission to a lesser extent than in a motor unit with fewer receptors. This mechanism is, of course, the likely explanation of the relative resistance of the diaphragm to non-depolarizing muscle relaxants.
4. Acetylcholinesterase activity may be diminished in the injured muscle, permitting acetylcholine molecules to exist for longer within the neuromuscular junction and minimizing the time for which the antagonist drug molecules can bind with the receptor. This is unlikely to be a major effect.

Acid-base disorders

Metabolic acidosis is a common feature in chronic renal failure. Metabolic acidosis and respiratory acidosis appear to have different effects on neuromuscular blockade in some situations. For example, tubocurarine is apparently potentiated by respiratory acidosis and antagonized by metabolic acidosis.[64] Respiratory acidosis potentiates pancuronium and vecuronium in man,[65] but the effects of metabolic acidosis have not been studied in man. Experimental metabolic acidosis produced by the deliberate infusion of an acid antagonizes tubocurarine but it may be unwise to extrapolate to the clinical situation in which acidosis is often associated with decreased tissue and organ perfusion, anaerobic metabolism and hyperkalaemia: experimental and clinical data are conflicting. The rate of Hofmann elimination is retarded at low pH and the duration of atracurium might be expected to be prolonged by acidosis. However, the magnitude of this effect is too small to be of clinical significance. In summary, it is probably safe in clinical practice to assume that non-depolarizing relaxant drugs (with the exception of atracurium) may be potentiated in clinical situations where there is acidosis and particular attention should be given to monitoring the process of maintenance and reversal of neuromuscular blockade. The phenomenon of re-curarization is discussed in Chapter 8. Respiratory alkalosis induced by deliberate hyperventilation ($PaCO_2 = 25$ mmHg) decreases the requirement for pancuronium and vecuronium[65] and it is probable that the newer neuromuscular blocking drugs are similarly affected. The effect of derangements in acid-base balance on the reversal of neuromuscular blockade are discussed in Chapter 8.

Acquired neuromuscular disorders

Myasthenia gravis

Neuromuscular function is reduced in myasthenia as a result of impaired synaptic transmission at the neuromuscular junction. There is a reduction in the number of acetylcholine receptors on the postjunctional membrane which appears to be the result of an autoimmune process. Autoantibodies against cholinoceptors are detectable in approximately 80% of patients and it is possible to identify a specific region on the alpha subunit to which they bind. The loss of functional receptors encroaches on the 'margin of safety'. There is classically exercise-related fatigue and a progressive reduction in the twitch response to repetitive stimulation.

The effect of muscle relaxant drugs in the myasthenic patient depends on concurrent drug therapy as well as the duration and severity of the disease. Many authors have commented on the variability of the response to neuromuscular blocking drugs and it is preferable to avoid these agents if possible by using local or regional anaesthesia. Inhalational agents impair neuromuscular transmission to a greater than normal extent in myasthenia gravis and the force of the twitch response may fall by 50–60% in the absence of a neuromuscular blocking drug. There is a theoretical risk that chronic anticholinesterase therapy may inhibit plasma cholinesterase in addition to the expected inhibition of acetylcholinesterase at the neuromuscular junction with the result that the hydrolysis of the ester local anaesthetics may be impeded.[66] If it is imperative that a muscle relaxant is used, facilities should be available for the continuation of mechanical ventilation in the postoperative period if full neuromuscular function cannot be restored at the end of surgery.

Preoperative preparation should be directed towards optimizing the therapeutic regimen if time permits. It has been suggested that the dose of anticholinesterase should be reduced preoperatively so that additional muscle relaxation is unnecessary during the surgical procedure,[67] but this approach would seem to predispose to postoperative ventilatory insufficiency. In severe cases, plasmaphoresis may produce an improvement in neuromuscular function that persists into the postoperative period.[68]

Preoperative tests of pulmonary function have been disappointing in predicting the need for postoperative mechanical ventilation[69] but estimation of FVC and FEV1 at least provides baseline data with which to compare postoperative values. A history of myasthenia longer than 6 years, coexisting respiratory disease and a requirement for pyridostigmine greater than 750 mg a day have been demonstrated to be predictive of the need for postoperative mechanical ventilation.[70] Myasthenic patients may be treated with long-term corticosteroid therapy and consequent adrenal suppression must be considered when planning the anaesthetic technique.

Myasthenic patients are characteristically sensitive to non-depolarizing muscle relaxants and an initial dose of approximately one-fifth of the usual dose may be adequate.[67] Monitoring of neuromuscular transmission throughout anaesthesia is mandatory. Reversal of residual non-depolarizing blockade may paradoxically increase the extent of Train-of-Four fade. This phenomenon may be the result of a rapid increase in the first response of the Train-of-Four to greater than the pre-relaxant value.[67] It is probably desirable to avoid the use of reversal agents by using a non-depolarizing muscle relaxant with a rapid recovery rate.

Because the number of functional receptors is reduced in myasthenia gravis, there is resistance to agonist agents, including suxamethonium.[71] The dose of suxamethonium necessary to facilitate endotracheal intubation may be increased to 1.5 mg kg^{-1}. The duration of a single dose of suxamethonium appears to be relatively unaffected by myasthenia in isolation. Repeated doses demonstrate a progressive increase in duration; in contrast to the non-myasthenic patient. This may be the result of a further reduction of the margin of safety consequent on progressive desensitization of the end-plate receptors.[72] The duration of a single dose of suxamethonium may be prolonged in myasthenia gravis if plasma cholinesterase is inhibited by recent, chronic therapy with anticholinesterases, e.g. pyridostigmine. The duration of suxamethonium 1.5 mg kg^{-1} may be increased to 35 minutes in these circumstances.[73]

The choice of muscle relaxant during anaesthesia is governed by the need to ensure recovery of neuromuscular function to as near normal as possible at the end of the surgical procedure. A drug which is normally short acting would appear to have considerable advantages. Suxamethonium in an increased dose is satisfactory for intubation but repeated doses demonstrate bradyphylaxis (see above). The short duration of mivacurium suggests its use in myasthenia gravis but inhibition of plasma cholinesterase by oral anticholinesterase therapy might be expected to prolong the duration of blockade to an unacceptable extent. There is considerable clinical experience of atracurium in myasthenia gravis patients and this drug appears to be satisfactory if the dose is reduced by approximately 80%.[67] A short-acting non-depolarizing muscle relaxant which is independent of plasma cholinesterase for its metabolism would appear to be desirable in myasthenia gravis.

Myasthenic syndrome

First described by Eaton and Lambert, this syndrome is characterized by proximal limb weakness which is ameliorated by exercise, and autonomic dysfunction. These clinical observations distinguish myasthenic syndrome from myasthenia gravis. There is commonly an association with bronchial carcinoma, usually small cell.

The neuromuscular defect is at the prejunctional membrane and the motor nerve terminals may be atrophic on microscopy. The number of postjunctional cholinoceptors is normal. In myasthenic syndrome there is evidence that autoantibodies bind to the voltage-gated calcium channels at the motor nerve terminal and a reduced number of quanta is released in response to nerve stimulation. The quantal content and, therefore, the MEPP amplitude, is normal. In contrast to myasthenia gravis which is characterized by tetanic fade, patients with myasthenic syndrome demonstrate an increase in the muscle twitch in response to rapid nerve stimulation.

In common with myasthenia gravis, these patients are sensitive to non-depolarizing muscle relaxants which should be used in a much reduced dose. The response to anticholinesterase drugs is unreliable and it is suggested that, at the end of surgery, mechanical ventilation is continued until spontaneous recovery from the non-depolarizing neuromuscular blocking drug has reached a maximum.

Inherited muscle disease

The inherited muscle diseases may by grouped into the muscular dystrophies, the myotonias, glycogen storage diseases and the myopathies[74] (Table 9.4). In general, severe cardiac impairment may occur in all inherited muscle diseases. Respiratory dysfunction is common and may be due to muscle weakness or severe deformity of the thoracic spine and ribs.

Suxamethonium and anticholinesterases should be avoided in the myotonias because the resulting, severe muscle contracture may render ventilation impossible. In the glycogen storage diseases suxamethonium may lead to the production of myoglobinuria with the associated risk of renal failure. Severe hyperkalaemia after suxamethonium has been described in Duchenne muscular dystrophy.[75] Malignant hyperpyrexia (MH) is discussed in Chapter 4. Not all myopathies are associated with a risk of MH but it is wise to avoid suxamethonium in these patients in the absence of published data supporting its safety. There is little published information concerning the use of non-depolarizing muscle relaxants in inherited muscle disease. Patients with myotonia or Duchenne muscular dystrophy do not appear to be abnormally sensitive.[76-78] Nevertheless, neuromuscular monitoring is mandatory and anticholinesterases should probably be avoided.

Table 9.4 Potential anaesthetic hazards associated with muscle disorders

	Myotonia	Contracture with cholinergics	Cardiac impairment	Respiratory impairment	MH susceptibility	Cardiac arrhythmias
Muscular dystrophy	—	+/−[a]	++	++	+/−	+
Dystrophia myotonica	+	++	++[b]	+	—	++
Myotonia congenita	++	++	+/−	+/−	+	—
Glycogen storage diseases	—	[a]	++/−	+/−	—	+/−
Myopathies	—	++	—	+/−	assume ++	+/−

[a] Suxamethonium may produce myoglobinuria and renal failure.
[b] Mitral valve defects common.

References

1. Vanlinthout LEH, Van Egmond J, De Boo T, et al. Factors influencing magnitude and time course of neuromuscular block produced by suxamethonium. *British Journal of Anaesthesia* 1992; **69:** 29–35.
2. Kent AP, Parker CJR, Hunter JM. Pharmacokinetics of atracurium and laudanosine in the elderly. *British Journal of Anaesthesia* 1989; **63:** 661–666.
3. Parker CJR, Hunter JM, Snowdon SC. The effect of age, gender and anaesthetic technique on the pharmacodynamics of atracurium. *British Journal of Anaesthesia* 1993; **70:** 38–41.
4. d'Hollander AA, Luyckx C, Barvais L, De Ville A. Clinical evaluation of atracurium besylate requirement for a stable muscle relaxation during surgery: lack of age-related effects. *Anesthesiology* 1983; **59:** 237–240.
5. d'Hollander A, Massaux F, Nevelsteen M, Agoston S. Age-dependent dose–response relationship of ORG NC45 in anaesthetized patients. *British Journal of Anaesthesia* 1982; **54:** 653–657.
6. Rupp SM, Fisher DM, Miller RD, Castagnoli K. Pharmacokinetics and pharmacokinetics of vecuronium in the elderly. *Anesthesiology* 1987; **67:** 45–49.
7. Lien CA, Matteo RS, Ornstein E, et al. Distribution, elimination and action of vecuronium in the elderly. *Anesthesia and Analgesia* 1991; **73:** 39–42.
8. Basta SJ, Dresner DL, Shaff LP, et al. Neuromuscular effects and pharmacokinetics of mivacurium in elderly patients under isoflurane anaesthesia. *Anesthesia and Analgesia* 1989; **68:** S18.
9. Fiset P, Balendran P, Bevan DR. Onset, duration and recovery from ORG 9426 in the elderly. *Anesthesiology* 1990; **73:** A881.
10. Matteo RS, Ornstein E, Schwartz AE, et al. Pharmacokinetics and pharmacodynamics of ORG 9426 in elderly surgical patients. *Anesthesiology* 1991; **75:** A1065.
11. McLeod K, Hull CJ, Watson MJ. Effects of ageing on the pharmacokinetics of pancuronium. *British Journal of Anaesthesia* 1979; **57:** 435–438.
12. Duvaldestin P, Saada J, Berger JL, et al. Pharmacokinetics, pharmacodynamics, and dose–response relationships of pancuronium in control and elderly subjects. *Anesthesiology* 1982; **56:** 36–40.
13. Azad SS, Larijani GE, Goldberg ME, et al. A dose–response evaluation of pipecuronium bromide in elderly patients under balanced anaesthesia. *Journal of Clinical Pharmacology* 1989; **29:** 657–659.
14. Matteo RS, Schwartz AE, Ornstein E, et al. Pharmacokinetics and pharmacodynamics of pipecuronium in elderly surgical patients. *Anesthesia and Analgesia* 1991; **72:** S172.
15. Koscielniak-Nielsen ZJ, Law-Min JC, Donati F, et al. Dose–response relations of doxacurium and its reversal with neostigmine in young adults and healthy elderly patients. *Anesthesia and Analgesia* 1992; **74:** 845–850.
16. Dresner DL, Basta SJ, Ali HH, et al. Pharmacokinetics and pharmacodynamics of doxacurium in young and elderly patients during isoflurane anaesthesia. *Anesthesia and Analgesia* 1990; **71:** 498–502.
17. Martlew RA, Harper NJN. The pharmacodynamics of doxacurium in elderly patients. *Anaesthesia* (in press).
18. Stephens ID, Ho PC, Holloway DWA, et al. Pharmacokinetics of alcuronium in elderly patients undergoing total hip replacement or aortic reconstruction surgery. *British Journal of Anaesthesia* 1984; **56:** 465–471.
19. Walker J, Shanks CA, Triggs EJ. Clinical pharmacokinetics of alcuronium chloride in man. *European Journal of Clinical Pharmacology* 1980; **17:** 449–454.
20. Matteo RS, Backus WW, McDaniel DD, et al. Pharmacokinetics and pharmacodynamics of d-tubocurarine and metocurine in the elderly. *Anesthesia and Analgesia* 1985; **64:** 23–29.

21. Hunter JM. Resistance to non-depolarizing neuromuscular blocking agents. Editorial. *British Journal of Anaesthesia* 1991; **67:** 511–514.
22. Tatman AJ, Wrigley SR, Jones RM. Resistance to atracurium in a patient with an increase in alpha 1 globulins. *British Journal of Anaesthesia* 1991; **67:** 623–625.
23. Eston AC, Bayliss MK, Park GR. Effect of renal failure on drug metabolism by the liver. *British Journal of Anaesthesia* 1993; **71:** 282–290.
24. Miller RD, Way WL, Hamilton WK, et al. Succinylcholine-induced hyperkalemia in patients with renal failure? *Anesthesiology* 1972; **36:** 138–141.
25. Hunter JM, Jones RS, Utting JE. Comparison of vecuronium, atracurium and tubocurarine in normal patients and in patients with no renal function. *British Journal of Anaesthesia* 1984; **56:** 941–951
26. Fahey MR, Rupp SM, Canfell SM, et al. Effect of renal failure on laudanosine excretion in man. *British Journal of Anaesthesia* 1985; **57:** 1049–1051.
27. Smith CL, Hunter TM, Jones RS. Vecuronium infusions in patients with renal failure in an ITU. *Anaesthesia* 1987; **42:** 387–393.
28. Slater RM, Pollard BJ, Doran BRH. Prolonged neuromuscular blockade with vecuronium in renal failure. *Anaesthesia* 1988; **43:** 250–251.
29. Ali HH, Savarese JJ, Embree PB, et al. Clinical pharmacology of mivacurium chloride BW 1090U infusion: comparison with vecuronium and atracurium. *British Journal of Anaesthesia* 1988; **61:** 541–546.
30. Savarese JJ, Ali HH, Basta SJ, et al. Clinical neuromuscular pharmacology of mivacurium chloride BW1090U. *Anesthesiology* 1984; **68:** 723–732.
31. Cook DR, Freeman JA, Lai AA, et al. Pharmacokinetics of mivacurium in normal patients and in those with hepatic or renal failure. *British Journal of Anaesthesia* 1992; **69:** 580–583.
32. Ryan RW. Preoperative serum cholinesterase concentration in chronic renal failure. *British Journal of Anaesthesia* 1977; **49:** 945–949.
33. Phillips BJ, Hunter JM. Use of mivacurium chloride by constant infusion in the anephric patient. *British Journal of Anaesthesia* 1992; **68:** 492–498.
34. Wierda JMKH, Kleef UW, Lambalk LM, et al. The pharmacodynamics and pharmacokinetics of ORG 9426, a new non-depolarizing neuromuscular blocking agent, in patients anaesthetized with nitrous oxide, halothane and fentanyl. *Canadian Journal of Anaesthesia* 1991; **38:** 430–435.
35. Khuenl-Brady KS, Pomaroli A, Pühringer F, et al. The use of rocuronium ORG 9426 in patients with chronic renal failure. *Anaesthesia* 1993; **48:** 873–875.
36. Cooper RA, Maddineni VR, Mirakhur RK, et al. Time course and pharmacokinetics of rocuronium bromide ORG 9426 during isoflurane anaesthesia in patients with and without renal failure. *British Journal of Anaesthesia* 1993; **71:** 222–226.
37. McLeod K, Watson MJ, Rawlins MD. Pharmacokinetics of pancuronium in patients with normal and impaired renal function. *British Journal of Anaesthesia* 1976; **48:** 341–345.
38. d'Hollander AA, Camu F, Sanders M. Comparative evaluation of neuromuscular blockade after pancuronium administration in patients with and without renal failure. *Acta Anaesthesiologica Scandinavica* 1978; **22:** 21–26.
39. Caldwell JE, Canfell PC, Castagnoli KP, et al. The influence of renal failure on the pharmacokinetics and duration of pipecuronium bromide in patients anesthetized with halothane and nitrous oxide. *Anesthesiology* 1989; **70:** 7–12.
40. Cook DR, Freeman JA, Lai AA, et al. Pharmacokinetics and pharmacokinetics of doxacurium in normal patients and in those with hepatic or renal failure. *Anesthesia and Analgesia* 1991; **72:** 145–150.
41. Feldman SA, Levi JA. Prolonged paresis following gallamine. *British Journal of Anaesthesia* 1963; **35:** 804–806.

42. White RD, de Weerd JH, Dawson B. Gallamine in anaesthesia for patients with chronic renal failure undergoing bilateral nephrectomy. *Anesthesia and Analgesia* 1971; **50:** 11–16.
43. Duvaldestin P, Bertrand P, Concina D, *et al*. Pharmacokinetics of fazadinium in patients with renal failure. *British Journal of Anaesthesia* 1979; **51:** 943–947.
44. Miller RD, Cullen DJ. Renal failure and postoperative respiratory failure: recurarization. *British Journal of Anaesthesia* 1976; **48:** 253–256.
45. Somogyi AA, Shanks CA, Triggs EJ. Disposition kinetics of pancuronium bromide in patients with total biliary obstruction. *British Journal of Anaesthesia* 1977; **49:** 1103–1108.
46. Lebrault C, Duvaldestin P, Henzel D, Chauvin M, Guesnon P. Pharmacokinetics and pharmacodynamics of vecuronium in patients with cholestasis. *British Journal of Anaesthesia* 1986; **58:** 983–987.
47. Westra P, Keulemans GTP, Houwertjes MC, *et al*. Mechanisms underlying the prolonged duration of action of muscle relaxants caused by extrahepatic cholestasis. *British Journal of Anaesthesia* 1981; **53:** 217–226.
48. Bell CF, Hunter JM, Jones RS, Utting JE. Use of atracurium and vecuronium in patients with oesophageal varices. *British Journal of Anaesthesia* 1985; **57:** 160–168.
49. Bion JF, Bowden MI, Chow B, Honisberger L, Weatherley BC. Atracurium infusions in patients with fulminant hepatic failure awaiting liver transplantation. *Intensive Care Medicine* 1993; **19:** S94–S98.
50. Nigrovic V, Klaunig JE, Smith SL. Potentiation of atracurium toxicity in isolated rat hepatocytes by inhibition of its hydrolytic degradation pathway. *Anesthesia and Analgesia* 1987; **66:** 512–516.
51. Reckendorfer H, Burgmann H, Sperlich M, *et al*. Hepatotoxicity testing of atracurium and laudanosine in the isolated, perfused rat liver. *British Journal of Anaesthesia* 1992; **69:** 288–291.
52. Duvaldestin P, Berger JL, Videcoq M, Desmonts JM. Pharmacokinetics and pharmacodynamics of ORG NC45 in patients with cirrhosis. *Anesthesiology* 1982 **57:** A238.
53. Devlin JC, Head-Rapson AG, Parker CJR, Hunter JM. Pharmacodynamics of mivacurium chloride in patients with hepatic cirrhosis. *British Journal of Anaesthesia* 1993; **71:** 227–231.
54. Khuenl-Brady K, Castagnoli KP, Canfell PC. The neuromuscular blocking effects and pharmacokinetics of ORG 9426 and ORG 9616 in the cat. *Anesthesiology* 1990; **72:** 669–674.
55. Servin F, Lavault E, Desmonts JM. Pharmacokinetics of repeated doses of rocuronium in cirrhotic and control patients. *Anesthesiology* 1993; **79:** A962.
56. Duvaldestin P, Agoston S, Henzel D, *et al*. Pancuronium pharmacokinetics in patients with hepatic cirrhosis. *British Journal of Anaesthesia* 1978; **50:** 1131–1136.
57. D'Honneur GD, Khalil M, Dominique C, *et al*. Pharmacokinetics and pharmacodynamics of pipecuronium in patients with cirrhosis. *Anesthesia and Analgesia* 1993; **77:** 1203–1206.
58. Baraka A, Gabaldi F. Correlation between tubocurarine requirements and plasma protein pattern. *British Journal of Anaesthesia* 1968; **40:** 89–93.
59. Duvaldestin P, Henzel D. Binding of tubocurarine, fazadinium, pancuronium and ORG NC45 to serum proteins in normal man and in patients with cirrhosis. *British Journal of Anaesthesia* 1982; **54:** 513–516.
60. Iwasaki H, Namika A, Omote K, *et al*. Response differences of paretic and healthy extremities to pancuronium and neostigmine in hemiplegic patients. *Anesthesia and Analgesia* 1985; **64:** 864–866.
61. Gronert GA. Disuse atrophy with resistance to pancuronium. *Anesthesiology* 1981; **55:** 547–549.

62. Mills AK, Martin JAJ. Evaluation of atracurium neuromuscular blockade in paediatric patients with burn injury. *British Journal of Anaesthesia* 1988; **60:** 450–455.
63. Martyn J, Goldhill DR, Goudsouzian NG. Clinical pharmacology of muscle relaxants in patients with burns. *Journal of Clinical Pharmacology* 1986; **26:** 680–685.
64. Miller RD, Van Nyhuis LS, Eger II EI, Way WL. The effect of acid-base balance on neostigmine antagonism and on d-tubocurarine-induced neuromuscular blockade. *Anesthesiology* 1975; **42:** 377–381.
65. Gencarelli PJ, Swen J, Koot HWJ, Miller RD. The effects of hypercarbia and hypocarbia on pancuronium and vecuronium neuromuscular blockades in anesthetized humans. *Anesthesiology* 1983; **59:** 376–380.
66. Baraka A. Anaesthesia and myasthenia gravis. *Canadian Journal of Anaesthesia* 1992; **39:** 476–486.
67. Bell CF, Florence AM, Hunter JM, Jones RS, Utting JE. Atracurium in the myasthenic patient. *Anaesthesia* 1984; **39:** 961–968.
68. d'Empaire G, Hoaglin DC, Perlo VP, Pontoppidian H. Effect of prethymectomy plasma exchange on postoperative respiratory function in myasthenia gravis. *Journal of Thoracic and Cardiovascular Surgery* 1985; **89:** 592–596.
69. Loach AB, Young AC, Spalding JMK, *et al.* Postoperative management after thymectomy. *British Medical Journal* 1975; **1:** 309–312.
70. Leventhal S, Orkin FK, Hirsh RA. Prediction of the need for postoperative mechanical ventilation in myasthenia gravis. *Anesthesiology* 1980; **53:** 26–30.
71. Eisencraft JB, Book WJ, Mann SM, Papatestias AE, Hubbard M. Resistance to succinylcholine in myasthenia gravis: a dose–response study. *Anesthesiology* 1988; **69:** 760–763.
72. Baraka A, Baroody M, Yazbeck V. Repeated doses of suxamethonium in the myasthenic patient. *Anaesthesia* 1993; **48:** 782–784.
73. Baraka A. Suxamethonium block in a myasthenic patient. *Anaesthesia* 1992; **47:** 217–219.
74. Ellis FR. Inherited muscle disease. *British Journal of Anaesthesia* 1980; **52:** 153–164.
75. Smith CL, Bush GH. Anaesthesia and progressive muscular dystrophy. *British Journal of Anaesthesia* 1985; **57:** 1113–1118.
76. Nightingale P, Healy TEJ, McGuiness K. Dystrophia myotonica and atracurium. *British Journal of Anaesthesia* 1985; **57:** 1131–1135.
77. Rosewarne FA. Anaesthesia, atracurium and Duchenne muscular dystrophy. *Canadian Anaesthetists Society Journal* 1986; **33:** 250–251.
78. Mitchell MM, Ali HH, Savarese JJ. Myotonia and neuromuscular blocking agents. *Anesthesiology* 1978; **49:** 44–48.

10

Drug interactions

Brian J Pollard

When more than one drug is in use at the same time, there is the potential for one to be affected by the other(s). Few anaesthetics require the use of only one drug and if a relaxant is in use there will also already be other drugs in use – the anaesthetic agents. It is common practice to use several drugs during the course of an anaesthetic, e.g. systemic analgesics, local anaesthetics, antibiotics. In addition, the patient may be receiving therapy for one or more pre-existing medical disorders, with the clear potential for interactions. Furthermore, as the number of drugs being used increases, so the potential for interactions increases. Because this book is devoted to the muscle relaxant family only interactions concerning muscle relaxants are considered.

Sites of drug interactions

It is appropriate to consider first where possible interactions concerning the neuromuscular blocking agents might take place, and the neuromuscular junction must clearly be considered as a primary location for such interactions. Acetylcholine released from the prejunctional nerve ending crosses the synaptic cleft to interact with receptors on the postjunctional membrane. Any action to increase or decrease the release, mobilization, storage, synthesis or breakdown of acetylcholine will alter neuromuscular block.

Released acetylcholine crosses the synaptic cleft to combine with the receptors on the postjunctional membrane. The two alpha subunits lie in different environments, surrounded by different protein molecules, and drugs may act differentially on one or other alpha subunit. Binding to other parts of the receptor/ion channel complex may also occur.

There are additional sites of interaction of drugs which do not lie within or near the neuromuscular junction. These include pharmacokinetic factors, changes in protein binding, binding to other non-specific acceptor sites, alterations in the volume of distribution and changes in the rate of metabolism and/or elimination.

178 *Drug interactions*

Table 10.1 Drug interactions involving the muscle relaxants

Drugs	Interaction	Extent
Alfentanil	No effect	0
Aminophylline	Antagonism	++
Antibiotics	Potentiation (some only)	+++
Anticonvulsants	Antagonism	+++
Aprotinin	{ Potentiation	+
	{ Prolongs suxamethonium	++
Aspirin	No effect	0
Azathioprine	Antagonism	+
Benzodiazepines	No effect	0
Beta blockers	Potentiation	+
Bretylium	Potentiation	+
Bupivacaine	Potentiation	++
Butyrophenones	No effect	0
Calcium antagonists	Potentiation	++
Carbamazepine	Antagonism	++
Cephalosporins	No effect	0
Chlorpromazine	No effect	0
Clindamycin	Potentiation	+++
Colistin	Potentiation	+++
Cyclo-oxygenous inhibitors	No effect	0
Cyclosporin	Potentiation	+
Dantrolene	Potentiation	++
Dexamethasone	Antagonism	++
Diazepam	No effect	0
Disopyramide	Potentiation	+
Doxapram	Potentiation	++
Droperidol	No effect	0
Ecothiopate	Prolongs suxamethonium	+++
Enflurane	Potentiation	+++
Erythromycin	No effect	0
Esmolol	{ Potentiation	+
	{ Prolongs suxamethonium	++
Etomidate	Weak potentiation	+
Fentanyl	No effect	0
Frusemide	Potentiation	+
Ganglion blockers	Potentiation	+++
Gentamicin	Potentiation	+++
Glyceryl trinitrate	No effect	0
Haloperidol	No effect	0
Halothane	Potentiation	+++
Hexamethonium	Potentiation	+++
Hydrocortisone	Antagonism	++
Immunosuppressants	Prolongs suxamethonium	++
Isoflurane	Potentiation	+++
Kanamycin	Potentiation	+++
Ketamine	Weak potentiation	+
Ketorolac	No effect	0
Lignocaine	Potentiation	++
Lincomycin	Potentiation	+++

Drugs	Interaction	Extent
Lincosamines	Potentiation	+++
Local analgesics	Potentiation	++
Lorazepam	No effect	0
Mannitol	No effect	0
Meptazinol	Prolongs suxamethonium	+++
Methohexitone	Weak potentiation	+
Metronidazole	No effect	0
Midazolam	No effect	0
Morphine	No effect	0
Neomycin	Potentiation	+++
Netilmicin	No effect	0
Nifedipine	Potentiation	++
Nitrates	No effect	0
Nitrous oxide	No effect	0
Opioids	No effect	0
Penicillamine	No effect	0
Penicillin	No effect	0
Pentolinium	Potentiation	+++
Pethidine	No effect	0
Phenothiazines	No effect	0
Phenytoin	Antagonism	+++
Phosphodiesterase inhibitors	Antagonism	++
Polymyxin	Potentiation	+++
Prilocaine	Potentiation	++
Primidone	Antagonism	++
Procainamide	Potentiation	++
	Prolongs suxamethonium	+++
Procaine	Potentiation	++
	Prolongs suxamethonium	+++
Propofol	Weak potentiation	+
Propranolol	Potentiation	+
	Prolongs suxamethonium	++
Quinidine	Potentiation	++
Sodium nitroprusside	No effect	0
Steroids	Antagonism	++
Streptomycin	Potentiation	+++
Sufentanil	No effect	0
Tetracyclines	Potentiation	+
Thiopentone	Weak potentiation	+
Tobramycin	Potentiation	+++
Trimetaphan	Potentiation	+++
Valproate sodium	Antagonism	++
Verapamil	Potentiation	++
Volatile agents	Potentiation	+++

0 No effect.
+ Very little effect which is of no clinical significance.
++ Weak effect which may be of clinical significance under certain circumstances.
+++ Clinically significant effect.

180 Drug interactions

Having considered the possible sites where drugs can theoretically interact, it is now appropriate to examine those interactions which are known and which are of importance concerning the neuromuscular blocking agents. The drugs will be considered in their principal families. A summary table is also included (Table 10.1). Interactions between the neuromuscular blocking agents and ions, including acid-base changes, are also considered in this chapter, although these are not drug interactions in the strict sense of the definition. A summary table for electrolytes and acid-base balance is included (Table 10.2).

Table 10.2 The effects of acid base and electrolyte change on the muscle relaxants

	H^+	K^+	Ca^{2+}	Mg^{2+}	Li^+
Increased plasma ion concentration	+	−	−	+	+
Decreased plasma ion concentration	−	+	+	−	

− Antagonism of relaxant effect.
+ Potentation of relaxant effect.

Volatile agents

Every anaesthetist is aware from an early stage in training that an interaction exists between the volatile anaesthetic agents and the muscle relaxants. This interaction is useful clinically and it is likely that most anaesthetists make use of it almost every day. The extent of the interaction depends upon the volatile agent and relaxant combination.

At high concentrations volatile agents alone will depress neuromuscular transmission.[1] The existence of potentiation between volatile anaesthetic agents and muscle relaxants is therefore hardly surprising. Numerous studies have demonstrated this interaction in both clinical and *in vitro* situations.[2, 3]

The extent of the potentiation differs depending upon the exact volatile agent muscle relaxant combination; for example, isoflurane and enflurane potentiate a tubocurarine and pancuronium neuromuscular block more than does halothane, whereas enflurane is more potent than either halothane or isoflurane on a vecuronium neuromuscular block.[4] All of the three volatile agents, enflurane, isoflurane and halothane, appear to potentiate an atracurium block to a similar degree.[4] The new volatile agents, desflurane and sevoflurane also potentiate a neuromuscular block. There may be changes in the extent of potentiation with time. That due to an enflurane block becomes greater with time, whereas the effects due to halothane appear to remain constant.[5] A suxamethonium neuromuscular block is also potentiated by volatile agents and it has been suggested that the characteristics of a depolarizing block may change with time in the presence of a volatile agent.[6]

The presence of enflurane has been reported to delay the reversal of a pancuronium block and isoflurane to delay the reversal of a vecuronium block.[7] The reversibility of atracurium or vecuronium is less affected by the presence of a volatile agent than are the longer-acting relaxants.[8]

The mechanism behind the volatile agents' potentiation of muscle relaxants is unknown. Proposed mechanisms have included an increase in muscle blood flow with increased delivery of relaxant to the neuromuscular junction,[9] a decrease in the release of acetylcholine from nerve endings[10] and an action on the postjunctional membrane.[5] The volatile agents still show potentiation *in vitro*, making the muscle blood flow hypothesis unlikely although some modification of the pharmacokinetics of muscle relaxants are produced by volatile agents, for example, the elimination half-life of pancuronium is increased following enflurane or halothane. There is no good evidence either in support of, or against, the interaction being prejunctional in origin.

Actions on the postjunctional membrane have received the most attention. Volatile agents depress agonist-induced depolarization of the end-plate.[11] The conductance of the ion channels at the end-plate is less in the presence of volatile anaesthetic agents. In view of the high lipid solubility of the volatile agents, it is tempting to speculate that there could be non-specific actions upon the lipid component of the cell membrane. A direct action on the muscle contraction system or through the medium of calcium flux are also possible. It is very likely that more than one mechanism is involved simultaneously, and different volatile agents are unlikely to all be acting in exactly the same manner.

Benzodiazepines

Studies reporting the effects of the benzodiazepines on neuromuscular transmission have shown conflicting results. Most studies in laboratory animals have shown potentiation of the action of muscle relaxants, an action which is probably not mediated through benzodiazepine receptors.[12,13] These effects have not, however, been demonstrated in the clinical situation and the benzodiazepines are usually regarded as without effect on a neuromuscular block.

Intravenous anaesthetic agents

The barbiturates have been reported to potentiate weakly the effects of muscle relaxants,[14] although no effect was reported by methohexitone on a pipecuronium neuromuscular block.[15] Etomidate may potentiate pancuronium[16] and also vecuronium[17] but not pipecuronium[15] or suxamethonium.[18] Ketamine has been examined in a number of studies and the overall conclusion lies in favour of potentiation, although there is not universal agreement. Propofol has been shown to potentiate a neuromuscular block resulting from atracurium, vecuronium, pancuronium and suxamethonium, but only in doses higher than the clinical range.

Althesin and propanidid potentiated the actions of muscle relaxants. One particular interaction between propanidid and suxamethonium followed the inhibition of plasma cholinesterase by propanidid which could lead to awareness in a paralysed patient if the action of the propanidid was terminated before that of the suxamethonium.

The mechanism of action of the interactions between muscle relaxants and intravenous anaesthetic agents appears likely to be due to reduction in the sensitivity of the postjunctional membrane. It must be remembered that these interactions are weak and not of clinical importance.

Local analgesics

The local analgesics can produce neuromuscular block alone and also potentiate a non-depolarizing neuromuscular block.[19] Tetanic fade can be produced by lignocaine alone and the onset of a tubocurarine block is faster following a dose of 2.5 mg kg^{-1} lignocaine.[20] The mechanism of this interaction appears likely to be due both to a decrease in responsiveness of the postjunctional membrane and a decrease in acetylcholine release although other mechanisms cannot be discounted.

The ester local analgesics also inhibit plasma cholinesterase. The action of any other drug which is broken down by plasma cholinesterase (suxamethonium and mivacurium) may be prolonged, particularly in the presence of an atypical cholinesterase.

Opioid analgesics

A number of studies have demonstrated that the opioids potentiate muscle relaxants but there is not universal agreement. In many studies any effect of the opioid on neuromuscular transmission was not antagonized by naloxone; furthermore, naloxone alone also potentiates a neuromuscular block.[21]

It is likely that the mechanism involved is not simple and does not involve opioid receptors. A biphasic effect has been reported with pethidine on an *in vitro* preparation where a low concentration augmented twitch height, which was followed by a slowly developing block at higher concentrations.[22] These concentrations were, however, considerably greater than those likely to be reached clinically.

Meptazinol demonstrates an interesting phenomenon. In higher doses, it is capable of inhibiting neuromuscular transmission, but in low doses it paradoxically antagonizes a tubocurarine block. This latter effect is due to an inhibitory action on cholinesterase activity.[23]

Diuretics

The diuretics are a broad family of drugs including agents with widely differing actions, e.g. osmotic diuretics (mannitol), carbonic anhydrase inhibitors (acetazolamide) and loop diuretics (frusemide). Frusemide potentiates tubocurarine although not to any clinically important extent.[24] Frusemide will also accelerate recovery from a pancuronium neuromuscular block.[25] *In vitro* studies with frusemide have confirmed the potentiation of tubocurarine at low concentrations and also demonstrated antagonism at higher concentrations. In animals, chlorthalidone, chlorothiazide and acetazolamide potentiate a neuromuscular block although this has not been demonstrated in man. Mannitol has no effect on a neuromuscular block.[24]

Antibiotics

Many of the antibiotics are known to interact with the neuromuscular blocking agents and a great many studies have been undertaken to determine and quantify these interactions.

Aminoglycosides

Interest has centred around the aminoglycoside antibiotics for some time, probably because the aminoglycosides were the antibiotics principally implicated in the original reports documenting an interaction. It is now known that most of the aminoglycosides are comparatively potent in their ability to potentiate muscle relaxants. A number of aminoglycoside antibiotics have also been shown to be capable of producing neuromuscular block alone.[26]

Neomycin, streptomycin, dihydrostreptomycin, tobramycin, gentamicin and kanamycin are all capable of potentiating a neuromuscular block. This has been shown to be the case for a tubocurarine, pancuronium, gallamine, suxamethonium or vecuronium neuromuscular block. It is wise, therefore, to assume that all aminoglycosides are capable of interacting in this way with all neuromuscular blocking agents.

It does not appear to be necessary to receive a particularly large dose of the antibiotic for problems to arise and it is easy for the anaesthetist to be caught unawares. The speed of onset and intensity of effect depend upon the dose absorbed and the route of administration, but enough can be absorbed from irrigation of the intrapleural space, peritoneal cavity or even a wound. Enough neomycin may be absorbed orally to result in a measurable effect.

It is interesting to note that there are as yet no reports of any interaction involving netilmicin. Furthermore, netilmicin has been reported to be devoid of any neuromuscular action.[27]

The mechanism of action of aminoglycoside antibiotics on a neuromuscular block is unknown. Similarities exist between the action of the aminoglycosides and that of magnesium.[28] The output of acetylcholine from the prejunctional nerve endings may be reduced in the presence of an aminoglycoside antibiotic and the sensitivity of the postjunctional end-plates may also be lower. Streptomycin has been reported to possess a local analgesic-like effect on nerves and also possibly to affect muscle contraction.

A block which results from the presence of an aminoglycoside antibiotic is not antagonized by an anticholinesterase although it may be partially reversed by calcium and 4-aminopyridine.

Polymyxins and colistin

These antibiotics also potentiate the actions of the neuromuscular blocking agents[29] and a combination of polymyxin with neomycin has been shown to be especially potent in this respect.[30] The mechanism of action is not understood. Decreases in acetylcholine output and in postjunctional receptor sensitivity have been proposed. A block involving a polymyxin may also be difficult to antagonize and there is a possibility that an anticholinesterase may even make it worse. Some recovery is produced by 4-aminopyridine.

Lincosamines

The principal agents in this family, lincomycin and clindamycin, both potentiate a neuromuscular block. The proposed mechanism of action is a mixture of prejunctional and postjunctional effects. This block is also difficult to reverse. Anticholinesterases either have a slight effect or accentuate the block, whereas the administration of 4-aminopyridine will produce some antagonism.

Penicillins, cephalosporins and erythromycin

These are the only families of antibiotic drugs which have no effect on a neuromuscular block.

Metronidazole

There has been one report implicating metronidazole in the potentiation of a vecuronium neuromuscular block, although this has been disputed. It is likely that metronidazole is without significant effect.

Tetracyclines

The tetracyclines exhibit weak potentiation of a neuromuscular block, although it is unlikely that this interaction has any clinical significance.

Anticonvulsants

Resistance to the non-depolarizing muscle relaxants is seen in many patients receiving long-term treatment with phenytoin. This phenomenon has so far been observed with pancuronium, vecuronium, tubocurarine and metocurine, but not with atracurium.[31, 32] There is an increase in requirements of relaxant to achieve a given degree of block, an increase in the infusion rate requirements to maintain a steady state block and reduction in the duration of action of a bolus dose. Similar effects are seen with the other anticonvulsants carbamazepine and sodium valproate.[33] Phenytoin has been reported to exacerbate pre-existing myasthenia gravis (Chapter 9).

Resistance, as described above, is a feature of chronic phenytoin therapy. When phenytoin is administered acutely to a patient with a steady state block, however, potentiation may be seen.[34] This effect is unlikely to be of any clinical significance.

The mechanism underlying these effects of the anticonvulsants is unknown. Phenytoin therapy increases the metabolism of pancuronium and there is an increased binding of metocurine to plasma proteins in the presence of phenytoin. It is thought most likely, however, that the interactions are taking place within the neuromuscular junction. Phenytoin and carbamazepine both decrease acetylcholine release; primidone increases acetylcholine release; carbamazepine, phenobarbitone, primidone, trimethadione and ethosuximide all decrease the sensitivity of the postjunctional membrane to acetylcholine.

Beta adrenergic blocking agents

It has been known for over 70 years that catecholamines have an effect on the neuromuscular junction. It would be natural therefore to expect adrenergic antagonists to also have an effect on neuromuscular transmission. The alpha adrenergic antagonists have received little attention. The beta adrenergic antagonists, which have a more widespread use in medicine, have been investigated.

Although the beta adrenergic blocking agents potentiate a nondepolarizing neuromuscular block both clinically and *in vitro*, it seems unlikely that beta blockers have any significant effects in normal clinical practice.[35, 36] Treatment with a beta adrenergic blocker has been shown to worsen the symptoms of myasthenia gravis (Chapter 9).

An interesting interaction exists with suxamethonium, where its duration of action may be prolonged by propranolol and esmolol due to inhibition of breakdown by plasma cholinesterase. The clinical significance of this interaction, if any, is unclear.

Calcium antagonists

Calcium and calmodulin play essential roles in the release of neurotransmitters and an effect of the calcium antagonists on neuromuscular transmission has been confirmed.[37, 38] Verapamil and nifedipine potentiate a nondepolarizing neuromuscular block *in vitro* and *in vivo*. This potentiating effect is accentuated by antibiotics and also by enflurane. Verapamil may increase weakness in patients with Duchenne muscular dystrophy (Chapter 9). In the presence of verapamil, a neuromuscular block may be easier to reverse with edrophonium than with neostigmine.

The mechanism of action of calcium antagonists on a neuromuscular block is unknown. Verapamil possesses a local analgesic-like effect on excitable membranes, may block ion channels at the neuromuscular junction and may have a direct effect on skeletal muscle. It is possible that there is a combination of prejunctional and postjunctional mechanisms.

Ganglionic blockers

The ganglionic blocking agents, hexamethonium, pentolinium and trimetaphan have a neuromuscular blocking action alone *in vitro* and also prolong the action of the non-depolarizing muscle relaxants.[39] This potentiating action appears to exist for all non-depolarizing muscle relaxants and also for suxamethonium. In the case of suxamethonium and trimetaphan the prolongation of effect is partly due to a reduction in the breakdown of suxamethonium by plasma cholinesterase because trimetaphan is also broken down by that enzyme.

The mechanism underlying the interaction between the muscle relaxants and the ganglion blocking agents is unclear. A prejunctional mechanism is possible because the prejunctional receptors at the neuromuscular junction resemble those nicotinic cholinergic receptors on autonomic ganglia more than they do the cholinergic receptors on the postjunctional membrane at the neuromuscular junction.[40] Evidence has also been advanced, however, to support a postjunctional action of hexamethonium. It is likely that the interaction is not a simple one because antagonism is seen with low concentrations of ganglionic blocker *in vitro* followed by potentiation at higher concentrations.[39]

Anti-arrhythmic agents

The anti-arrhythmic agents belong to several different families and include the beta adrenergic blockers, calcium channel blockers, phenytoin, lignocaine, disopyramide, bretylium, quinidine and procainamide. For interactions concerning phenytoin, lignocaine, the beta blockers and calcium channel blockers the reader is referred to their separate sections.

Quinidine and procainamide

These two agents potentiate both a depolarizing and a non-depolarizing neuromuscular block.[41] Re-curarization has been noted in the presence of quinidine. Both quinidine and procainamide may precipitate an episode of myasthenia gravis (Chapter 9) in susceptible patients. The mechanism is not known, but it seems likely that both a prejunctional and a postjunctional action are involved.

Disopyramide and bretylium

Both of these drugs have been shown to potentiate a non-depolarizing neuromuscular block *in vitro*.[42, 43] No effect has been described in the clinical situation, however, and it seems likely that if there is an effect, it is of no importance.

Steroids

The steroids comprise a large family of compounds with a common structure but which exhibit a variety of pharmacological actions. Although some muscle relaxants are steroids, there are a number of reports of an effect of steroids which are not muscle relaxants on neuromuscular transmission.

A tubocurarine neuromuscular block is antagonized by low concentrations of dexamethasone and potentiated by high concentrations of dexamethasone, *in vitro*.[44] This interaction with tubocurarine has been both confirmed and refuted in the clinical situation. A similar antagonism exists between corticosteroids and pancuronium and resistance to the effect of vecuronium has been described in a patient receiving long-term testosterone therapy.

The mechanism of these interactions is unknown. An increase in acetylcholine release, depression of postjunctional receptor excitability, inhibition of phosphodiesterase, an increase in prejunctional choline uptake and the inhibition of cholinesterase have all been proposed. Whatever the mechanism, the interaction seems unlikely to be of any clinical significance.

Hypotensive agents

The hypotensive agents are a heterogeneous group of drugs with representatives from a number of different families. Included under this heading would be the ganglionic blockers and adrenergic blockers (both covered in separate sections above), sodium nitroprusside, glyceryl trinitrate and various nitrate derivatives.

Glyceryl trinitrate has been reported both to increase the duration of action of pancuronium and to have no effect on a pancuronium, tubocurarine, gallamine, vecuronium or suxamethonium block. Sodium nitroprusside has no effect on a neuromuscular block.

Immunosuppressants

Cyclosporin potentiates an atracurium and vecuronium neuromuscular block. Azathioprine both antagonizes a tubocurarine and a pancuronium neuromuscular block although it has also been shown to have no effect on a tubocurarine block. The immunosuppressants may prolong the action of suxamethonium by an inhibitory effect on plasma cholinesterase.

Phosphodiesterase inhibitors

Aminophylline alone, in the absence of a muscle relaxant, enhances diaphragmatic contractility. In the presence of a pre-existing pancuronium neuromuscular block, aminophylline produces partial antagonism. Theophylline, caffeine and a number of other xanthines may facilitate neuromuscular transmission *in vitro*, an action which might be expected to reduce the muscle relaxant requirement. The mechanism for these interactions may be related to either changes in cyclic AMP levels or to an action involving adenosine receptors.

Miscellaneous drugs

There are a number of drugs which do not fit into any of the above categories, yet have been shown to affect a neuromuscular block. Many of these actions have been noticed in an accidental fashion.

Aprotinin

Aprotinin weakly potentiates a neuromuscular block. Aprotinin also inhibits the activity of plasma cholinesterase and therefore may prolong a suxamethonium neuromuscular block.

Penicillamine

Penicillamine has been reported to worsen or precipitate a myasthenic episode (Chapter 9), although there have been no incidents described of any interaction between penicillamine and the muscle relaxants.

Dantrolene

Dantrolene potentiates both a tubocurarine and a vecuronium neuromuscular block. It is uncertain what mechanism is involved, because there has not been any effect of dantrolene described on the neuromuscular junction.

Phenothiazines

An increase in weakness in a myasthenic patient (Chapter 9) has been reported, although no direct effect on a neuromuscular block has been described.

Cyclo-oxygenase inhibitors

All members of this family of drugs appear to be devoid of any action at the neuromuscular junction.

Ecothiopate

Eye drops containing the organophosphate anticholinesterase ecothiopate are occasionally used in ophthalmology. Sufficient may be absorbed for a parenteral effect to be seen. The duration of action of suxamethonium is prolonged due to inhibition of plasma cholinesterase and it is possible that mivacurium may be similarly affected.

Doxapram

This respiratory stimulant has been shown to retard the rate of recovery of a neuromuscular block. It is unlikely that this interaction has any clinical relevance in normal doses.

Acid-base balance

Acute changes in acid-base balance are common during anaesthesia and Intensive Care management. It is common practice to deliberately hyperventilate patients (usually to a $PaCO_2$ of approximately 4.5 kPa, although occasionally to a $PaCO_2$ between 3.5 and 4.0 kPa), resulting in an acute respiratory alkalosis. At the completion of surgery, there may be hypoventilation due to residual narcosis or paralysis with an associated change to an acute respiratory acidosis. An acidosis or alkalosis (metabolic or respiratory) may already be present resulting from a pre-existing medical condition upon which these acute changes will be superimposed. It is clearly important, therefore, to know the effect of such changes in acid-base balance on the action of the muscle relaxants (see Table 10.2).

It is generally accepted that a neuromuscular block is potentiated by acidosis. The implication is that acute respiratory acidosis in the early postoperative period may give rise to difficulties in reversal, or even to the late reappearance of a block. Not all of the muscle relaxants appear to be equally affected. A pancuronium or vecuronium neuromuscular block is affected very little by changes in acid-base status while tubocurarine is quite markedly affected. Metocurine, gallamine and suxamethonium are also affected but to a lesser extent than tubocurarine.

Potassium

The resting membrane potential is largely dependent upon differences in potassium concentrations between the inside and outside of a cell. Changes in the extracellular potassium concentration will therefore alter cellular excitability, and may also modify acetylcholine release. An increase in extracellular potassium concentration increases acetylcholine release. An increase in extracellular potassium will also decrease the sensitivity to a tubocurarine or pancuronium neuromuscular block which is reflected in an increased requirement for tubocurarine. These effects are present within the clinical range – an increase in potassium from 3.5 to 5.0 mmol l^{-1} would be expected to increase the requirement of relaxant by approximately one-third.

The converse also holds true and a decrease in plasma potassium concentration reduces the requirement for non-depolarizing muscle relaxants. The amount of neostigmine required for reversal may be higher and there may be a risk of re-curarization (Chapter 8).

In the condition familial periodic paralysis, the potassium characteristically undergoes marked swings. Great care should be taken with the use of muscle relaxants in these patients.

Lithium

Lithium potentiates a tubocurarine, pancuronium or suxamethonium neuromuscular block. Lithium may also produce muscular weakness when given alone in the absence of a muscle relaxant. The mechanism underlying these effects of lithium appears to be due to a decrease in acetylcholine release from the nerve endings.

Calcium

Alterations in the plasma concentrations of calcium ions and also other divalent cations affect neuromuscular transmission. The actions of calcium at the neuromuscular junction are, however, many, and do not all affect a neuromuscular block in an identical fashion. Some of the actions of calcium oppose one another and it is therefore difficult to predict the exact effect on theoretical grounds alone. To illustrate these problems, both potentiation and antagonism of a pancuronium and a tubocurarine neuromuscular block have been described in the presence of an increase in calcium concentration.[45] The duration of action of suxamethonium is prolonged but that of atracurium decreases in the presence of a raised serum calcium secondary to hyperparathyroidism.

The mechanisms behind these actions are complex. An increase in calcium ion concentration will increase acetylcholine release, decrease the sensitivity of the postjunctional membrane and enhance the excitation contraction coupling process in skeletal muscle. The result may be partial antagonism of a non-depolarizing neuromuscular block; and also antagonism of a block which is partly due to antibiotics.

Magnesium

Magnesium has an opposing action to calcium at the neuromuscular junction and potentiates muscle relaxants. When suxamethonium is administered in the presence of raised magnesium levels, potentiation of the block can be expected and it has been suggested that a phase I block may more readily change to a phase II block (Chapter 4) in the presence of a raised magnesium concentration.[46] Knowledge of the interaction with magnesium is important because of its widespread use in medicine. It is routinely used in pre-eclampsia and eclampsia in many obstetric units and also during open heart surgery. Hypomagnesaemia is common in patients with malnutrition and also in critically ill patients.

The mechanism by which magnesium affects neuromuscular transmission is probably principally related to an opposing effect on the actions of calcium. An increased magnesium concentration leads to a reduction in acetylcholine release and it is possible that the main effect is prejunctional in origin.

There are a number of other divalent cations which may behave in a manner similar to magnesium. Manganese, beryllium, lead and cadmium all share similar properties at the neuromuscular junction.

Combinations of muscle relaxants

If two drugs act in exactly the same way on the same receptor system then, when they are given simultaneously, their effects should be additive. There are two different types of muscle relaxants, depolarizing and non-depolarizing. These clearly might not be simply additive because of the different nature of their actions. It would, however, be logical to assume that the muscle relaxants within one family were all additive in their actions, but is this the case?

The non-depolarizing relaxants

Riker and Wescoe in 1951[47] were the first to examine combinations of two non-depolarizing muscle relaxants. They showed that, in cats, the effect of consecutive doses of either gallamine or tubocurarine was additive because the effect was the same whether the second dose was the same agent, or the other agent.

Wong in 1969[48] also studied tubocurarine and gallamine. In view of the different side-effects of tubocurarine and gallamine, he argued that mixing the two would, by allowing the use of less of each, lead to a reduction in dose-dependent side-effects. He was surprised to note that less was needed for surgical relaxation than he predicted. Further experiments led him to conclude that these two agents were not additive, but synergistic, i.e. they potentiated each other's action.

As more muscle relaxants became available, they were investigated for interactions. The initial acceptance of the existence of synergism then had to be modified, because certain of the two-component combinations were found not to be synergistic, but to be additive. The present state of our knowledge is summarized in Table 10.3.[39] The important clinical lesson to note here is that if two different non-depolarizing relaxants are used in the same patient, less will be needed for the same effect if they are a synergistic combination. If the dose is not reduced, or the duration of surgery is not as long as anticipated, then a prolonged block may result, or reversal may be delayed.

An explanation for these observations is, of course, required. There are a number of variables to be considered in the clinical situation, which include uptake, binding, metabolism, redistribution and regional blood flow patterns.

Table 10.3 Interactions between the non-depolarizing relaxants

	Tubocurarine	Pancuronium	Alcuronium	Gallamine	Metocurine	Atracurium	Vecuronium	Mivacurium	Doxacurium	Pipecuronium
Pancuronium	+++									
Alcuronium	+++	0								
Gallamine	0	0	(0)							
Metocurine	0	+++	(++)	+						
Atracurium	0	0	+	?	(0)					
Vecuronium	++	0	0	?	(+)	+				
Mivacurium	(0)	(+)	(+)	?	(0)	(0)	(+)			
Doxacurium	(0)	(+)	(+)	?	(+)	(0)	(0)	(0)		
Pipecuronium	(+)	(0)	(0)	?	(+)	(+)	(0)	(+)	(+)	
Rocuronium	(+)	(0)	(0)	?	(+)			(+)	(+)	(0)

+++ Marked potentiation at clinical doses.
++ Potentiation at clinical doses.
+ Weak potentiation, may not be noticeable at clinical doses.
0 Addition.
(+) Not studied but potentiation might be expected.
(0) Not studied but addition might be expected.
? Unknown.

Which of these might be involved in these interactions? The majority of those do not, however, apply when using an *in vitro* preparation yet synergism is still seen. This strongly suggests that the interaction is taking place at the neuromuscular junction.

The reasons for the different interactions between the non-depolarizing muscle relaxants are, as yet, not clear. Possible mechanisms include actions upon the prejunctional acetylcholine receptors, actions upon the postjunctional acetylcholine receptors, differences in the sensitivity of the two alpha subunit acetylcholine recognition sites and effects on cholinesterase. Pharmacokinetic effects outside the neuromuscular junction are unlikely to be the sole explanation.

Extending a neuromuscular block

As the end of surgery approaches it may be necessary to extend the block using an additional dose of muscle relaxant in order to assist surgical closure. This can be achieved using either increments of the same muscle relaxant that has been used for the main part of the procedure or using a second, shorter-acting agent. It must be noted, however, that if this technique is used, only a very small dose will be required (atracurium 1–2 mg or vecuronium 0.25 mg) in order to achieve an extension of block of 10–15 minutes.[49] If much larger doses are administered there may be a profound block lasting much longer than anticipated with associated difficulties in reversal.

Suxamethonium and non-depolarizing relaxants

Suxamethonium is the drug of choice when the airway has to be secured without delay because it has the fastest rate of onset of any of the muscle relaxants in current use. It is then common practice to continue with a non-depolarizing relaxant to maintain paralysis. It is recommended to await recovery from the suxamethonium before administering a second relaxant. This second agent is, however, very likely to be given before there is full recovery from suxamethonium. The potential for interactions between suxamethonium and the non-depolarizing agents therefore exists (see also Chapter 4).

Suxamethonium before a non-depolarizing agent

The requirement for a non-depolarizing neuromuscular blocking agent appears to be reduced following the prior administration of an intubating dose of suxamethonium.[50] This matter has been addressed in a number of studies and there is disagreement as to whether all non-depolarizing relaxants are similarly affected and, if so, to what extent. There is no clear reason for these differences. Although the evidence is conflicting it is generally believed by most clinicians that a prior dose of suxamethonium allows the use of a smaller dose of non-depolarizing relaxant for continuation of the block.

Pre-curarization before suxamethonium

Suxamethonium has several unwanted and potentially hazardous side-effects, many of which are probably related to the muscular fasciculations which follow its administration. When a small dose of a non-depolarizing muscle relaxant is administered before suxamethonium the fasciculations are less and those side-effects related to the fasciculations are reduced. This is the technique of pre-curarization.

The presence of a subparalysing dose of a non-depolarizing muscle relaxant produces a delay in the onset of the suxamethonium and also shortens its duration of action. It is therefore necessary to administer a slightly increased dose of suxamethonium when following a small dose of a non-depolarizing agent. Not all of the non-depolarizing agents are suitable for pre-curarization. Those most commonly used are tubocurarine (3–5 mg) or gallamine (5–20 mg). Pancuronium is not as effective as either tubocurarine or gallamine. It is likely that the mechanism responsible for the benefit of pre-curarization in decreasing the side-effects of suxamethonium is simply due to the partial subclinical block which reduces all of the responses to suxamethonium. In the presence of the partial non-depolarizing block it is not possible for suxamethonium to achieve its maximum agonist effect which limits both the required effects and the side-effects.

Suxamethonium following a non-depolarizing agent

It is occasionally noted that as the end of a surgical procedure approaches, the surgeon requests a temporary increase in the neuromuscular block. The block, which was adequate for surgery, is not quite deep enough to facilitate closure of the surgical incision. The administration of a small dose of suxamethonium at this time may help and has been recommended. The effect of a dose of suxamethonium at this time may, however, be unpredictable, depending upon the non-depolarizing agent, the degree of block present, the total dose administered and the dose of suxamethonium. In view of the uncertainty, this technique is best avoided. The use of an additional increment of the same, or a different non-depolarizing agent is more logical.

Conclusion

It is clear that there are many possible interactions concerning the relaxants and these may be difficult to predict. Some may be advantageous and some disadvantageous. It is recommended that neuromuscular function be monitored every time a muscle relaxant is used, so that much of the unpredictability can be removed from the equation. When a muscle relaxant is in use, the possibility of interactions must be constantly remembered because of the potentially hazardous consequences of a partial block continuing into the postoperative period.

References

1. Pollard BJ, Miller RA. Potentiating and depressant effects of inhalation anaesthetics on the rat phrenic nerve – diaphragm preparation. *British Journal of Anaesthesia* 1973; **45:** 404–415.
2. Katz RL, Gissen AJ. Neuromuscular and electromyographic effects of halothane and its interaction with d-tubocurarine in man. *Anesthesiology* 1967; **28:** 564–567.
3. Padfield A. The effects of general anaesthetic agents on the neuro-muscular junction. In *Proceedings of the Fourth World Congress of Anaesthesiologists, London* 1968, pp. 900–903.
4. Rupp SM, McChristian JW, Miller RD. Atracurium neuromuscular blockade during halothane/N_2O and enflurane/N_2O anesthesia in humans. *Anesthesiology* 1984; **61:** A288.
5. Stanski DR, Ham J, Miller RD, Sheiner LB. Pharmacokinetics and pharmacodynamics of d-tubocurarine during nitrous oxide-narcotic and halothane anesthesia in man. *Anesthesiology* 1979; **51:** 235–241.
6. Hilgenberg JC, Stoelting RK. Characteristics of succinylcholine-produced phase II neuromuscular block during enflurane, halothane and fentanyl anesthesia. *Anesthesia and Analgesia* 1981; **60:** 192–196.
7. Baurain MJ, d'Hollander AA, Melot C, Dernovoi BS, Bervais L. Effects of residual concentrations of isoflurane on the reversal of vecuronium-induced neuromuscular blockade. *Anesthesiology* 1991; **74:** 474–478.
8. Engbaek J, Ording H, Viby-Mogensen J. Neuro-muscular blocking effects of vecuronium and pancuronium during halothane anaesthesia. *British Journal of Anaesthesia* 1983; **55:** 497–500.
9. Miller RD, Crique M, Eger EI. Duration of halothane anesthesia and neuromuscular blockade with d-tubocurarine. *Anesthesiology* 1976; **44:** 206–210.
10. Hughes R, Payne JP. Interactions of halothane with nondepolarizing neuromuscular blocking drugs in man. *British Journal of Clinical Pharmacology* 1976; **7:** 485–490.
11. Waud BE, Waud DR. Comparison of the effects of general anesthetics on the end plate of skeletal muscle. *Anesthesiology* 1975; **43:** 540–547.
12. Driessen JJ, Vree TB, Booij LHDJ, Van der Pol FM, Crul JF. Effects of some benzodiazepines on peripheral neuromuscular function in the rat in-vitro hemidiaphragm preparation. *Journal of Pharmacy and Pharmacology* 1983; **36:** 244–247.
13. Driessen JJ, Vree TB, van Egmond J, Booij LHDJ, Crul JF. *In vitro* interaction of diazepam and oxazepam with pancuronium and suxamethonium. *British Journal of Anaesthesia* 1984; **56:** 1131–1137.
14. Cronnelly R, Morris RB, Miller RD. Comparison of thiopental and midazolam on the neuromuscular responses to succinylcholine or pancuronium in humans. *Anesthesia and Analgesia* 1983; **62:** 75–77.
15. Dutre P, Rolly G, Vermeulen H. Effect of intravenous hypnotics on the actions of pipecuronium. *European Journal of Anaesthesiology* 1992; **9:** 313–317.
16. Booij LHDJ, Crul JF. The comparative influence of gamma-hydroxy butyric acid, althesin and etomidate on the neuromuscular blocking potency of pancuronium in man. *Acta Anaesthesiologica Belgica* 1979; **30:** 219–223.
17. McIndewar IC, Marshall RJ. Interactions between the neuromuscular blocking drug ORG NC 45 and some anaesthetic, analgesic and antimicrobial agents. *British Journal of Anaesthesia* 1981; **53:** 785–792.
18. Doenicke A, Dittmann-Kessler I, Sramota A, Beyer E. Etomidate and suxamethonium. The duration of relaxation and pseudocholinesterase activity. A clinical experimental study. *Anaesthesist* 1980; **29:** 120–124.

19. Chapple DJ, Clark JS, Hughes R. Interaction between atracurium and drugs used in anaesthesia. *British Journal of Anaesthesia* 1983; **55:** 17S–22S.
20. Zukaitis MG, Hoech GP. Train of 4 measurement of potentiation of curare by lidocaine. *Anesthesiology* 1979; **51:** S288.
21. Soterpoulos GC, Standaert FG. Neuromuscular effects of morphine and naloxone. *Journal of Pharmacology and Experimental Therapeutics* 1973; **184:** 136–184.
22. Boros M, Chaudhry IA, Nagashima H, Duncalf RM, Sherman EH, Foldes FF. Myoneural effects of pethidine and droperidol. *British Journal of Anaesthesia* 1984; **56:** 195–201.
23. Strahan SK, Pleuvry BJ, Modla CY. Effect of meptazinol on neuromuscular transmission in the isolated rat phrenic nerve-diaphragm preparation. *British Journal of Anaesthesia* 1985; **57:** 1095–1099.
24. Miller RD, Sohn YJ, Matteo RS. Enhancement of d-tubocurarine neuromuscular blockade by diuretics in man. *Anesthesiology* 1976; **45:** 442–445.
25. Azar I, Cottrell J, Gupta B, Turndorf H. Furosemide facilitates recovery of evoked twitch response after pancuronium. *Anesthesia and Analgesia* 1980; **59:** 55–57.
26. De Rosayro M, Healy TEJ. Tobramycin and neuromuscular transmission in the rat isolated phrenic nerve-diaphragm preparation. *British Journal of Anaesthesia* 1978; **50:** 251–254.
27. Bendtsen A, Engbæk J, Lehnsbo J. The interaction between pancuronium and netilmicin on neuromuscular function. *Acta Anaesthesiologica Scandinavica* 1983; **27** (Suppl. 78): 83.
28. Singh YN, Harvey AL, Marshall IG. Antibiotic-induced paralysis of the mouse phrenic nerve-hemidiaphragm preparation, and reversibility by calcium and by neostigmine. *Anesthesiology* 1978; **48:** 418–424.
29. Kronenfeld MA, Thomas SJ, Turndorf H. Recurrence of neuromuscular blockade after reversal of vecuronium in a patient receiving polymyxin/amikacin sternal irrigation. *Anesthesiology* 1986; **65:** 93–94.
30. Lee C, De Silva AJC. Interaction of neuro-muscular blocking effects of neomycin and polymyxin B. *Anesthesiology* 1979; **50:** 218–220.
31. Hickey DR, Sangwan S, Bevan JC. Phenytoin-induced resistance to pancuronium. *Anaesthesia* 1988; **43:** 757–759.
32. Ornstein E, Matteo RS, Schwartz AE, Silverberg PA, Young WL, Diaz J. The effect of phenytoin on the magnitude and duration of neuromuscular block following atracurium and vecuronium. *Anesthesiology* 1987; **67:** 191–196.
33. Blanc-Bimar MC, Jadot G, Bruguerolle B. Modifications of the curarizing action of two short acting curare like agents after the administration of two antiepileptic agents. *Annales Anesthesiologie Français* 1979; **20:** 685–690.
34. Gray HStJ, Slater RM, Pollard BJ. The effect of acutely administered phenytoin on vecuronium-induced neuromuscular blockade. *Anaesthesia* 1989; **44:** 379–381.
35. Rozen MS, Whan FMcK. Prolonged curarization associated with propranolol. *Medical Journal of Australia* 1972; **1:** 467–469.
36. Usubiaga JE. Neuromuscular effects of beta-adrenergic blockers and their interaction with skeletal muscle relaxants. *Anesthesiology* 1968; **29:** 485–492.
37. Kraynack BJ, Lawson NW, Gintautas J. Neuromuscular blocking action of verapamil in cats. *Canadian Anaesthetists Society Journal* 1983; **30:** 242–247.
38. Durant NN, Nguyen N, Katz RL. Potentiation of neuromuscular blockade by verapamil. *Anesthesiology* 1984; **60:** 298–303.
39. Pollard BJ. Studies concerning the interactions between nondepolarising neuromuscular blocking agents. 1991, MD Thesis, University of Sheffield, pp. 211–250.
40. Bowman WC. Prejunctional and postjunctional cholinoceptors at the neuromuscular junction. *Anesthesia and Analgesia* 1980; **59:** 935–943.

41. Harrah MD, Way WL, Katzung BG. The interaction of d-tubocurarine with anti-arrhythmic drugs. *Anesthesiology* 1970; **33**: 406–410.
42. Healy TEJ, O'Shea M, Massey J. Disopyramide and neuromuscular transmission. *British Journal of Anaesthesia* 1981; **53**: 495–498.
43. Welch GW, Waud BE. Effect of bretylium on neuromuscular transmission. *Anesthesia and Analgesia* 1982; **61**: 442–444.
44. Leeuwin RS, Veldsema-Currie RD, Van Wilgenburg H, Ottenhof M. Effects of corticosteroids on neuromuscular blocking actions of d-tubocurarine. *European Journal of Pharmacology* 1981; **69**: 165–173.
45. Waud BE, Waud DR. Interaction of calcium and potassium with neuromuscular blocking agents. *British Journal of Anaesthesia* 1980; **52**: 863–866.
46. Crul JF, Long GJ, Brunner EA, Coolen JMW. The changing pattern of neuromuscular blockade caused by succinylcholine in man. *Anesthesiology* 1966; **27**: 729–735.
47. Riker WF, Wescoe WC. The pharmacology of Flaxedil with observations on certain analogs. *Annals of New York Academy of Science* 1951; **54**: 373–394.
48. Wong KC. Some synergistic effects of curare and gallamine. *Federation Proceedings* 1969; **28**: 420.
49. Middleton CM, Pollard BJ, Kay B, Healy TEJ. Use of atracurium or vecuronium to prolong the action of tubocurarine. *British Journal of Anaesthesia* 1988; **62**: 659–663.
50. Stirt JA, Katz RL, Murray AL, Schehl DL, Lee C. Modification of atracurium blockade by halothane and by suxamethonium. A review of clinical experience. *British Journal of Anaesthesia* 1983; **55**: 71S–75S.

11

Muscle Relaxants in Paediatric Anaesthesia

Cormac C McLoughlin and **Rajinder K Mirakhur**

The introduction of new and 'cleaner' muscle relaxant drugs within the last 10 years has produced many improvements, not least of which has been the provision of short-acting alternatives to suxamethonium appropriate to the shorter duration of paediatric surgery. The widespread use of peripheral neuromuscular monitoring has confirmed the previously held clinical impressions that children do not show the same responses to muscle relaxant drugs as adults, and the improvement in laboratory analysis techniques has improved our understanding of the factors controlling these differences. This chapter describes the current knowledge with respect to the pharmacokinetics and pharmacodynamics of muscle relaxants in children and deals with the current controversies regarding their administration.

Physiological factors modifying neuromuscular transmission

From birth to adolescence and adulthood the nature of neuromuscular transmission and consequently the dose–response relationships of muscle relaxant drugs is modified by changes in body physiology associated with growth. These influences may act in opposite directions and the net effect may vary between individual drugs. Broadly speaking the clinical characteristics of muscle relaxant drugs may be modified by the following:

1. Maturation of the neuromuscular junction.
2. Changes in the proportion of fast- and slow-contracting muscle fibres.
3. Changes in the relative size of the skeletal muscle compartment.
4. Changes in the size of body compartments and specifically the volume of the extracellular fluid. This will have obvious effects on the volume of distribution of water-soluble muscle relaxant drugs.
5. Possible changes in the metabolism and clearance of drugs.

The first three points are essentially pharmacodynamic alterations while the last two represent pharmacokinetic effects.

Maturation of the neuromuscular junction

The neuromuscular transmission system is immature in humans at birth and for the first 2 months of life.[1] The maturation process has been studied in rats and mice and calves (representing mammalian muscle) in order to demonstrate the changes in the neuromuscular junction. These changes include a reduction in polyneuronal innervation of muscle fibres in favour of the adult pattern of focal innervation and a gradual loss of extrajunctional receptors which are present in large numbers at birth.[2] Additionally the gamma subunit of in the foetal calf postjunctional receptor is replaced by the epsilon subunit soon after birth.[3] This maturation process is thought to be responsible for the shorter channel opening times and faster channel gating observed in adult muscle.[4] Neonatal rats, 1–2 weeks old, do not show repetitive neuronal firing and twitch potentiation when anticholinesterase agents are administered in contrast to the normal adult rat pattern. As this phenomenon is thought to require myelination between the distal node of Ranvier and the origin of the unmyelinated terminal, its absence is taken to indicate incomplete nerve myelination in the neonatal rat.[5] For the same reason fasciculations caused by either suxamethonium or acetylcholine are absent in the rats 1–2 weeks old. These become apparent from the age of about 3 weeks following myelination.

In the first month of life human neonates show a small degree of fade in response to a Train-of-Four stimulation. This is particularly marked in the premature infant (less than 33 weeks gestation) in whom the Train-of-Four ratio may be as low as 0.83.[6] The same tendency to fade is manifested in response to tetanic stimulation and once again the premature infant is particularly sensitive and shows post-tetanic exhaustion at stimulation frequencies as low as 20 Hz.[7] The frequency sweep electromyogram which tests the integrity of muscle by increasing the stimulus frequency from 1 Hz to 100 Hz over a stimulation period of 10 seconds shows that there is significantly greater fade at higher frequencies in infants less than 12 weeks of age than in those older than 12 weeks whose responses are similar to those seen in adults.[8] These observations may be explained by a shortfall in acetylcholine release at the neuromuscular junction and it has been observed in rat phrenic nerve-diaphragm preparations that the acetylcholine quantum output increases with age in young rats such that the rate of release at 30 days of age is only 75% of the value at 110 days.[9] As less acetylcholine is released from the motor nerve terminal, one would expect the immature neuromuscular junction to be relatively sensitive to the effects of competitive muscle relaxant drugs.

The sensitivity of the neuromuscular junction to relaxant drugs can also be expressed in terms of $Cp_{ss(50)}$ which is the steady state plasma concentration that results in 50% reduction of neuromuscular function. Fisher and colleagues describe an age dependence of $Cp_{ss(50)}$ for tubocurarine in children with values in neonates of only about one-third of those observed in children 12 years or older.[10] At the same time the $t_{1/2}k_{eo}$, the half-time for equilibration between neuromuscular junction and plasma and a measure of neuromuscular perfusion, was similar in all age groups. However, others have found no significant differences with age in respect to plasma concentration–response curves for tubocurarine.[11] However, the greater sensitivity of the neuromuscular junction of neonates and infants to non-depolarizing relaxant drugs is only one factor which affects the response of infants and children to them. Other factors such as the larger volume of the central compartment and the volume of distribution of water-soluble drugs, as we shall see later, also exerts an influence by reducing the plasma concentration when the drugs are administered on a dose per weight basis.

Changes in the proportion of fast and slow muscle fibres

Goudsouzian has shown that the contraction time of the adductor pollicis muscle in response to a 0.25 Hz stimulation was significantly slower in neonates.[1] However any difference from older children in the contraction time had disappeared by 2 months of age. This finding was explained on the basis of conversion of some fibres in peripheral muscles from slow-contracting (type 1) to fast-contracting (type 2) variety. There is experimental evidence for this observation and conversion to fast muscle fibres has been observed in the extensor digitorum longus muscle in rats.[12] The opposite effect however is seen in the respiratory muscles in humans. The proportion of slow-contracting fibres at birth in the diaphragm and intercostal muscles is 25% and 46% respectively; these proportions increase by 8 months of age to 55% and 65% respectively.[13] These changes in fibre types are important for a number of reasons. Firstly, it is known that fast-contracting fibres are intrinsically more sensitive to the effects of non-depolarizing relaxant drugs and will therefore be paralysed more readily; secondly, if the direction of change from one fibre type to another is different for different muscles then one must be guarded in the extrapolation of dose-response characteristics from one muscle to another. Whereas the masseter muscle may have similar sensitivity as the adductor pollicis,[14] the same is not true of the laryngeal musculature.[15]

Changes in relative size of the skeletal muscle compartment

Anaesthetists in general administer most drugs on a weight-related basis with the expectation that for most individuals this delivers a relatively constant proportion of a drug to each body compartment. While for most adults of lean build this may be largely true, the body compartments of children are different from adults and also may vary considerably during growth. During the first year of life, for example, the body fat compartment increases markedly and may reach 30% of body weight before diminishing in the succeeding years towards puberty. At the same time, the proportion of skeletal muscle decreases during the first year of life and thereafter increases to reach a maximum of 40% of body weight by the end of the active growth period.[16] As the muscle compartment represents the important site of action for non-depolarizing neuromuscular blocking drugs, it follows that an increase in the size of this compartment will tend to increase the dose requirements of these agents. In fact a number of workers have reported that children need higher doses of some non-depolarizing relaxants when given on an mg kg^{-1} basis in comparison to adults, neonates and infants.[17,18] This may be the reason for 'resistance' to some non-depolarizing relaxant drugs in this age group and will be dealt with later together with its clinical significance.

Changes in the size of body compartments

An additional pharmacokinetic variable which affects the clinical action of muscle relaxant drugs is the size of the extracellular compartment. As these drugs are predominantly ionized molecules they are distributed mainly in the plasma and extracellular fluid (ECF).[19] ECF volume varies with age and accounts for about 44% of body weight in the newborn.[16] This decreases to approximately 23% in the adult. Thus we would expect that the volume of distribution at steady state (Vd_{ss}) will show an age dependence. Fisher and co-workers found that the Vd_{ss} for tubocurarine decreased with age to the extent that the adult values were less than half those in neonates.[10] Thus a fixed per kg dose will produce lower plasma levels in neonates as compared with older children which may offset the supposedly increased sensitivity to non-depolarizing relaxants of the immature neuromuscular junction.[2] As the volume of the ECF is more closely related to the body surface area ($6-8 l m^{-2}$) than body weight,[20] some workers have suggested that age-related differences may be reduced if the dose of drug is calculated on the basis of surface area. Walts and Dillon observed no difference between infants and adults when suxamethonium was administered in equal doses on a surface area basis.[21] Others, however, have not found the same correction factor true for non-depolarizing drugs.[22,23]

Possible changes in metabolism and clearance of drugs

While the maturity of the neuromuscular junction determines its sensitivity to relaxants, the rate of clearance determines the mean residence time and, to some extent, their duration of action.

Neonates exhibit immature renal function and the creatinine clearance reaches normal levels only after 2 years of age.[24] As most non-depolarizing relaxants rely to a considerable extent on renal excretion, it is possible that the duration of action may be altered in the very young child. Fisher et al. studied the pharmacokinetics of tubocurarine and found its elimination half-life to be significantly prolonged in neonates in comparison with adults (174 vs 89 minutes).[10] Matteo and colleagues found that while renal excretion of tubocurarine was reduced in neonates and the elimination half-life increased in both neonates and children, recovery times were not significantly different from older children and adults.[11] Similarly, the rate of spontaneous recovery from 95% blockade with pancuronium does not differ in neonates, infants or children.[23] In this study however, the recovery was from equipotent rather than equal doses of the drug. Despite the large volume of distribution of atracurium in infants, a significantly greater plasma clearance produces a shorter elimination half-life in comparison with older children.[25] This finding may explain the shorter duration of neuromuscular blockade with atracurium observed in this age group.[26]

Suxamethonium

Suxamethonium is the only depolarizing muscle relaxant drug in common clinical usage and is still the drug with the most rapid onset and shortest duration of action. These characteristics have been felt to be advantageous in paediatric anaesthesia with a shorter duration of surgery. In fact the use of suxamethonium has diminished in recent years due to the greater availability of intermediate-acting non-depolarizing drugs and a greater unease about its side-effects. Despite this, suxamethonium remains a very valuable drug in paediatric practice.

Neuromuscular effects

Suxamethonium essentially consists of two acetylcholine molecules joined together and because of its small size and water solubility it is rapidly distributed throughout the extracellular space. As discussed earlier, the volume of the extracellular space as a percentage of body weight shows a considerable reduction with growth in childhood. As the volume of distribution of a water-soluble drug parallels the volume of this body compartment it is not surprising that neonates and small children are relatively 'resistant' to suxamethonium when the drug is administered in a weight-related manner. It was recognized within a few years of its introduction that young children required larger doses to depress respiration and produce apnoea than adults.[27] Several workers have quantified these differences and have shown that neonates in particular require much larger increases in dosage to produce similar neuromuscular blockade. Infants in the first 10 weeks of life required a dose of 1 mg kg^{-1} to produce the same twitch depression as 0.5 mg kg^{-1} in older children.[28] In spite of this increased requirement, recovery from suxamethonium-induced paralysis occurs more rapidly in younger children.

The dose–response curves for suxamethonium in different age groups in children have been recently described by various workers.[29,30] There is general agreement that the ED95 for suxamethonium, i.e. the dose required to produce 95% depression of adductor pollicis twitch strength, is age dependent and tends to decrease with increasing age (Table 11.1). If this difference in dose requirements is due to differences in volume of distribution relating to the size of the extracellular fluid compartment, then these differences should largely disappear if suxamethonium were administered on the basis of body surface area, as the latter parallels more closely the ECF volume. Meakin and colleagues in fact observed that differences between the age groups were largely eliminated when the ED90 for suxamethonium was calculated on the basis of surface area.[29] This finding supports previous observations by Walts and Dillon who reported that there were no differences in potency or duration of action if the dosage of suxamethonium was calculated on the basis of body surface area.[21] Such an approach, while being pharmacokinetically more precise, is nonetheless cumbersome and it is easier to consider neonates and small children relatively 'resistant' to suxamethonium. Whereas adults usually receive $1–1.5 \text{ mg kg}^{-1}$ for intubation; on the basis of ED90 values, neonates and infants require approximately twice as much suxamethonium on a weight basis, while older children require about 20% more than adults.[29]

The duration of action is also inversely related to age (Table 11.2). Brown and colleagues showed that following a dose of 1 mg kg^{-1}, children of 1–4

Table 11.1 Potency estimates of various relaxants in different age groups

	\multicolumn{4}{c}{Potency (ED95) ($\mu g\,kg^{-1}$)}
	Neonates	Infants	Children	Adults
Suxamethonium	620	729	423	310
Mivacurium	—	—	110	80
Atracurium	220	230	320	210
Vecuronium	48	47	81	43
Rocuronium	—	—	303[a]	305
Tubocurarine	—	410	500	480
Pancuronium	72	66	93	67
Doxacurium	—	—	27[a]	30
Pipecuronium	—	33	79	59

All values obtained under balanced anaesthesia except those marked [a] which are under halothane anaesthesia; neonates = up to 1 month old; infants = 1 month to 1 year; children = 1–12 years old; data derived from various sources.

years showed a recovery time of 3.5 minutes as compared with 4.6 and 8.2 minutes in children aged 5–10 and 11–15 years respectively.[30] These values for smaller children are in broad agreement with the observations of Cook and Fischer.[28] Although the plasma cholinesterase activity may be lower in children less than 6 months of age, this has little influence on the duration of action of suxamethonium which is not prolonged.[31] In another study the duration of suxamethonium-induced apnoea was observed to be shorter in children in comparison to adults without any significant difference in the plasma cholinesterase activity between the two groups.[32] This study also showed that the recovery of respiration and neuromuscular transmission in peripheral muscles (adductor pollicis) occurred almost simultaneously in children while the recovery of peripheral muscles was delayed in adults. These findings may be due to termination of its action by redistribution from a small muscle compartment into a relatively large extracellular fluid compartment in children.

The inverse relationship between age and dose requirement for suxamethonium is also evident when the drug is administered by continuous infusion for neuromuscular blockade. Infants of around 2 months of age require an infusion rate which is on average four times greater than for older children.[33] In addition, whereas the dose requirement in older children is relatively uniform there is considerable diversity in younger infants. Neonates less than 10 days of age are particularly unpredictable in their dose requirements and this effect is presumably related to an immature myoneural junction. Some neonates may show marked resistance whereas others are sensitive to suxamethonium.[33] The development of tachyphylaxis and phase II block during suxamethonium infusion seems to reflect the higher requirements for maintenance in children. Thus while phase II block may be seen at an earlier time during infusion it tends to occur after larger doses in small infants in comparison with adults.[33,34]

204 Muscle relaxants in paediatric anaesthesia

Table 11.2 Duration of action of different relaxants in the dosages as indicated

	Duration of action (min)			
	Neonates	Infants	Children	Adults
Suxamethonium	—	3.5 (50% rec, 1 mg kg^{-1})	4.6 (50% rec, 1 mg kg^{-1})	9.1 (90% rec, 1 mg kg^{-1})
Mivacurium	—	—	6.2 (25% rec, 0.12 mg kg^{-1})	15.9 (25% rec, 0.15 mg kg^{-1})
Atracurium	28.7 (25% rec, 0.5 mg kg^{-1})	35.9 (25% rec, 0.5 mg kg^{-1})	33.7 (25% rec, 0.5 mg kg^{-1})	44 (25% rec, 0.5 mg kg^{-1})
Vecuronium	60 (90% rec, 0.1 mg kg^{-1})	39 (90% rec, 0.1 mg kg^{-1})	18 (90% rec, 0.1 mg kg^{-1})	34 (90% rec, 0.07 mg kg^{-1})
Rocuronium	—	—	27 (25% rec, 0.6 mg kg^{-1})	21 (25% rec, 0.5 mg kg^{-1})
Tubocurarine	72 (50% rec, 0.3 mg kg^{-1})	66 (50% rec, 0.3 mg kg^{-1})	46 (50% rec, 0.3 mg kg^{-1})	55 (50% rec, 0.3 mg kg^{-1})
Pancuronium	—	25 (10% rec, 0.07 mg kg^{-1})	26 (10% rec, 0.07 mg kg^{-1})	46 (10% rec, 0.07 mg kg^{-1})
Doxacurium	—	—	28 (25% rec, 0.028 mg kg^{-1})	76 (25% rec, 0.039 mg kg^{-1})
Pipecuronium	—	20 (25% rec, 0.038 mg kg^{-1})	39 (25% rec, 0.07 mg kg^{-1})	45 (25% rec, 0.058 mg kg^{-1})

All values obtained under balanced anaesthesia; data derived from various sources.
rec = recovery.

The use of the intramuscular (i.m.) route for administration of suxamethonium was first reported by Beldavs as an aid to induction of anaesthesia and as an alternative method of producing relaxation in the absence of venous access.[35] He reported that flaccid paralysis occurred within 3 minutes with a dose of 2 mg lb^{-1}. This method has since been more rigorously investigated with generally similar results.[36, 37] Thus, a dose of 4 mg kg^{-1} is required i.m. for reliable muscle relaxation but, disappointingly, the onset of maximum block at the adductor pollicis muscle takes about 4 minutes. Thus it is debatable whether this method of administration of suxamethonium is useful in the emergency management of the paediatric airway. In addition, the pattern of onset at different muscles and particularly those of the larynx following i.m. suxamethonium is not known. Furthermore, Cook and colleagues have reported three cases of pulmonary oedema in infants following i.m. suxamethonium administration considered to be rare examples of a suxamethonium-induced non-cardiogenic pulmonary oedema.[38] The exact mechanism of such a reaction, however, remains unclear.

Side-effects

Cardiac effects

The intravenous administration of suxamethonium to infants and small children is frequently associated with bradycardia within 20 seconds of injection. Although bradycardia and cardiac arrhythmias can occur on first administration, the bradycardia is more pronounced when repeated doses are administered.[39, 40] In contrast, adults are more susceptible to a second dose of suxamethonium, especially if administered soon after the first injection.[41] Bradycardia is effectively attenuated with atropine 0.02 mg kg^{-1} or glycopyrrolate 0.01 mg kg^{-1}.[42]

Muscle damage

Muscle fasciculations which herald the onset of action of suxamethonium are said to be uncommon in neonates and infants, but in a recent study mild or moderate fasciculations were noted in over 50% of children aged 3–12 years.[43] Most studies in adults have shown little correlation between severity of fasciculations and subsequent muscle pains[44-46] or changes in creatine kinase (CK) and myoglobin.[47, 48] Whereas children frequently develop large increases in CK, myalgia is rarely a feature.[43, 49] Electrolyte and biochemical changes suggestive of muscle damage following suxamethonium administration in children are dependent on the anaesthetic technique. Increases in CK are significantly greater when inhalation of halothane precedes administration of suxamethonium.[43] In addition there is an increase in serum potassium of between 0.25 and 0.6 mmol l^{-1} in contrast to minimal changes in these biochemical variables when thiopentone precedes suxamethonium.[43, 50, 51] Despite previous reports that children with strabismus show exaggerated increases in CK following suxamethonium administration,[52] recent studies have failed to substantiate these findings.[43, 53]

Suxamethonium and muscle disease

There have been a number of case reports of cardiac arrest following the administration of suxamethonium to children with Duchenne muscular dystrophy.[54, 55] The aetiology of this reaction is not known but, as some of these patients have associated cardiomyopathy, it has been suggested that this may represent an adverse reaction to hyperkalaemia.[56] Alternatively, an association between Duchenne muscular dystrophy and malignant hyperpyrexia has been described but at present the consensus view as expressed by Brownell is that an association between these two conditions is considered unproven but 'possible'.[57] Although cardiac arrest following suxamethonium is thankfully rare, there is concern in some quarters about the use of suxamethonium in paediatric practice as the diagnosis of Duchenne muscular dystrophy may be unsuspected.[58] As the clinical syndrome develops during childhood it is conceivable that the first presentation during infancy may be an adverse reaction to suxamethonium. However, the situation must be put in proper perspective in view of an incidence of only about 30 per 100 000 live births.[59]

The congenital myotonias represent another uncommon group of patients who often present in childhood and in whom the use of suxamethonium is best avoided. Suxamethonium administration is associated with a sustained contracture of skeletal muscle or a localized or generalized myotonic response.[60] The contracture may be so extreme as to render ventilation of the lungs and tracheal intubation almost impossible.

Suxamethonium is capable of producing a contracture response in a variety of other circumstances. A sustained increase in muscle tension is observed in denervated muscles and the extraocular muscles in humans.[61, 62] A transient dose-related increase in muscle tension also occurs in the adductor pollicis and masseter muscles in children and adults but the effect is five times greater at the masseter muscle and lasts longer.[63, 64] The increase in muscle tone at the adductor pollicis muscle is also greater in neonates and infants compared with older children.[29] A similar contracture response is observed in the biventer cervicis muscle of the chick which contains muscle fibres innervated by multiple nerve endings. It responds to suxamethonium with a sustained increase in muscle tension.[65]

Food and Drug Administration in the USA has recently advised avoiding the use of suxamethonium in situations where another muscle relaxant can be used. This is based on the possibility of the presence of undiagnosed underlying muscle disease.

Masseter muscle rigidity

The tonic effect of suxamethonium on various muscles, in particular the masseter, has stimulated controversy concerning the significance of the clinical diagnosis of masseter muscle rigidity (MMR) following suxamethonium. MMR is a contracture of the masseter muscles in the presence of suxamethonium to such a degree that it interferes with tracheal intubation. Traditionally it has been taken as an early sign of the onset of a malignant hyperpyrexic reaction and the discontinuation of anaesthesia and cancellation of surgery has been recommended in such cases. There is little doubt that muscle rigidity is a common sign of malignant hyperpyrexia (MH) occurring in about 75% of cases when suxamethonium is the triggering agent.[66] The incidence following anaesthesia induction with halothane followed by suxamethonium administration in children has been reported as 1.03%.[67] The supposed association of MMR with MH and musculoskeletal abnormalities may be supported by the claim that children with strabismus show a four-fold greater incidence of MMR.[68] There are difficulties, however, in interpreting much of these data, which are largely derived from retrospective studies, but, if accurate, then it follows that the incidence of susceptibility to MH in the population is considerably higher than the actual reported incidence of the syndrome. While such a possibility has been suggested by some, others have cast doubt on the relative value of MMR as a predictor of MH.[69, 70]

About 50% of children referred for *in vitro* skeletal muscle contracture testing following an episode of MMR were shown to exhibit some element of sensitivity to MH.[71, 72] Littleford and co-workers reviewed the records of 68 cases of MMR in their institution over 10 years and found no cases of long-term morbidity when anaesthesia was continued.[73] This was in spite of the occurrence of intraoperative arrhythmias, hypercarbia and metabolic acidosis. They concluded that provided careful monitoring was employed anaesthesia and surgery could be safely continued in isolated cases of MMR. Increasingly, an expectant approach is advocated in the management of MMR with the continuation of anaesthesia and monitoring for signs of hypermetabolism. The contrary viewpoint, however, has been compellingly put by Allen and Rosenberg who found that no clinical sign occurring in association with MMR appeared predictive of malignant hyperpyrexia susceptibility on subsequent *in vitro* testing.[74] Consequently, they advised that the increased risk of a hyperpyrexic reaction following MMR cannot justify the continuation of anaesthesia in an elective surgical procedure.

Non-depolarizing relaxant drugs in paediatric anaesthesia

General characteristics

The onset of action following administration of non-depolarizing relaxants tends to occur more rapidly in neonates and infants in comparison with older children and adults. Fisher and Miller observed that maximum blockade following vecuronium $70\,\mu g\,kg^{-1}$ occurred in 1.5 minutes in infants less than 1 year of age as compared with 2.4 and 2.9 minutes for 1–8-year-old children and adults respectively.[75] Similar findings have been reported with pancuronium,[76] atracurium[17] and alcuronium and tubocurarine.[18] In the last report the onset time to maximum blockade was significantly faster even when the drugs were administered on predetermined equipotent dosages as distinct from a dose per weight basis. As the speed of onset does not parallel the ED95 of the individual drugs it is probably not due to age-related differences in the sensitivity of the neuromuscular junction but more likely to be related to a greater cardiac output and faster circulation in infants and a faster delivery of the drug to the neuromuscular junction.

In common with adults, the dose requirements for non-depolarizing relaxants in children are reduced when anaesthesia is supplemented with volatile agents.[77, 78] Although halothane is less effective in this respect, Brandom *et al.* showed, in children aged 2–10 years, a similar decrease of about 30% in atracurium infusion requirements during anaesthesia with 1 MAC of halothane and isoflurane in comparison with opioid-supplemented anaesthesia.[77]

As discussed earlier, changes in body composition and neuromuscular physiology during growth give rise to a variation in drug requirements which may be generally predictable on a theoretical basis. It has been suggested that calculation of drug dosage on the basis of body surface area may minimize age-related differences in the requirement of relaxants.[20, 26] Fisher and colleagues studying the pharmacokinetics and pharmacodynamics of tubocurarine in infants, children and adults suggested that the increased sensitivity of the neonatal neuromuscular junction was offset by the larger volume of distribution of the drug and consequently lower plasma concentrations.[10] They maintained, therefore, that dose requirements did not differ with age. This is disputed by others who have observed age-related differences particularly in young infants.[21, 22, 79] In practice, however, the most pertinent point concerning the administration of muscle relaxants to very young infants is the variation in response of the neonates particularly in the first 10 days of life such that some may appear resistant and others sensitive. Goudsouzian and colleagues, for example, described the ED95 of tubocurarine in a group of 12 neonates to vary four-fold from 0.15 to 0.62 mg kg^{-1}.[22]

The change in the size of the muscle compartment during growth as described earlier may explain another principle of dose requirement of non-depolarizing relaxant drugs in childhood. Thus one might expect children as distinct from adolescents and neonates or infants to have larger requirements and clinical evidence supports this prediction as shown by higher ED95s in this age group (Table 11.1). In general, the dose requirements are slightly smaller in infants than in adolescents while the latter group approximate to adult dosages.[17, 22, 23, 80] As an example, the ED95 for vecuronium during balanced anaesthesia varies from 81 μg kg^{-1} in 5–7 year-olds to 48 μg kg^{-1} for neonates, with 13–16-year-olds in between with an ED95 of 55 μg kg^{-1}.[81] While some of these differences are reduced by calculation of the ED95 on the basis of body surface area, the general trends remain. An additional factor may be a greater degree of protein or tissue binding in rapidly growing children. This will reduce the free portion of the drug available for diffusion into the effect compartment, thus increasing the overall requirement. Fisher and colleagues have indeed found that the steady state plasma concentration of vecuronium required for 50% neuromuscular blockade ($Cp_{ss(50)}$) is higher in children in comparison with infants.[82] The relationship between neuromuscular sensitivity, protein binding and $Cp_{ss(50)}$ is therefore complex.

The response of the paediatric patient to non-depolarizing muscle relaxant drugs may also be altered by coexistent pathology. Brown and co-workers have suggested that children with malignant tumours may show resistance to non-depolarizing drugs.[83] There was considerable prolongation of onset with tubocurarine and, to a lesser extent, alcuronium in association with malignant liver, kidney and bone tumours, but the observed changes did not correlate with the degree of malignancy. The reasons for such an abnormal response are not known but the response reverted to normal when the tumours were treated with chemotherapy and/or surgery.

Non-depolarizing relaxants in clinical use

Longer-duration relaxant drugs

While the popularity of this group of drugs is declining in general and in paediatric practice in particular, due largely to the development of intermediate-duration muscle relaxant drugs, there is a considerable accumulated knowledge on their clinical characteristics. Although the use of drugs such as tubocurarine is decreasing, new longer-acting agents such as pipecuronium and doxacurium have undergone investigation in children as well as adults.

Tubocurarine

As in adults, tubocurarine was the first relaxant to be used in children. The ED95s in different paediatric age groups vary between 0.32 and 0.5 mg kg^{-1} with those under 10 days old tending to be slightly higher.[22] An earlier report, however, suggested that neonates were more sensitive to tubocurarine if the drug was administered in a dose based on the body surface area.[21]

The onset of block is generally more rapid in children and so is the recovery.[84] Commonly used doses of 0.5–0.6 mg kg^{-1} result in maximum blockade in 3–4 minutes and provide a duration of clinical relaxation of 30–40 minutes. Addition of halothane not only results in the prolongation of neuromuscular block but also in marked decrease in arterial pressure due to the added effect of depression of cardiac output by halothane.

Paediatric patients tend to liberate less histamine than adults. For this reason tubocurarine still retains some popularity in paediatric use.

Pancuronium

Pancuronium has largely replaced the use of tubocurarine in paediatric anaesthesia because it is associated with very little histamine release and hypotension.[85] Changes in heart rate and blood pressure tend to be small and when observed, as in adults, tend to show an increase.[86] The rate of onset of neuromuscular blockade with pancuronium, as with most relaxants, decreases with age.

Meretoja and Luosto examined the potency of pancuronium in four different age groups of children and observed that the ED95 values in children (3–9 years old) were higher when compared with neonates, infants and adolescents (Table 11.1).[23] The difference was as large as 33% between neonates and children. Bevan et al. reported that a 90% block after a standard intubating dose of pancuronium occurred in 1.3 minutes in infants under 1 year of age in comparison to 2.1 and 3.4 minutes in children aged 3–10 years and adults respectively.[76] They also observed a significant age-related prolongation of the time to 10% recovery of twitch height.

Pipecuronium

Pipecuronium bromide is a relatively new steroidal relaxant which has been in use in Hungary for many years but has been introduced in Europe and the USA only recently. Studies in adults have shown its clinical duration to be similar to pancuronium but without any significant cardiovascular effects.[87, 88]

Sarner and colleagues have reported that children (1–6 years) require larger doses to achieve comparable neuromuscular blockade than infants (3–6 months) with ED95s of 49 and 33 µg kg^{-1} respectively during anaesthesia with halothane.[89] Similar trends have been reported by Pittet and co-workers using balanced anaesthesia.[90] The effects of isoflurane on pipecuronium requirements are more pronounced than those of halothane with a 30% reduction in the ED95.[91] These workers also reported the speed of onset of neuromuscular block with pipecuronium to be slow and similar in children and adults and that it does not appear to be influenced by volatile agents.

The time to recovery of the twitch to 25% of control (duration of clinical relaxation) is significantly shorter in children receiving halothane supplementation as compared with isoflurane.[91] The clinical duration is shorter in infants as compared with older children.[89, 90] After cumulative dosing the twitch height returns to 25% of control in about 40 minutes in children and about 15 minutes in infants. Thus, while pipecuronium is a long-acting agent in children and adults, it has only an intermediate duration in infants. The dosage is usually in the range of 50–75 $\mu g\,kg^{-1}$.

Doxacurium

Doxacurium chloride is the most potent muscle relaxant currently available and in many respects is similar to pipecuronium with negligible cardiovascular effects both in adults and in children.[92, 93] In common with pipecuronium there is a tendency for children to require larger doses than adults to achieve comparable degrees of muscle relaxation. The ED95 for doxacurium in children aged 2–12 years is about 30 $\mu g\,kg^{-1}$ during halothane anaesthesia.[93, 94] This is comparable to what has been reported for adults under balanced anaesthesia.[95, 96]

The onset of neuromuscular block with equipotent doses of doxacurium is slightly faster in children when compared to adults, as is the recovery to 25% neuromuscular function.[94] A dose of 28 $\mu g\,kg^{-1}$ (approximately 1 × ED95) in children during halothane anaesthesia results in a clinical duration of about 28 minutes, while the comparable duration with approximately double this dose is 50 minutes. The onset of action, however, is relatively slow. Both pipecuronium and doxacurium are now licensed for use in the USA.

Intermediate-duration muscle relaxant drugs

As in adult practice, the use of atracurium and vecuronium is widespread in children. At this time these are the only agents with an intermediate duration of action, although rocuronium, a new rapidly acting non-depolarizing drug, has been undergoing extensive clinical trials.

Atracurium

Atracurium is a bisquaternary benzylisoquinolinium compound which undergoes spontaneous breakdown in the body by a process known as Hofmann elimination in addition to organ uptake.

While Goudsouzian and colleagues[78] have claimed that the dose requirements for atracurium are unaffected by age, the more conventional pattern of age dependence for the drug has been described by other workers.[17, 26] However, the speed of recovery from atracurium blockade would appear to be slightly more rapid in neonates compared with other children.[17] The ED95 in neonates and infants is not much different from that in adults, but it is higher in children (Table 11.1). The duration of action in all the paediatric groups is shorter in comparison with adults (Table 11.2). A commonly used dose of 0.5 mg kg^{-1} gives a duration of clinical relaxation of 20–30 minutes.[97]

The dosage of atracurium when it is administered as an infusion for maintenance of relaxation is also age related (Table 11.3). Under balanced anaesthesia the maintenance requirements for children receiving atracurium infusions are about 0.5 mg kg^{-1} h^{-1} with neonates requiring about 25% less.[98, 99] The addition of a volatile agent reduces requirements by about 30%.[77]

Clinically useful doses of atracurium exert minimal cardiovascular effects, but if it is administered there may be an increase in heart rate and decrease in arterial pressure due to histamine release.[100] Although actual histamine levels have not been measured in children receiving intubating doses of atracurium, histamine liberation does not appear to be as common a problem in children as in adults and less likely to be associated with significant cardiovascular changes.[26, 97, 101] Bronchospasm has, however, been reported in paediatric patients following administration of atracurium.[102]

Although there is little firm evidence for harmful effects of long-term muscle relaxant infusions, concern over possible accumulation of laudanosine, one of the breakdown products of atracurium, has made some wary of its use in critical care situations over long periods. Reported blood levels of laudanosine following prolonged infusion in children, however, are small and reach only a fraction of those reported to be epileptogenic in dogs.[103]

Vecuronium

Vecuronium is the monoquaternary analogue of pancuronium and has been widely used in paediatric practice. The age-related variation in estimated potency values seen with other muscle relaxant drugs is also observed with vecuronium, with the usual pattern of higher ED95s in children as compared to neonates, infants and adolescents ranging from 48 to 80 µg kg^{-1}.[81]

Unlike atracurium, vecuronium may show a prolonged duration of action in neonates and infants. In recent work reported by Meretoja, 90% recovery

Table 11.3 Dosage of relaxants for maintaining 90–95% block

	Infusion requirements (µg kg^{-1} h^{-1})			
	Neonates	Infants	Children	Adults
Vecuronium	—	62	144	102
Atracurium	400	456[a]	558	383
Mivacurium	—	—	780[a]	360

All dosages are under balanced anaesthesia except [a] which is under halothane anaesthesia; data derived from various sources.

following 0.1 mg kg^{-1} of vecuronium took 18 minutes in 3–10-year-old children, while recovery from the same dose in infants less than 3 months of age took 60 minutes.[80] This feature of recovery would appear to be unique to vecuronium and is apparent until about 12 months of age. The reason is not entirely clear, but Fisher and colleagues have suggested that the longer recovery from vecuronium in infants is probably due to age-related changes in volume of distribution rather than changes in plasma clearance.[82] While the Vd$_{ss}$ decreases with increasing age, the clearance does not, leading to a shorter mean residence time and faster recovery in children compared with infants. A 0.1 mg kg^{-1} dose of vecuronium is associated with 25% recovery in 20–30 minutes.[104] The maintenance requirements in children are about 30% more than in adults, but reduce after 30 minutes if a volatile agent is being used.[105]

Vecuronium is virtually free of any cardiovascular effects in adults.[106, 107] It rarely liberates histamine even in high doses and therefore larger doses can be administered to shorten the onset time. Sloan and colleagues showed that by increasing the dose of vecuronium in children from 0.1 to 0.4 mg kg^{-1} the onset to 95% blockade could be reduced from 83 to 39 seconds.[108] There was, however, a corresponding increase in clinical duration from 24 to 75 minutes. This is significantly shorter than the reported duration of 115 minutes by one group using a similar dose in adults.[109]

Vecuronium is also suitable for use as a continuous infusion, the dosage in older children being higher than in adults (Table 11.3). It has been used by infusion in the paediatric Intensive Care situation and seems to have relatively few side-effects and relatively rapid recovery in those with normal renal and hepatic function.[110]

Newer muscle relaxant drugs

Mivacurium

Mivacurium is a bisquaternary benzylisoquinolinium compound resembling atracurium and doxacurium in structure. It has a short duration of action due to metabolism by plasma cholinesterase.[111, 112]

The ED95 in children aged 2–10 years under balanced anaesthesia is about 0.1 mg kg^{-1}.[113] This is in comparison with a value of about 0.07 mg kg^{-1} in adults under similar conditions.[114] The addition of halothane lowers the ED95 by about 15%. When the potency values for mivacurium are calculated on the basis of surface area, the differences between children and adults become insignificant. It has been suggested that because of this the age-related differences for mivacurium may simply be related to volume of distribution of the drug.[113]

The usual clinical dose of mivacurium is 0.15–0.2 mg kg^{-1} which provides a duration of clinical relaxation of 8–10 minutes in children. Following 250 μg kg^{-1} during halothane anaesthesia 25% recovery occurs in 11 minutes, while 95% recovery takes 18 minutes.[113] Antagonism of neuromuscular block is unnecessary in most situations but the block may be prolonged in those with very low or abnormal plasma cholinesterase.

Dose requirements of mivacurium by infusion are higher in children than in adults (Table 11.3). This is irrespective of whether it is calculated on the basis of body weight (13 µg kg^{-1} min^{-1}) or body surface area (375 µg m^{-2} min^{-1}).[115, 116]

Evidence of histamine liberation in the form of cutaneous erythema and associated hypotension is seen in adults when the dose exceeds 150 µg kg^{-1}.[117] Sarner and colleagues noted cutaneous flushing in 3 out of 18 children receiving 250 µg kg^{-1} of mivacurium and one of these showed a fall in mean arterial pressure of 32%.[113] While histamine liberation may be less frequent in children following mivacurium, it would appear that the larger drug requirements may lead to occasional hypotension.

Rocuronium

Rocuronium (Org 9426) is a new steroidal relaxant drug which is a desacetoxy derivative of vecuronium. The ED95 in adults has been reported to be approximately 300 µg kg^{-1}.[118] The main feature in adults has been shown to be a rapid onset of action but with an intermediate duration similar to that of vecuronium.[119-121]

There are to date few data available about the use of rocuronium in children, but as in adults the drug appears to show a very rapid onset of action. Under halothane anaesthesia the ED95 in children is about 300 µg kg^{-1} which is similar to that reported in adults under balanced anaesthesia.[118, 122] Complete neuromuscular block following a dose of 600 µg kg^{-1} has been reported to occur in 60–90 seconds in both adults and children with 25% recovery occurring in 27–30 minutes.[120-122]

Rocuronium in doses of 600–900 µg kg^{-1} appears to be relatively free of significant cardiovascular effects in adults except for a small and clinically insignificant increase in heart rate.[121, 123] A transient increase in heart rate within 1 minute of administration has also been reported in children receiving twice the ED95 dose.[122] Rocuronium offers an advantage over vecuronium not only in the rapidity of effect but also in being dispensed as a solution. One can achieve a faster onset of action without having to increase the dose unduly. Although not fully assessed, the drug could potentially be used as an alternative to suxamethonium in rapid sequence induction. As with vecuronium, reliable antagonism by anticholinesterases occurs provided some spontaneous recovery has taken place.

Antagonism of neuromuscular blockade

Reversal of neuromuscular blockade in children as in adults is commonly achieved with neostigmine although the use of pyridostigmine and edrophonium has also been described. Dosage recommendations for anticholinesterases in children have generally been empirical and, presumably based on occasional episodes of prolonged block, have been larger than those suggested in adult practice. Infants and children have been reported to require smaller doses of neostigmine, the equivalent ED50 doses being 13.1 µg kg^{-1} in infants, 15.5 µg kg^{-1} in children and 22.9 µg kg^{-1} in adults.[124] The commonly used dose of neostigmine is 40–50 µg kg^{-1} but actual dosage depends upon the circumstances at the time of antagonism of block.

Dose-response studies with edrophonium in infants and children and have suggested the requirement of slightly larger doses than in adults.[125] These workers concluded that paediatric patients should receive edrophonium in a dose of about 1 mg kg^{-1} to antagonize neuromuscular blockade consistently. They attributed this apparent paradox to a difference in the mechanism of action of the two drugs. These effects notwithstanding, the speed of reversal of neuromuscular blockade in children would appear to be more rapid with both neostigmine and edrophonium.[126] The requirements of atropine and glycopyrrolate are similar to those in adults (20–25 µg kg^{-1} for atropine and 10 µg kg^{-1} for glycopyrrolate).[127]

Postoperative neuromuscular function in children appears to be better maintained following reversal than in adults. Neuromuscular monitoring in the recovery ward suggests that 30–40% of adult patients may show some residual curarization.[128] This has not been observed in children older than 2 years.[129] Possible reasons for this are reduced potency of relaxants and a more rapid spontaneous recovery in this group of patients.

References

1. Goudsouzian NG. Maturation of neuromuscular transmission in the infant. *British Journal of Anaesthesia* 1980; **52:** 205–213.
2. Goudsouzian NG, Standaert FG. The infant and the myoneural junction. *Anesthesia and Analgesia* 1986; **65:** 1208–1217.
3. Bowman WC. Neuromuscular transmission: postjunctional events. In *Pharmacology of Neuromuscular Function* 1990, London: Wright, pp. 100–133.
4. Michler A, Sakmann B. Receptor stability and channel conversion in the subsynaptic membrane of the developing mammalian neuromuscular junction. *Developmental Biology* 1980; **80:** 1–17.
5. Okamoto M, Waleski JL, Artusio JF, Riker WF. Neuromuscular pharmacology in rat neonates: development of responsiveness to prototypic blocking and reversal drugs. *Anesthesia and Analgesia* 1992; **75:** 361–371.
6. Goudsouzian NG, Crone RK, Todres ID. Recovery from pancuronium blockade in the neonatal intensive care unit. *British Journal of Anaesthesia* 1981; **53:** 1303–1309.
7. Koenigsberger MR, Pattern B, Lovelace RED. Studies of neuromuscular function in the newborn: 1. A comparison of myoneural function in the full term and the premature infant. *Neuropediatrics* 1973; **4:** 350–361.
8. Crumrine RS, Yodlowski EY. Assessment of neuromuscular function in infants. *Anesthesiology* 1981; **54:** 29–32.
9. Kelly SS, Roberts DV. The effect of age on the safety factor in neuromuscular transmission in the isolated diaphragm of the rat. *British Journal of Anaesthesia* 1977; **49:** 217–221.
10. Fisher DM, O'Keefe C, Stanski D, Cronnelly R, Miller RD, Gregory GA. Pharmacokinetics and pharmacodynamics of d-tubocurarine in infants, children and adults. *Anesthesiology* 1982; **57:** 203–208.
11. Matteo RS, Lieberman IG, Salanitre E, McDaniel DD, Diaz J. Distribution, elimination, and action of d-tubocurarine in neonates, infants, children, and adult. *Anesthesia and Analgesia* 1984; **63:** 799–804.

12. Close R. Dynamic properties of fast and slow skeletal muscles of rat during development. *Journal of Physiology* (Lond) 1964; **173:** 74–95.
13. Keens TG, Bryan AC, Levison H, Iannuzo CP. Developmental pattern of muscle fibre types in human ventilatory muscles. *Applied Physiology* 1978; **44:** 909–915.
14. Saddler JM, Bevan JC, Plumley MH, Donati F, Bevan DR. Potency of atracurium on masseter and adductor pollicis muscles in children. *Canadian Journal of Anaesthesia* 1990; **37:** 26–30.
15. Donati F, Meistelman C, Plaud B. Vecuronium neuromuscular blockade at the adductor muscles of the larynx and adductor pollicis. *Anesthesiology* 1991; **74:** 833–837.
16. Friis-Hansen B. Body composition during growth. *In vivo* measurements and biochemical data correlated to differential anatomical growth. *Pediatrics* 1971; **47:** 264–274.
17. Meakin G, Shaw EA, Baker RD, Morris P. Comparison of atracurium-induced neuromuscular blockade in neonates, infants and children. *British Journal of Anaesthesia* 1988; **60:** 171–175.
18. Meretoja OA, Brown TCK, Clare D. Dose response of alcuronium and d-tubocurarine in infants, children and adolescents. *Anaesthesia and Intensive Care* 1990; **18:** 449–451.
19. Crankshaw DP, Cohen EN. Uptake, distribution and elimination of skeletal muscle relaxants. In Katz RL ed. *Muscle Relaxants*, 1975, New York: American Elsevier, pp. 125–141.
20. Cook DR. Muscle relaxants in infants and children. *Anesthesia and Analgesia* 1981; **60:** 335–343.
21. Walts LF, Dillon JB. The response of newborns to succinylcholine and d-tubocurarine. *Anesthesiology* 1969; **31:** 35–38.
22. Goudsouzian NG, Donlon JV, Saverese JJ, Ryan JF. Re-evaluation of dosage and duration of action of d-tubocurarine in the pediatric age group. *Anesthesiology* 1975; **43:** 416–426.
23. Meretoja OA, Luosto T. Dose-response characteristics of pancuronium in neonates, infants and children. *Anaesthesia and Intensive Care* 1990; **18:** 455–459.
24. McCance RA. The role of the developing kidney in the maintenance of internal stability. *Journal of the Royal College of Physicians* (Lond) 1972; **6:** 235–245.
25. Brandom BW, Stiller RL, Cook DR, Woelfel SK, Chakravorti S, Lai A. Pharmacokinetics of atracurium in anaesthetized infants and children. *British Journal of Anaesthesia* 1986; **58:** 1210–1213.
26. Brandom BW, Rudd GD, Cook DR. Clinical pharmacology of atracurium in pediatric patients. *British Journal of Anaesthesia* 1983; **55:** 117S–121S.
27. Telford J, Keats AS. Succinylcholine in cardiovascular surgery of infants and children. *Anesthesiology* 1957; **18:** 841–848.
28. Cook DR, Fischer CG. Neuromuscular blocking effects of succinylcholine in infants and children. *Anesthesiology* 1975; **42:** 662–665.
29. Meakin G, McKiernan EP, Morris P, Baker RD. Dose–response curves for suxamethonium in neonates, infants and children. *British Journal of Anaesthesia* 1989; **62:** 655–658.
30. Brown TCK, Meretoja OA, Bell B, Clare D. Suxamethonium-electromyographic studies in children. *Anaesthesia and Intensive Care* 1990; **18:** 473–476.
31. Zsigmond EK, Downs JR. Plasma cholinesterase activity in newborns and infants. *Canadian Anaesthetists Society Journal* 1971; **18:** 278–285.
32. Mirakhur RK, Elliott P, Lavery TD. Plasma cholinesterase activity and the duration of suxamethonium apnoea in children. *Annals of the Royal College of Surgeons of England* 1984; **66:** 43–45.
33. Goudsouzian NG, Liu LMP. The neuromuscular response of infants to a continuous infusion of succinylcholine. *Anesthesiology* 1984; **60:** 97–101.

34. Bevan JC, Donati F, Bevan DR. Prolonged infusion of suxamethonium in infants and children. *British Journal of Anaesthesia* 1986; **58:** 839–843.
35. Beldavs J. Intramuscular succinylcholine for endotracheal intubation in infants and children. *Canadian Anaesthetists Society Journal* 1959; **6:** 141–147.
36. Liu LMP, DeCook TH, Goudsouzian NG, Ryan JF, Liu PL. Dose response to succinylcholine in children. *Anesthesiology* 1981; **55:** 599–602.
37. Sutherland GA, Bevan JC, Bevan DR. Neuromuscular blockade in infants following intramuscular succinylcholine in two or five per cent concentration. *Canadian Anaesthetists Society Journal* 1983; **30:** 342–346.
38. Cook DR, Westman HR, Rosenfeld L, Hendershot RJ. Pulmonary edema in infants: possible association with intramuscular succinylcholine. *Anesthesia and Analgesia* 1981; **60:** 220–223.
39. Leigh MD, McCoy DD, Belton MK, Lewis GB. Bradycardia following intravenous administration of succinylcholine chloride to infants and children. *Anesthesiology* 1957; **18:** 698–702.
40. Craythorne NWB, Turndorf H, Dripps RD. Changes in pulse rate and rhythm associated with the use of succinylcholine in anesthetized children. *Anesthesiology* 1960; **21:** 465–470.
41. Mathias JA, Evans-Prosser CDG, Churchill-Davidson HC. The role of non-depolarizing drugs in the prevention of suxamethonium bradycardia. *British Journal of Anaesthesia* 1970; **42:** 609–613.
42. Lerman J, Chinyanga HM. The heart rate response to succinylcholine in children: a comparison of atropine and glycopyrrolate. *Canadian Anaesthetists Society Journal* 1983; **30:** 377–381.
43. McLoughlin CC, Mirakhur RK, Elliott P, Craig HJL, Trimble ER. Changes in serum potassium, calcium and creatine kinase following suxamethonium administration in children with and without strabismus. *Paediatric Anaesthesia* 1991; **1:** 101–106.
44. Ferres CJ, Mirakhur RK, Craig HJL, Browne ES, Clarke RSJ. Pre-treatment with vecuronium as a prophylactic against post-suxamethonium muscle pain: comparison with other non-depolarizing neuromuscular blocking drugs. *British Journal of Anaesthesia* 1983; **55:** 735–741.
45. Sosis M, Broad T, Larijani GE, Marr AT. Comparison of atracurium and d-tubocurarine for prevention of succinylcholine myalgia. *Anesthesia and Analgesia* 1987; **66:** 657–659.
46. O'Sullivan EP, Williams NE, Calvey TN. Differential effects of neuromuscular blocking agents on suxamethonium-induced fasciculations and myalgia. *British Journal of Anaesthesia* 1988; **60:** 367–371.
47. Plotz J, Braun J. Failure of 'self-taming' doses of succinylcholine to inhibit increases in postoperative serum creatine kinase activity in children. *Anesthesiology* 1982; **56:** 207–209.
48. Laurence AS. Myalgia and biochemical changes following intermittent suxamethonium administration. Effects of alcuronium, lignocaine, midazolam and suxamethonium pretreatments on serum myoglobin, creatine kinase and myalgia. *Anaesthesia* 1987; **42:** 503–510.
49. Bush GH, Roth F. Muscle pains after suxamethonium in children. *British Journal of Anaesthesia* 1961; **33:** 151–155.
50. Keneally JP, Bush GH. Changes in serum potassium after suxamethonium in children. *Anaesthesia and Intensive Care* 1974; **2:** 147–150.
51. Henning RD, Bush GH. Plasma potassium after halothane-suxamethonium induction in children. *Anaesthesia* 1982; **37:** 802–805.
52. Tammisto T, Brander P, Airaksinen MM, Tommila V, Listola V. Strabismus as a possible sign of latent muscular disease predisposing to suxamethonium-induced muscular injury. *Annals of Clinical Research* 1970; **2:** 126–130.

53. Saddler JM, Bevan JC, Plumley MH, Polomena RC, Donati F, Bevan DR. Jaw muscle tension after succinylcholine in children undergoing strabismus surgery. *Canadian Journal of Anaesthesia* 1990; **37:** 21–25.
54. Henderson WAV. Succinylcholine-induced cardiac arrest in unsuspected Duchenne muscular dystrophy. *Canadian Anaesthetists Society Journal* 1984; **31:** 444–446.
55. Solares G, Herranz JL, Sanz MD. Suxamethonium-induced cardiac arrest as an initial manifestation of Duchenne muscular dystrophy. *British Journal of Anaesthesia* 1986; **58:** 576.
56. Genever EE. Suxamethonium-induced cardiac arrest in unsuspected pseudohypertrophic muscular dystrophy. *British Journal of Anaesthesia* 1971; **43:** 984–986.
57. Brownell AKW. Malignant hyperthermia: relationship to other diseases. *British Journal of Anaesthesia* 1988; **60:** 303–308.
58. Delphin E, Jackson D, Rothstein P. Use of succinylcholine during elective paediatric anesthesia should be re-evaluated. *Anesthesia and Analgesia* 1987; **66:** 1190–1192.
59. Smith CL, Bush GH. Anaesthesia and progressive muscular dystrophy. *British Journal of Anaesthesia* 1985; **57:** 1113–1118.
60. Cody JR. Muscle rigidity following administration of succinylcholine. *Anesthesiology* 1968; **29:** 159–162.
61. Gronert GA, Lambert E, Theye RA. The response of denervated muscle to succinylcholine. *Anesthesiology* 1973; **39:** 13–22.
62. Eakins KE, Katz RL. The action of succinylcholine on the tension of extraocular muscles. *British Journal of Pharmacology* 1966; **26:** 205–211.
63. Plumley MH, Bevan JC, Saddler JM, Donati F, Bevan DR. Dose-related effects of succinylcholine on the adductor pollicis and masseter muscles in children. *Canadian Journal of Anaesthesia* 1990; **37:** 15–20.
64. Van der Spek AFL, Fang WB, Ashton-Miller JA, Stohler CS, Carlson DS, Schork MA. The effects of succinylcholine on mouth opening. *Anesthesiology* 1987; **67:** 459–465.
65. Bowman WC, Rand MJ. Striated muscle and neuromuscular transmission. In *Textbook of Pharmacology* 2nd edn, 1980, Oxford, London, Edinburgh, Melbourne: Blackwell, pp. 17.1–17.53.
66. Hackl W, Mauritz W, Schemper M, Winkler M, Sporn P, Steinbereithner K. Prediction of malignant hyperthermia susceptibility; statistical evaluation of clinical signs. *British Journal of Anaesthesia* 1990; **64:** 425–429.
67. Schwartz L, Rockoff MA. Masseter muscle spasm with anesthesia; incidence and implications. *Anesthesiology* 1984; **61:** 772–775.
68. Carroll JB. Increased incidence of masseter spasm in children with strabismus anesthetized with halothane and succinylcholine. *Anesthesiology* 1987; **67:** 559–561.
69. Kalow W, Britt B, Chan F-Y. Epidemiology and inheritance of malignant hyperthermia. *International Anesthesiology Clinics* 1979; **17:** 114–140.
70. Leary NP, Ellis FR. Masseteric muscle spasm as a normal response to suxamethonium. *British Journal of Anaesthesia* 1990; **64:** 488–492.
71. Rosenberg H, Reed S. *In vitro* contracture tests for susceptibility to malignant hyperthermia. *Anesthesia and Analgesia* 1983; **62:** 415–420.
72. Christian AS, Ellis FR, Halsall PJ. Is there a relationship between masseteric muscle spasm and malignant hyperpyrexia? *British Journal of Anaesthesia* 1989; **62:** 540–544.
73. Littleford JA, Patel LR, Bose D, Careron CB, McKillop C. Masseter muscle spasm in children: implications of continuing the triggering anesthetic. *Anesthesia and Analgesia* 1991; **72:** 151–160.

74. Allen GC, Rosenberg H. Malignant hyperthermia susceptibility in adult patients with masseter muscle rigidity. *Canadian Journal of Anaesthesia* 1990; **37:** 31–35.
75. Fisher DM, Miller RD. Neuromuscular effects of vecuronium (ORG NC45) in infants and children during N_2O, halothane anesthesia. *Anesthesiology* 1983; **58:** 519–523.
76. Bevan JC, Donati F, Bevan DR. Attempted acceleration of the onset of action of pancuronium. Effects of divided doses in infants and children. *British Journal of Anaesthesia* 1985; **57:** 1204–1208.
77. Brandom B, Cook DR, Woelfel SK, Rudd D, Fehr B, Lineberry CG. Atracurium infusion requirements in children during halothane, isoflurane, and narcotic anesthesia. *Anesthesia and Analgesia* 1985; **64:** 471–476.
78. Goudsouzian N, Liu LMP, Gionfriddo M, Rudd GD. Neuromuscular effects of atracurium in infants and children. *Anesthesiology* 1985; **62:** 75–79.
79. Cook DR. Sensitivity of the newborn to tubocurarine. *British Journal of Anaesthesia* 1981; **53:** 319–320.
80. Meretoja OA. Is vecuronium a long acting neuromuscular agent in neonates and infants? *British Journal of Anaesthesia* 1989; **62:** 184–187.
81. Meretoja OA, Wirtavuori K, Neuvonen PJ. Age-dependence of the dose-response curve of vecuronium in pediatric patients during balanced anaesthesia. *Anesthesia and Analgesia* 1988; **67:** 21–26.
82. Fisher DM, Castagnoli K, Miller RD. Vecuronium kinetics and dynamics in anesthetized infants and children. *Clinical Pharmacology and Therapeutics* 1985; **37:** 402–406.
83. Brown TCK, Gregory M, Bell B, Clare D. Response to non-depolarizing muscle relaxants in children with tumours. *Anaesthesia and Intensive Care* 1990; **18:** 460–465.
84. Baxter M, Bevan JC, Donati F, Bevan DR. d-Tubocurarine priming in children. *Canadian Anaesthetists Society Journal* 1986; **33:** S86–S87.
85. Nightingale DA, Bush GH. A clinical comparison between tubocurarine and pancuronium in children. *British Journal of Anaesthesia* 1973; **45:** 63–70.
86. Yamamoto T, Baba H, Shiratsuchi T. Clinical experience with pancuronium bromide in infants and children. *Anesthesia and Analgesia* 1972; **51:** 919–924.
87. Stanley JC, Mirakhur RK, Bell PF, Sharpe TDE, Clarke RSJ. Neuromuscular effects of pipecuronium bromide. *European Journal of Anaesthesiology* 1991; **8:** 151–156.
88. Stanley JC, Carson IW, Gibson FM, McMurray TJ, Elliott P, Lyons SM, *et al.* Comparison of the haemodynamic effects of pipecuronium and pancuronium during fentanyl anaesthesia. *Acta Anaesthesiologica Scandinavica* 1991; **35:** 262–266.
89. Sarner JB, Brandom BW, Dong M-L, Pickle D, Cook DR, Weinberger MJ. Clinical pharmacology of pipecuronium in infants and children during halothane anesthesia. *Anesthesia and Analgesia* 1990; **71:** 362–366.
90. Pittet JF, Tassonyi E, Morel DR, Gemperle G, Rouge JC. Neuromuscular effect of pipecuronium bromide in infants and children during nitrous oxide-alfentanil anesthesia. *Anesthesiology* 1990; **72:** 432–435.
91. Pittet JF, Tassonyi E, Morel DR, Gemperle G, Richter M, Rouge JC. Pipecuronium-induced neuromuscular blockade during nitrous oxide-fentanyl, isoflurane, and halothane anesthesia in adults and children. *Anesthesiology* 1989; **71:** 210–213.
92. Stoops CM, Curtis CA, Kovach DA, McCammon RL, Stoelting RK, Warren TM, *et al.* Hemodynamic effects of doxacurium chloride in patients receiving oxygen sufentanil anesthesia for coronary artery bypass grafting or value replacement. *Anesthesiology* 1988; **69:** 365–370.

93. Goudsouzian NG, Alifimoff JK, Liu LMP, Foster V, McNulty B, Savarese JJ. Neuromuscular and cardiovascular effects of doxacurium in children anaesthetized with halothane. *British Journal of Anaesthesia* 1989; **62:** 263–268.
94. Sarner JB, Brandom BW, Cook DR, Dong M-L, Horn MC, Woelfel SK, *et al*. Clinical pharmacology of doxacurium chloride (BW A938U) in children. *Anesthesia and Analgesia* 1988; **67:** 303–306.
95. Basta SJ, Savarese JJ, Ali HH, Embree PB, Schwartz AF, Rudd GD, *et al*. Clinical pharmacology of doxacurium chloride. A long acting non-depolarizing relaxant. *Anesthesiology* 1988; **69:** 478–486.
96. Maddineni VR, Cooper R, Stanley JC, Mirakhur RK, Clarke RSJ. Clinical evaluation of doxacurium chloride. *Anaesthesia* 1992; **47:** 554–557.
97. Lavery GG, Mirakhur RK. Atracurium besylate in paediatric anaesthesia. *Anaesthesia* 1984; **39:** 1243–1246.
98. Kalli I, Meretoja OA. Infusion of atracurium in neonates, infants and children. *British Journal of Anaesthesia* 1988; **60:** 651–654.
99. Goudsouzian NG. Atracurium infusion in infants. *Anesthesiology* 1988; **68:** 267–269.
100. Scott RPF, Savarese JJ, Basta SJ, Sunder N, Ali HH, Gargarian M, *et al*. Atracurium: clinical strategies for preventing histamine release and attenuating the haemodynamic response. *British Journal of Anaesthesia* 1985; **57:** 550–553.
101. Goudsouzian NG, Young ET, Moss J, Liu LMP. Histamine release during the administration of atracurium or vecuronium in children. *British Journal of Anaesthesia* 1986; **58:** 1229–1233.
102. Woods I, Morris P, Meakin G. Severe bronchospasm following the use of atracurium in children. *Anaesthesia* 1985; **40:** 207–208.
103. Tabardel Y, Paquay T, Santerre S. Prolonged infusion of atracurium in infants. *Developmental Pharmacology and Therapeutics* 1990; **15:** 52–56.
104. Ferres CJ, Crean PM, Mirakhur RK. An evaluation of Org NC45 (vecuronium) in paediatric anaesthesia. *Anaesthesia* 1983; **38:** 943–947.
105. Woelfel SK, Dong ML, Brandom BW, Sarner JB, Cook DR. Vecuronium infusion requirements in children during halothane-narcotic-nitrous oxide, isoflurane-narcotic-nitrous oxide, and narcotic-nitrous oxide anesthesia. *Anesthesia and Analgesia* 1991; **73:** 33–38.
106. Robertson EN, Booij LHDJ, Fragen RJ, Crul JF. Clinical comparison of atracurium and vecuronium (ORG NC45). *British Journal of Anaesthesia* 1983; **55:** 125–130.
107. Lavery GG, Mirakhur RK, Clarke RSJ, Gibson FM. The effect of atracurium, vecuronium and pancuronium on heart rate and arterial pressure in normal individuals. *European Journal of Anaesthesiology* 1986; **4:** 143–147.
108. Sloan MH, Lerman J, Bissonnette B. Pharmacodynamics of high dose vecuronium in children during balanced anesthesia. *Anesthesiology* 1991; **74:** 656–659.
109. Tullock WC, Dianna P, Cook DR, Wilks DH, Brandom BW, Stiller RL, *et al*. Neuromuscular and cardiovascular effects of high-dose vecuronium. *Anesthesia and Analgesia* 1990; **70:** 86–89.
110. Fitzpatrick KTJ, Black GW, Crean PM, Mirakhur RK. Continuous vecuronium infusion for prolonged muscle relaxation in children. *Canadian Journal of Anaesthesia* 1991; **38:** 169–174.
111. Cook DR, Stiller RL, Weakly JN, Chakravorti S, Brandom BW, Welch RM. In vitro metabolism of mivacurium chloride (BW B1090U) and succinylcholine. *Anesthesia and Analgesia* 1989; **68:** 452–456.
112. Savarese JJ, Ali HH, Basta SJ, Embree PB, Scott RPF, Sunder N, *et al*. The clinical neuromuscular pharmacology of mivacurium chloride (BW B1090U): a short-acting non-depolarizing ester neuromuscular blocking drug. *Anesthesiology* 1988; **68:** 723–732.

113. Sarner JB, Brandom BW, Woelfel SK, Dong M-L Horn MC, Cook DR, et al. Clinical pharmacology of mivacurium chloride (BW B1090U) in children during nitrous oxide-halothane and nitrous oxide-narcotic anesthesia. *Anesthesia and Analgesia* 1989; **68:** 116–121.
114. Weber S, Brandom BW, Powers DM, Sarner JB, Woelfel SK, Cook DR, et al. Mivacurium chloride (BW B1090U)-induced neuromuscular blockade during nitrous oxide-isoflurane and nitrous oxide-narcotic anesthesia in adult surgical patients. *Anesthesia and Analgesia* 1988; **67:** 495–499.
115. Alifimoff JK, Goudsouzian NG. Continuous infusion of mivacurium in children. *British Journal of Anaesthesia* 1989; **63:** 520–524.
116. Brandom BW, Sarner JB, Woelfel SK, Dong ML, Horn MC, Borland LM, et al. Mivacurium infusion requirements in pediatric surgical patients during nitrous oxide-halothane and nitrous oxide-narcotic anesthesia. *Anesthesia and Analgesia* 1990; **71:** 16–22.
117. Caldwell JE, Heier T, Kitts JB, Lynam DP, Fahey MR, Miller RD. Comparison of the neuromuscular block induced by mivacurium, suxamethonium or atracurium during nitrous oxide-fentanyl anaesthesia. *British Journal of Anaesthesia* 1989; **63:** 393–399.
118. Cooper AR, Mirakhur RK, Elliott P, McCarthy G. Estimation of the potency of ORG 9426 using two different modes of nerve stimulation. *Canadian Journal of Anaesthesia* 1992; **39:** 139–142.
119. Wierda JMKH, Kleef UW, Lambalk LM, Kloppenburg WD, Agoston S. The pharmacodynamics and pharmacokinetics of Org 9426, a new non-depolarizing neuromuscular blocking agent, in patients anaesthetized with nitrous oxide, halothane and fenlanyl. *Canadian Journal of Anaesthesia* 1991; **38:** 430–435.
120. Cooper R, Mirakhur RK, Clarke RSJ, Boules Z. Comparison of intubating conditions after administration of Org 9426 (rocuronium) and suxamethonium. *British Journal of Anaesthesia* 1992; **69:** 269–273.
121. Cooper RA, Mirakhur RK, Maddineni VR. Neuromuscular effects of rocuronium bromide (Org 9426) during fentanyl and halothane anaesthesia. *Anaesthesia* 1993; **48:** 103–105.
122. Woelfel SK, Brandom BW, Cook DR, Sarner JB. Effects of administration of ORG-9426 in children during nitrous oxide-halothane anesthesia. *Anesthesiology* 1992; **76:** 939–942.
123. Booij LHDJ, Knape HDA. The neuromuscular blocking effect of Org 9426, a new intermediate acting non-depolarizing muscle relaxant in man. *Anaesthesia* 1991; **46:** 341–343.
124. Fisher DM, Cronnelly R, Miller RD, Sharma M. The neuromuscular pharmacology of neostigmine in infants and children. *Anesthesiology* 1983; **59:** 220–225.
125. Fisher DM, Cronnelly R, Sharma M, Miller RD. Clinical pharmacology of edrophonium in infants and children. *Anesthesiology* 1984; **61:** 428–433.
126. Meakin G, Sweet PT, Bevan JC, Bevan DR. Neostigmine and edrophonium as antagonists of pancuronium in infants and children. *Anesthesiology* 1983; **59:** 316–321.
127. Black GW, Mirakhur RK, Keilty SR, Love SHS. Reversal of non-depolarizing neuromuscular block in children: comparison of atropine and glycopyrrolate in a mixture with neostigmine. *Anaesthesia* 1980; **35:** 913–916.
128. Viby-Mogensen J, Jorgensen BC, Ording H. Residual curarization in the recovery room. *Anesthesiology* 1979; **50:** 539–541.
129. Baxter MRN, Bevan JC, Samuel J, Donati F, Bevan DR. Postoperative neuromuscular function in pediatric day-care patients. *Anesthesia and Analgesia* 1991; **72:** 504–508.

12

Muscle Relaxants in the Intensive Care Unit

Brian J Pollard

A patient admitted to an Intensive Care Unit often requires a period of artificial ventilation of the lungs as one component of their overall management. To have a tracheal tube passed and then to be subjected to external control of ventilation is a traumatic experience and sedation is required. The sedative agent(s) used depend on a number of factors and this needs to be determined on an individual basis for each patient. Members of the different drug families possess a variety of pharmacological properties certain of which may be useful at specific occasions. For example, benzodiazepines reduce anxiety and promote sleep; opioids provide analgesia, respiratory depression and sedation. The use of one or a combination of these or other agents is important and the actual choice, a matter of individual preference, patient requirements and unit policy.

It is often possible to achieve continued tolerance of intermittent positive pressure ventilation (IPPV) with the use of sedatives and analgesics alone. On occasions when it proves difficult to achieve adequate oxygenation, however, a non-depolarizing neuromuscular blocking agent is often introduced. When a neuromuscular blocking agent is introduced it is extremely important to ensure that the patient is adequately sedated.

The use of the neuromuscular blocking drugs in Intensive Care practice has undergone a marked change over the last 15 years. At the end of the 1970s, over 90% of Intensive Care Units were using muscle relaxants (usually pancuronium) in most patients.[1] By 1986, it appeared that this number had fallen to about 16%.[2] A recent unpublished survey (AJ Stone, personal communication) suggests that the proportion is increasing again. While the exact reasons for these changes in practice are unclear, several factors are known to have exerted an influence.

In the rapidly developing speciality of Intensive Care, the muscle relaxants acquired a spurious place in the 'sedation' of patients. A common regimen was to administer a combination of pancuronium and an opioid by intermittent bolus injection. Pancuronium was an important component because a paralysed patient would remain immobile, allowing nursing and other procedures to be more easily undertaken. This did not seem to cause any concern because many people held the mistaken view that neuromuscular paralysis was a calm and pleasant state. This erroneous assumption was highlighted by an editorial in the *Lancet* in 1981 which drew attention to the problems of paralysis in inadequately sedated patients.[3] Further reinforcement was supplied by the appearance of published recollections of medical colleagues who had been patients in an Intensive Care Unit. This led to an increase in attempts to manage patients completely without the use of a muscle relaxant.

The original ventilators used on the Intensive Care Unit were modified anaesthetic ventilators and were therefore very rigid in their actions. This meant that the patient could not breathe through the ventilator. The result was that they were difficult to use unless the patient was heavily sedated or paralysed. The introduction of intermittent mandatory ventilation and other improved modes subsequently made it possible to use sedation alone for many patients.

Attempts to limit the routine use of muscle relaxants in Intensive Care patients and to rely solely on sedative agents, anaesthetic agents and opioids have not been uniformly successful. Many studies have noted that despite attempting to rely principally on a continuous infusion of sedative agents, a considerable proportion of patients additionally require the administration of a muscle relaxant to facilitate controlled ventilation. To attempt to discontinue the routine use of neuromuscular blocking agents is not therefore the most logical solution. The drugs should be retained and used more sensibly by improving education.

If we are going to take this path and improve education, it is necessary to begin with some indications for where and when muscle relaxants are most useful in Intensive Care.

Use of neuromuscular blocking agents in Intensive Care

There are a number of situations where the use of a neuromuscular blocking agent is useful or even essential (Table 12.1).

Tracheal intubation

Intubation of the trachea may be performed without the use of a muscle relaxant. In situations where the patient is already weak or has received a

Table 12.1 Indications for the use of a neuromuscular blocking agent in Intensive Care patients

Tracheal intubation
Facilitation of procedures
To assist mechanical ventilation
Critical gas exchange
Multiple trauma
Tetanus

generous dose of an anaesthetic or sedative agent, no other assistance will be required much of the time. In a number of Intensive Care Units, it is commonplace to intubate the trachea following local analgesia to the pharynx and larynx. The use of a muscle relaxant does, however, allow intubation to be performed more easily and with less potential for trauma. A muscle relaxant is also particularly useful when the airway has to be secured without delay or when there is a risk of regurgitation of gastric contents. In these situations suxamethonium is the agent of choice unless there are any specific contraindications to its use (Chapter 4).[4]

Facilitation of procedures

A completely immobile patient is highly desirable or even essential for a number of procedures. This includes many surgical or endoscopic procedures, e.g. bronchoscopy (especially rigid bronchoscopy) and tracheostomy. The use of a muscle relaxant is also of considerable benefit in order to obtain high quality CT and NMR images.[5]

To assist ventilation

There are a number of patients who are unable to tolerate IPPV even though they appear to be adequately sedated. The result is marked swings in intrathoracic pressure, intracranial pressure and intra-abdominal pressure. The maintenance of a stable blood pressure and cardiac output is likely to be compromised and excess strain put on abdominal wall suture lines which may lead to wound dehiscence. The use of a muscle relaxant in patients with a raised intracranial pressure is considered by many to be mandatory.

Critical gas exchange

The more critically ill patient often has greatly reduced lung compliance which results in difficulties in achieving adequate gas exchange during IPPV when sedative agents are used alone. The addition of a muscle relaxant in these patients may result in a small increase in compliance following the abolition of tone in the thoracic musculature and this may improve ventilation. The patient's oxygen requirement and carbon dioxide output will also fall following a reduction in the activity of striated muscle.

Multiple trauma

A number of multiple trauma patients with unstable fractures may benefit from paralysis. Movement of fracture sites is minimized which should assist in promoting healing. It is unlikely, however, that long-term benefit will be significant because it would be unusual for neuromuscular block to be continued for the several weeks which would be needed for a fracture to heal. There is a potential problem resulting from the abolition of all muscle tone. The presence of the muscle tone may be holding an unstable fracture in an acceptable position and the fracture be more easily displaced by moving the patient when paralysed. These factors must, of course, be assessed for each individual situation.

Tetanus

The use of a neuromuscular blocking agent is positively indicated in patients with tetanus in order to reduce muscle spasms and permit IPPV to be undertaken. Tetanus is accompanied by periods of gross cardiovascular instability and so the choice of a relaxant with minimal cardiovascular action is beneficial.

Disadvantages of using relaxants

There are no absolute contraindications to the non-depolarizing muscle relaxants except for known or suspected allergy to the drug. There are, however, a number of relative contraindications, or situations where the use of a muscle relaxant may complicate management (Table 12.2). Absolute contraindications exist to suxamethonium and these include hyperkalaemia, raised intracranial pressure, myotonia and certain muscle disorders (Chapter 4). Each patient must therefore be considered on individual merits and the advantages weighed against the disadvantages for that patient.

Difficulties with neurological assessment

When a patient is pharmacologically paralysed it is difficult or impossible to assess fully the neurological state. Spontaneous movement or movement to command is not possible. One particularly important problem is that focal or localizing neurological signs may be missed in the paralysed patient, leading to delay in treatment. This could have potentially serious consequences in some circumstances, e.g. expansion of a subdural haematoma. Bearing in mind the advantages of muscle relaxants in patients with raised intracranial pressure, the clinician is faced with a dilemma which can only be resolved by considering each case on an individual basis. If it is decided to use a muscle relaxant, it should be discontinued at suitable intervals to allow the assessment of the patient. Atracurium, vecuronium or mivacurium would seem, therefore, to be the agents of choice.

Disconnection

In the event of any patient becoming disconnected from the ventilator, spontaneous ventilation will not be possible in the presence of a neuromuscular blocking agent. The critical nature of gas exchange in most Intensive Care patients, however, makes it doubtful whether much benefit would be

Table 12.2 Disadvantages to the use of muscle relaxants in Intensive Care

Difficulties with neurological assessment
Disconnection
Awareness
Increased risk of thromboembolism
Disuse atrophy
Pancuronium belly

obtained from attempts at spontaneous ventilation. In addition, with the allocation of a nurse to every patient and the presence of comprehensive alarms on all ventilators, this risk must be very small indeed.

Awareness

This problem has already been noted as one of the factors which has led to a reduction in the use of the muscle relaxants. The use of a muscle relaxant may mask inadequate sedation by preventing or limiting most of the currently employed measurements of the level of sedation. This problem will unfortunately remain until a more objective method of measuring sedation can be developed.[6,7]

The incidence of thromboembolism

It has been reported that the incidence of deep venous thrombosis and pulmonary emboli was more common in the paralysed Intensive Care patient. The evidence is conflicting, however, and this must remain uncertain at present.

Disuse atrophy

Patients in an Intensive Care Unit may suffer muscle weakness due to disuse atrophy (Chapter 9) as part of an extension of the disease process surrounding multiple organ failure. The suggestion has been made that patients who receive long-term therapy with neuromuscular blocking agents may be at particular risk of this problem.[8]

Pancuronium belly

This is an unusual condition which may be seen in the neonate. If a neonate is paralysed from birth it does not have the chance to swallow any air. The absence of air from the gastrointestinal tract results in an unusual appearance on an abdominal X-ray which is described as 'ground glass' in appearance.[9]

Properties of the ideal relaxant

The properties of the ideal muscle relaxant for use in Intensive Care patients are outlined in Table 12.3.

Table 12.3 The properties of the ideal muscle relaxant for Intensive Care

Good cardiovascular stability
Negligible potential for histamine release
Non-cumulative
Elimination independent of the kidneys
Elimination independent of the liver
Rapid and complete termination of effect
Suitable for administration by infusion

Cardiovascular stability

Intensive Care patients usually possess a variable degree of cardiovascular instability. This may be a feature of the disease process or result from aspects of the treatment. It is nevertheless important to select all drugs with a view to achieving as little further disturbance of cardiovascular stability as possible.

Histamine release

Severe histamine reactions are uncommon in Intensive Care patients, given the multiplicity of drugs received by most patients. It is appropriate, however, to consider the potential for histamine release when selecting all drugs in these patients.

Lack of cumulation

Patients often remain on an Intensive Care Unit for several days. During this time, drugs are administered in repeated doses or using an infusion. Any drug with the potential for accumulation is therefore clearly undesirable.

Renal and hepatic disease

The Intensive Care patient not infrequently has a degree of impairment of either liver function, kidney function, or both. A drug which is completely independent of either or both of these systems for its elimination will therefore be at an advantage.

Rapid termination of effect

A muscle relaxant whose effect is rapidly and completely terminated without the need for any reversal agents is clearly advantageous.

Suitable for infusion administration

When neuromuscular block is required, a constancy of effect is advantageous. The choice of a drug which easily lends itself to administration by infusion would therefore be a logical choice.

Choice of muscle relaxant

Depolarizing agents

Suxamethonium is the only member of this family which remains in regular clinical use. In view of the unique position of suxamethonium as the agent with the shortest onset time of any of the muscle relaxants, it finds its main use in situations where it is necessary to achieve rapid intubation. This is a clear advantage in a new admission to an Intensive Care Unit because the time of last food intake is frequently not known and it is known that stress, including trauma, infection and acute renal failure all delay gastric emptying. Once the trachea is intubated it is unusual to administer any further doses of suxamethonium in order to maintain neuromuscular block.

Suxamethonium possesses a number of unwanted actions which are of importance in the Intensive Care patient (Chapter 4).[10] It increases the serum potassium by approximately $0.5\,\text{mmol}\,\text{l}^{-1}$ in healthy patients. Although this increase is usually of a similar magnitude in patients with renal failure such cases may well already have a raised serum potassium prior to the dose of suxamethonium. A further rise in plasma potassium concentration may be sufficient to cause ventricular arrhythmias which may proceed to cardiac arrest. There are a number of acute conditions where the increase in serum potassium may be exaggerated after suxamethonium. These include burns, massive trauma to muscle and spinal cord injuries (Chapter 9).

Non-depolarizing agents

When the long-term use of a muscle relaxant is considered in Intensive Care it is a member of this family which is selected. Although all of the non-depolarizing relaxants have been used at some time in Intensive Care practice, there are only a small number which have found regular use. Patients vary widely in their response and sensitivity to the non-depolarizing agents; the critically ill patient with multiple organ failure especially so. For example, factors including reduced metabolism of the drug due to impairment of liver function, or delayed excretion of drug or its breakdown products due to impairment of renal function, are likely to lead to cumulation of muscle relaxant or of metabolites after repeated doses. Additional factors, which may include muscular disuse atrophy from immobilization (Chapter 9) or the effects of other drugs, may also modify the response to the muscle relaxant. It is appropriate to examine each available non-depolarizing muscle relaxant in order to assess its suitability for use in Intensive Care.

Pancuronium

This synthetic aminosteroid neuromuscular blocking agent has been popular in Intensive Care practice since its introduction into clinical practice. The reasons for this probably lie in its cardiovascular action. Pancuronium has both a vagolytic and also a direct sympathetic stimulant effect which may increase the pulse rate and assist in maintaining blood pressure in the critically ill patient. It appears to be devoid of any potential to release histamine. Pancuronium has negligible ganglionic blocking activity, a feature which is of considerable advantage in the critically ill patient.

Pancuronium is metabolized in the liver partly to 3-desacetyl pancuronium which has about 40% of the neuromuscular blocking activity of the parent compound. Both pancuronium and its metabolites are excreted through the kidney. In the critically ill patient with renal and liver dysfunction, therefore, there will be a delay in the clearance of pancuronium, leading to the potential for accumulation. It would appear, therefore, that it is not appropriate to use pancuronium by continuous infusion in Intensive Care, but to administer it by intermittent bolus doses.

Tubocurarine

This naturally occurring substance has been a popular relaxant for use in anaesthesia. Its use in Intensive Care has never been extensive principally due to its capacity to reduce the blood pressure. This is at least partly due to block of autonomic ganglia and to histamine release. It may cause quite marked falls in blood pressure in critically ill, hypovolaemic patients.

Alcuronium

This relaxant is a synthetic derivative of the naturally occurring alkaloid toxiferine. It enjoyed moderate popularity in Intensive Care for a number of years, rivalling pancuronium, although its cardiostability was not quite as good. Alcuronium may cause some histamine release.

The principal disadvantage of alcuronium is that it is primarily excreted through the kidneys and therefore may have a prolonged action in patients with renal impairment.[11] This situation is not uncommon in Intensive Care patients and it is more appropriate to administer alcuronium by bolus doses in the critically ill.

Gallamine

This synthetic relaxant is infrequently used in Intensive Care. The main side-effect of gallamine is tachycardia, resulting from its potent vagolytic action. Gallamine should therefore not be used in patients in whom a tachycardia might be hazardous, e.g. coronary artery disease. The principal reason for its lack of popularity in Intensive Care, however, is due to problems with cumulation. Gallamine is solely dependent upon the kidneys for excretion. Its action is therefore prolonged in patients with renal insufficiency (common in Intensive Care patients).

Atracurium

This intermediate-acting neuromuscular blocking agent is popular in Intensive Care. It has virtually no cardiovascular side-effects. Histamine release is rarely seen in the clinical dose range. Its short duration of action makes administration by intermittent bolus injection not practical in the Intensive Care setting but it is very suitable for administration by continuous infusion. It thus meets a number of the criteria advanced for the ideal muscle relaxant in Intensive Care.

Atracurium is a unique muscle relaxant in that it was designed to break down in the plasma at body temperature and pH by the Hofmann elimination reaction. In the normal healthy patient approximately 40% is destroyed in this way, the remainder being broken down by ester hydrolysis in the liver.[12] This makes the termination of action of atracurium independent of the liver or the kidneys. Thus, in the critically ill patient, who may have quite marked impairment of hepatic or renal function, there will be no prolongation of the action of atracurium. Recovery from neuromuscular block is predictable and reliable.

These properties make atracurium a particularly valuable agent for Intensive Care. Any assessment of the patient which may require the absence of a neuromuscular block can be easily planned in the certain knowledge that after 1 hour, the muscle relaxant will no longer be active. There is, furthermore, no delay in weaning, a problem which might result from the use of one of the longer-acting agents.

There is one potential problem with atracurium, namely the existence of metabolites with pharmacological activity. Laudanosine has been shown to possess cerebral excitatory activity in certain laboratory animals. Evidence exists for the accumulation of laudanosine in patients who have received a continuous infusion of atracurium and this is accentuated in the presence of renal failure.[13] The significance of this is unclear, however, because it is not known whether laudanosine is similarly toxic in humans and, if so, what is the toxic plasma concentration. It seems likely on present evidence that laudanosine will prove to be of little clinical relevance.[14] Interest has recently focused on another metabolite, the quaternary acrylate. Acrylates are highly reactive substances and might produce tissue damage. As yet, however, no side-effects have been attributed to the acrylates.

An interesting observation has been made when atracurium is used by infusion in Intensive Care. After a delay of usually about 2–4 days, resistance to the neuromuscular blocking effect may begin to develop. An increasing dose is therefore required to maintain the same degree of paralysis, and the required infusion rate slowly and steadily rises. The mechanism is thought to be related to an increase in the number of extrajunctional acetylcholine receptors. It is unclear whether this phenomenon might exist with any of the other muscle relaxants, because they are all cumulative to a greater or lesser extent which would mask the appearance of such an effect.

A new neuromuscular blocking agent, known at present only by the number 51W89, may have a promising future in Intensive Care. 51W89 is one of the component isomers of atracurium (which is a racemic mixture) and therefore shares the same pathways of metabolism. Its cardiostability equals that of atracurium and its potency is higher; less is therefore required to secure the same level of neuromuscular block. One result will be that metabolites will be present in a lower concentration. Early studies on the use of 51W89 by infusion in Intensive Care indicate that there may not be the same problem with increasing infusion requirements.[15]

Vecuronium

This synthetic steroid-based intermediate-acting agent possesses good cardiovascular stability. It is intermediate acting and therefore, like atracurium, requires to be administered by continuous intravenous infusion in Intensive Care patients. It appears to be devoid of histamine releasing properties.

The metabolism of vecuronium takes place in the liver and, like pancuronium, there are three possible metabolites, the 3-hydroxy, 17-hydroxy and 3,17-dihydroxy derivatives. The first of these is also a neuromuscular blocking agent with a potency about 50–70% of that of vecuronium.[16] Vecuronium and its metabolites are excreted in the urine and therefore accumulation is possible in patients with impairment of renal function. Delayed recovery from a vecuronium neuromuscular block has been reported in critically ill patients with renal failure,[17] although it is uncertain whether it is the parent compound or its 3-hydroxy metabolite which is implicated.[18] Accumulation does not seem to occur when vecuronium is used in less severely ill patients who have normal renal and hepatic function, e.g. patients admitted to the Intensive Care Unit for routine postoperative care.[19]

Recent evidence has drawn attention to a possible potential detrimental effect of the steroid-based muscle relaxants in Intensive Care patients. A myopathy has been reported which may be related to the use of these agents.[20] Critically ill patients on the Intensive Care Unit do, however, frequently develop a myopathy related to the disease process, or to other factors not yet clearly understood.[21] It will be difficult to determine whether there is a steroid-induced myopathy resulting from the use of these muscle relaxants and, if so, whether or not it is important.

Pipecuronium and doxacurium

Pipecuronium is a synthetic steroid and doxacurium is a member of the benzylisoquinolinium series. Both are long-acting relaxants which have excellent cardiovascular stability with minimal potential for histamine release. By virtue of their lack of side-effects, they might be expected to be of use in Intensive Care patients. Their long action, however, would make control of block difficult. They are also excreted in part through the kidney which introduces the potential for cumulation in multiple organ failure patients. There is currently little information concerning the use of these two new agents in Intensive Care and it is difficult to anticipate that they will prove to be useful adjuncts to management in Intensive Care.

Rocuronium

This steroid neuromuscular blocking agent is very similar to vecuronium with respect to cardiovascular side-effects and duration of action. It differs in having a much lower potency and a much more rapid onset of action. Rocuronium is metabolized in the liver and the metabolites are excreted in the urine. It might therefore be expected that its action would be prolonged with the potential for cumulation in multiple organ failure patients on the Intensive Care Unit. Rocuronium is likely to find its place in the rapid intubation of patients for surgery. There is little experience of the use of rocuronium in Intensive Care and it is difficult to see what role it might play in the future.

Mivacurium

Mivacurium is a member of the benzylisoquinolinium family which is short acting. It possesses good cardiovascular stability, but may release histamine in higher doses. The elimination half-life of mivacurium is approximately 17 minutes and this is due to its metabolism in the plasma by plasma cholinesterase. This enzyme is synthesized in the liver and its production may well be reduced in critically ill patients with liver dysfunction. The action of mivacurium is likely therefore to be prolonged in critically ill patients on the Intensive Care Unit. The action of mivacurium may also be affected by other drugs which interfere with the activity of plasma cholinesterase.

The short duration of action of mivacurium will mean that it will have to be administered by continuous infusion. The consumption of mivacurium may be considerable and this may make it inappropriate on grounds of convenience and finance. It will be interesting see whether mivacurium does find a role in the management of patients on the Intensive Care Unit.

Monitoring neuromuscular blockade in Intensive Care

In order to use a muscle relaxant at maximum efficiency in any situation, the effects should be monitored and recorded. This enables the dose or infusion rate to be adjusted accordingly which prevents inadequate block with its inherent disadvantages, or excessive block with the possibility of delayed recovery.

The techniques of monitoring neuromuscular block are described in Chapter 6. The optimum method of monitoring neuromuscular block in Intensive Care patients has still not been fully decided. The use of complex EMG systems requires regular replacement of electrodes and is subject to much drift. The use of a simple hand-held nerve stimulator with observation of response is simple and satisfactory and allows the muscular response to be assessed at convenient intervals by either visual or tactile means. Accelerometry is a relatively new technique and may also prove to be satisfactory. The most appropriate level of block depends upon clinical circumstances. A useful general rule is to maintain a level of block where the first, or first and second twitches of the Train-of-Four are present. If there is no response to Train-of-Four stimulation the infusion rate should be decreased and if there are three or more twitches then the infusion rate should be increased.

The use of a nerve stimulator on Intensive Care patients introduces two principal problems. The first problem concerns electrode type and position. Most ECG-type surface electrodes are suitable for use and it is advisable to change them daily although the positions should be kept as constant as possible. The second problem is that of variation in response. This appears to be due to the accumulation of tissue oedema which reduces the intensity of what was previously a supramaximal stimulus. It may be possible to reduce this problem by exerting pressure over the electrodes for 1–2 minutes before stimulating.

Conclusion

The pattern of the use of the muscle relaxants in Intensive Care patients has undergone considerable change. Intensive Care Units still vary considerably although it is clear that there are a number of indications where the use of a muscle relaxant is valuable in Intensive Care patients. With the availability of intermediate-acting non-depolarizing muscle relaxants with few side-effects, the use of muscle relaxants in the Intensive Care Unit by continuous infusion should present little problem. It is, however, surprising that despite the increasing use of invasive monitoring in many aspects of intensive therapy, the routine monitoring of neuromuscular function in the Intensive Care patient is not common.

It must be remembered that there are certain pathophysiological changes which occur commonly in Intensive Care patients which can potentiate a muscle relaxant. These factors include acid-base and electrolyte disturbances, drug interactions, and developing and pre-existing muscular and neuromuscular disease. The most logical agent would appear at present to be atracurium by continuous infusion. Whichever agent is used, it is recommended that neuromuscular function is regularly monitored. Finally, it is necessary to return to first principles and to remember that all patients who receive a muscle relaxant must be adequately sedated during this phase of their treatment.

References

1. Merriman HM. The techniques used to sedate ventilated patients: a survey of methods used in 34 ICUs in Great Britain. *Intensive Care Medicine* 1981; **7:** 217–224.
2. Bion JF, Ledingham IMcA. Sedation in intensive care – a postal survey. *Intensive Care Medicine* 1987; **13:** 215–216.
3. Editorial. Paralysed with fear. *Lancet* 1981; **1:** 427.
4. Natanson C, Shelhamer JH, Parrillo JE. Intubation of the trachea in the intensive care setting. *Journal of the American Medical Association* 1985; **253:** 1160–1165.
5. Raval B, Steinberg R, Rauschkolb E. Artifact free computed tomography of the chest and abdomen in the severely ill patient. *Journal of Computed Tomography* 1985; **9:** 9–11.
6. Shelly MP, Wang DY. The assessment of sedation. *British Journal of Intensive Care* 1992; **2:** 195–203.
7. Sneyd JR, Wang DY, Edwards D, Pomfrett CJD, Doran BRH, Healy TEJ, et al. Effect of physiotherapy on the auditory evoked response of paralysed, sedated patients in the intensive care unit. *British Journal of Anaesthesia* 1992; **68:** 349–351.
8. Wokke JHJ, Jennekens FGI, van den Oord CJM, Veldman H, van Gijn J. Histological investigations of muscle atrophy and end-plates in two critically ill patients with generalized weakness. *Journal of the Neurological Sciences* 1988; **88:** 95–106.
9. Thomas S, Sainsbury C, Murphy JF. Pancuronium belly. *Lancet* 1984; **2:** 870.
10. Coakley, J. Suxamethonium and intensive care. *Anaesthesia* 1991; **46:** 330.
11. Smith CL, Hunter JM, Jones RS. Prolonged paralysis following an infusion of alcuronium in a patient with renal dysfunction. *Anaesthesia* 1987; **42:** 522–525.
12. Fisher DM, Canfell PC, Fahey MR, Rosen JI, Rupp SM, Sheiner LB, et al. Elimination of atracurium in humans. Contribution of Hofmann elimination and ester hydrolysis versus organ based elimination. *Anesthesiology* 1986; **65:** 6–12.
13. Fahey MR, Rupp SM, Canfell PC, Fisher DM, Miller RD, Sharma M, et al. Effect of renal failure on laudanosine excretion in man. *British Journal of Anaesthesia* 1985; **57:** 1049–1051.

14. Gwinnutt CL, Eddleston JM, Edwards D, Pollard BJ. Concentrations of atracurium and laudanosine in cerebrospinal fluid and plasma in three intensive care patients. *British Journal of Anaesthesia* 1990; **65:** 829–832.
15. Stone AJ, Harper NJN, Pollard BJ. Infusion requirements of patients receiving 51W89 in the Intensive Care Unit. *British Journal of Anaesthesia* 1994; **72:** 486P–487P.
16. Bencini AF, Houwertjes MC, Agoston S. Effects of hepatic uptake of vecuronium bromide and its putative metabolites on their neuromuscular blocking actions in the cat. *British Journal of Anaesthesia* 1985; **57:** 789–795.
17. Cody MW, Dormon FM. Recurarization after vecuronium in a patient with renal failure. *Anaesthesia* 1987; **42:** 993–995.
18. Segredo V, Matthay MA, Sharma ML, Gruenke LD, Caldwell JE, Miller RD. Prolonged neuromuscular blockade after long-term administration of vecuronium in two critically ill patients. *Anesthesiology* 1990; **72:** 566–570.
19. Darrah WC, Johnston JR, Mirakhur RK. Vecuronium infusions for prolonged muscle relaxation in the intensive care unit. *Critical Care Medicine* 1987; **17:** 1297–1300.
20. Griffin D, Fairman N, Coursin D, Rawsthorne L, Grossman JE. Acute myopathy during treatment of status asthmaticus with corticosteroids and steroidal muscle relaxants. *Chest* 1992; **102:** 510–514.
21. Zochodne DW, Bolton CF, Wells GA, Gilbert JJ, Hahn AF, Brown JD, *et al.* Critical illness polyneuropathy. A complication of sepsis and multiple organ failure. *Brain* 1987; **110:** 819–842.

13

Adverse Reactions to Muscle Relaxants
The role of the laboratory in clinical investigations

John Watkins

Introduction

The complexity of modern surgery requires the use of several drugs since no single drug supplies all the required pharmacology. The effects of these drugs may be superimposed on drug regimens already prescribed for conditions unrelated to the necessity for surgery (e.g. oral contraceptives, beta-blockers, antibiotics). The more drugs a patient receives the greater the risk of an untoward reaction.[1] In one case investigated by the author, the patient became life threateningly hypotensive after she had received no fewer than 23 intravenously and orally administered drugs over a 12-hour period prior to and including induction of anaesthesia.

The predominant and necessary use of the intravenous route by the anaesthetist for drug administration increases the risk of adverse response[2] since pharmacologically active drugs are presented directly and in high concentration to sensitive cells (notably mast cells) and to complex blood enzyme systems including complement and coagulation cascades which are normally protected by the biological barriers, the skin and mucous membranes. The direct 'chemical' activation of these adds to problems of genuine drug allergy but similarly leads to the release of a plethora of small effector molecules of the inflammatory response, notably histamine. Depending upon the quantity of mediator released, the clinical response will manifest as an immediate hypersensitivity-type response.[2]

These mechanisms, together with genuine antibody-mediated response to specific drugs, will be described later but for the present it suffices to note that the anaesthetist, more than any other clinician, will experience not only a high incidence of serious adverse reactions but that these will predominantly be of the immediate hypersensitivity type.

The term 'anaphylactoid' has been widely used in the world literature to describe the clinical manifestations of immediate hypersensitivity-type reactions, whether or not these prove to be drug specific. Only after laboratory testing can the *mechanisms* be described as anaphylactic or anaphylactoid. Such decision making is important for the future care of the patient. Unfortunately, many anaesthetists consider that only anaphylactic reactions (i.e. antibody mediated) pose life-threatening situations.

The anaphylactoid reaction in anaesthesia
Definition
It is proposed to restrict the discussion of adverse reactions in anaesthesia to those presenting immediately, i.e. within a few minutes or even seconds of drug presentation, and presenting clinically with the expected manifestation of type 1, antibody-mediated anaphylaxis. These manifestations will usually include cutaneous response such as flushing or urticaria, localized or diffuse, and varying degrees of cardiovascular involvement ranging from tachycardia and mild depression of systolic blood pressure to cardiac arrest. The most hazardous manifestation requiring immediate action by the anaesthetist remains acute bronchospasm but pulmonary effects too will range from cough and perhaps initial pulmonary resistance to acute shutdown.

The vast majority of these 'adverse responses' will be very minor indeed, will not require the intervention of the anaesthetist and will have causes not specifically related to the drug(s) themselves. These minor adverse reactions confuse the clinical problems of genuine hazardous drug reactions, their frequency and their management.

In 1977 Ring, Messmer and their colleagues[3] addressed the problem of reaction severity with plasma substitutes by defining four grades of reaction, I–IV (Table 13.1). Grade I is a histaminoid reaction presenting with marked flushing or urticaria, tachycardia and the like. No action is necessary. Grade II reactions are again predominantly histaminoid but more severe and manifest in mild hypotension with a fall in systolic blood pressure

Table 13.1 Classification of anaphylactoid reactions according to clinical severity

Grade	Skin	Respiratory	Cardiovascular
Mild			
I	Flush	—	—
II	Urticaria, flush	Pulmonary resistance	Marked tachycardia Hypotension (≥ 20 mmHg systolic)
Life-threatening			
III	Urticaria, flush	Bronchospasm, cyanosis	Gross hypotension (≥ 60 mmHg systolic) Shock
IV	Urticaria, flush	Respiratory arrest	Shock Cardiac arrest

⩾20 mmHg and marked tachycardia. They may require intervention but this will be largely cosmetic (e.g. chlorpheniramine and/or steroids). The reaction should be noted in the patient's records. Grades III and IV are truly life threatening involving acute bronchospasm and cardiovascular collapse. Immediate action is necessary and, furthermore, laboratory investigation should be initiated.

Death is sometimes referred to as grade V, but death may simply represent a mismanaged grade III or IV reaction. We are now subdividing our adverse reactions reports in terms of grades I–IV for statistical purposes. In practice, few grade I reactions are reported since their control poses no problem for the anaesthetist, and this laboratory effectively monitors only reactions of grade II–IV severity.

Extent of the problem

Using the above criteria and observations based on more than 3000 life-threatening reactions (following induction and prior to surgery) reported and analysed since 1974, the author suggests that out of 3.5 million general anaesthetics administered each year, life-threatening morbidity may be as high as 5000–10 000 reactions, encompassing severity grades II–IV.[4,5]

While these figures are markedly reduced by the exclusion of grade II reactions it would be wrong to dismiss these as of an irrelevant nature. It is certainly true that grades III and IV include significantly more genuine immune-mediated reactions, but grade II reactions are not devoid of these and in some cases represent harbingers of real disasters if the reaction mechanisms are not properly investigated and the responsible drug excluded from future procedures.

Mortality

An unfortunate feature of the majority of the adverse reactions is that they involve apparently fit, healthy young people, presenting for minor elective procedures (e.g. dental, gynaecological) rather than the high risk surgical patient. Of the reactions reported to our laboratory each year 2 or 3% have terminated in death within hours, or infrequently days, of the initial reaction. On this basis we should expect a mortality of 100–300 deaths. This figure is remarkably similar to that reported in the 1982 survey of Lunn and Mushin.[6] These authors discuss some 300 deaths per annum directly implicating anaesthesia and a further 1800 deaths in which anaesthesia had played a part.

Causes of anaesthetic reactions

Based on some 20 years' investigation of life-threatening reactions in the UK the author suggests that the major causes of anaesthetic reactions are

Table 13.2 Causes of adverse reactions

Immune
1. Genuine drug-specific hypersensitivity

Non-immune
1. Direct action of drugs on cells rich in vasoactive substances
2. Indirect release of vasoactive substances through activation of intermediary biological systems
3. Drug interactions and drug overload
4. Aggregate (non-immune) anaphylaxis (a further complication of 3)
5. Unforeseen patient pathology
6. Surgical stimulation
7. Psychosomatic response
8. Technique and error

distributed amongst the following immune and non-immune mechanisms (Table 13.2) which apply equally to almost all 'anaesthetic drugs'.

None of these 'mechanisms' is mutually exclusive of the others but the inference is that genuine immune hypersensitivity response to a specific drug(s) is relatively uncommon, perhaps involved in 20–25% of all drug reactions, and even lower in response to plasma substitutes, contrast media and the like.

Reporting trends: a hypothesis

Since drug-specific response is the minority mechanism the probability arises that a given drug will be implicated (named, but not necessarily the cause) in adverse reports simply at a frequency related to its market share of usage.

The above statement receives support from our studies[5] on reporting patterns over the period 1983–1989 (Fig. 13.1). The latter study compared reporting patterns involving the principal hypnotic drugs over a period which saw the demise of the steroid anaesthetic Althesin® (Glaxo) in 1984 and the introduction and increasing usage of propofol (Diprivan®, ICI) from 1985 onwards. The increased incidence of reports naming thiopentone (with minor increases for etomidate and methohexitone) in 1984 following the demise of Althesin falls back continuously from 1986 to date following the increasing use of propofol but with a compensating increase in adverse reports for the latter. Overall, the total number of adverse reports remains approximately constant. One must presume that 'reactions' to propofol for the most part included patients who would have so reacted to any hypnotic.

Neuromuscular blocking drugs

If the above hypothesis is correct, then we should be able to predict the distribution of the neuromuscular blockers in reactions simply in terms of the

238 *Adverse reactions to muscle relaxants*

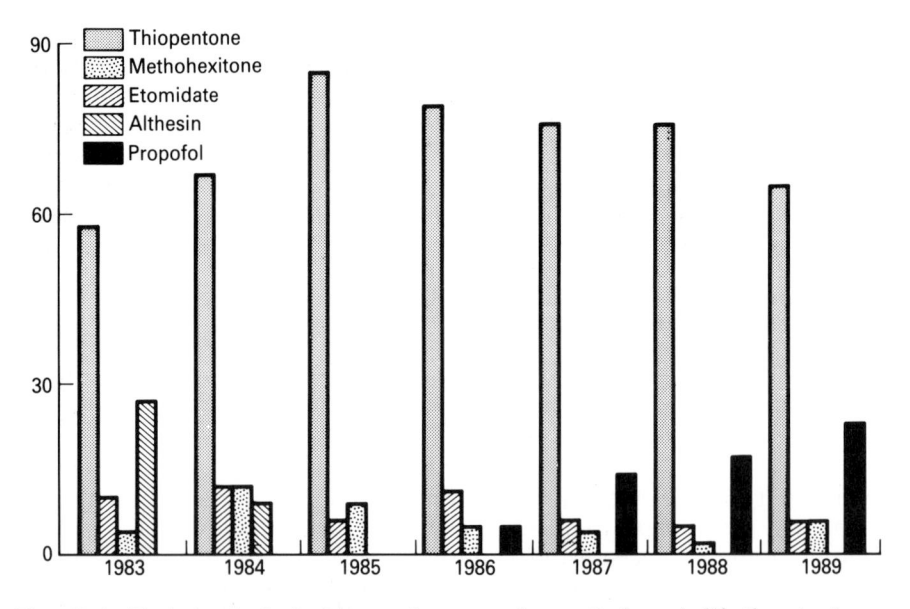

Fig. 13.1 Variation in the incidence of common hypnotic drugs in life-threatening reactions to general anaesthetics reported to the author's laboratory in the five years 1983–1989. More than 100 serious reactions are investigated each year such that the above may be seen as 'genuine percentage' figures. (From Watkins.[5])

hypnotic drug distributions. Table 13.3 compares the expected distribution with that actually observed for reactions investigated in 1990. Agreement with the hypothesis is good, although vecuronium appears to show an unusual distribution towards etomidate. This probably reflects the relatively small number of reports received involving these two drugs. Out of 139 adverse reports, vecuronium was mentioned 17 times and etomidate only 6 times. Nevertheless, the few etomidate-implicated reactions in 1989 showed

Table 13.3 Distribution of various neuromuscular blockers between co-administered hypnotic drugs in 139 life-threatening adverse drug reactions (ADRs)

Neuromuscular drug mentioned in report	Co-administered with					
	Thiopentone (67%)		Propofol (27%)		Etomidate (4%)	
	Actual	Expected	Actual	Expected	Actual	Expected
Suxamethonium	70%	67%	23%	27%	2%	4%
Atracurium	71%	67%	29%	27%	0%	4%
Vecuronium	53%	67%	24%	27%	23%	4%

Percentages in parenthesis for hypnotics indicate fraction of total ADRs involving that hypnotic; this is also the expected distribution of the neuromuscular drug. Total ADRs include three patients given methohexitone.
From Watkins.[5]

a similar bias towards co-involvement with vecuronium. A simple answer may be that the anaesthetist, fearing a life-threatening histamine-releasing reaction in his or her patient, for whatever reason, selects etomidate together with vecuronium as the drugs least likely to cause mast cell degranulation. Unfortunately, all that glisters is not gold or even histamine, and in the 1990 series an 18-year-old girl died following such induction for investigative laparoscopy, although nothing abnormal was discovered at autopsy.

The cynical reader may wish to reflect that in France where atracurium is little used and vecuronium use increasingly predominates, the latter is now considered second only to suxamethonium in its incidence of adverse reactions.[7]

Any of the drugs administered intravenously, orally and intramuscularly in general anaesthesia are capable of producing clinically significant anaphylactoid response. This includes drugs used for premedication and as antibiotics. For the most part these drugs bear little molecular similarity to one another even for the same pharmacological activity. However, the neuromuscular blocking drugs possess an identifying pharmacologically active, molecular group, the quaternary ammonium ion, and it is not surprising therefore that neuromuscular blockers have become the most closely studied.

A frequent question is that if a patient has been shown to be genuinely sensitive to one neuromuscular blocker can it be safe to prescribe an alternative? This is clearly not a problem with the hypnotic drugs whose molecular structures, ranging from steroid skeletons through phenols to simple heterocyclic molecules, provide adequate alternatives. Before we can answer the question, or indeed design a systematic investigation of the anaphylactoid reaction, we must consider the general mechanisms of adverse reactions to anaesthetic drugs.

Mechanisms of anaphylactoid reactions to anaesthetic drugs

Situations involving histamine release

Mast cells and basophils degranulate upon stimulation of their cell surfaces by specific IgE antibody–antigen combination and by a variety of non-immune factors (including high drug concentrations), releasing histamine and mediators into plasma.[8] Depending upon the level of such release, severe hypotension may occur as a result of peripheral blood pooling and increased vascular permeability, giving rise to severe hypovolaemia. Cutaneous vasodilation, as well as vasoconstriction, may occur alternatively. Lethal and sublethal shock is associated with malign arrhythmias or cardiac arrest. This is due to combinations of decreased venous return, myocardial ischaemia and hypoxaemia and direct action on H_2 receptors. While an acute hypotensive response is most frequently associated with histamine release the reader should remember it may also occur by other means.

Plasma histamine levels

Non-specific drug release of plasma histamine reaches 1 or $2\,\text{ng}\,\text{ml}^{-1}$ in some 30% of the population, but this is largely without significant clinical manifestation. However, rapid multiple drug administration through small veins may lead to overlapping waves of histamine which become additive or even synergistic.[9] Once a threshold of 10 ng ml^{-1} histamine is breached, clinical sequelae of increasing clinical severity will be observed.[10] Non-specific histamine release can be reduced by slow administration of the drugs into large veins, preferably into a fast-flowing carrier stream. If bolus administration is carried out into small veins, then the cannula should be flushed out with saline between drugs with preferably several minutes allowed between each addition. This reduces the overlap of histamine waves and also reduces pharmacological and physiochemical drug interactions, all potentially hazardous.[10] A potent source of histamine release is provided by the use of formulations containing morphine for premedication in patients with mast cell syndromes. Mild versions of mastocytosis are probably more common than generally accepted and while intravenous administration is immediate and dramatic, intramuscular administration may produce delayed response 1 hour or so later, at the time of anaesthetic induction.

Individual drugs and histamine release

General clinical considerations

Few anaesthetic drugs fail to liberate some measurable, but subclinical, histamine in a high proportion of patients in clinical trials. The most prominently used exceptions are etomidate, propofol and vecuronium. All three are associated with pain on injection and two, propofol and vecuronium, appear to exhibit bronchospasm as their major adverse response, i.e. the shock organ is the lung rather than the cardiovascular system. There is increasing evidence, at least for propofol, that this reaction is not associated with systemic mast cell, and possibly not exclusively with lung mast cell, activation. The author has tentatively suggested that these three drugs may predominantly release prostaglandins which would tend to suppress histamine release.

Of the neuromuscular agents themselves, atracurium and vecuronium have attracted particular attention as founder members of a new series of relaxants. Despite its benzylisoquinoline structure, atracurium is only a mild histamine release agent, with a potential slighty higher than that of thiopentone: such release may be regarded as of no consequence in healthy patients. Histamine release will achieve clinical significance only when doses >0.6 mg kg^{-1} are injected rapidly. However, atracurium does produce a high incidence of generalized flushing which, even under the most carefully controlled administration, is unlikely to fall below an incidence 8%.[11] A number of studies have shown that this cutaneous response is unconnected with *plasma* histamine release,[9] but may involve neurological stimulation and the release of substance-P, rather than histamine, at skin nerve endings.[12] The new benzylisoquinoline ester relaxant doxacurium shows no signs of histamine release or cardiovascular effects,[13] but mivacurium may release more *plasma* histamine than atracurium at similarly effective dose levels.[14]

Many clinical studies have documented the remarkable lack of cardiovascular effects of vecuronium. Cutaneous manifestations are rare and in trials there is no evidence for even subclinical histamine release. Pipecuronium shows comparable cardiovascular stability to vecuronium[15] and a significantly lower incidence of tachycardia than experienced with the 'parent' steroid pancuronium. Nevertheless, in an overall assessment of clinical safety atracurium and vecuronium emerged equally matched with a low and acceptable incidence of anaphylactoid response.[10, 16] Usage of the new analogues has not distorted this profile.

The high histamine release potential of tubocurarine[17, 18] has been recognized since 1946. The dose-related histamine release potential of both morphine and tubocurarine is indicative of chemically mediated histamine release rather than of an immunological release. The mechanisms giving rise to histamine release (as well as other mediators of the inflammatory and shock responses) via mast cell degranulation are illustrated[19] in Fig. 13.2, and elaborated below.

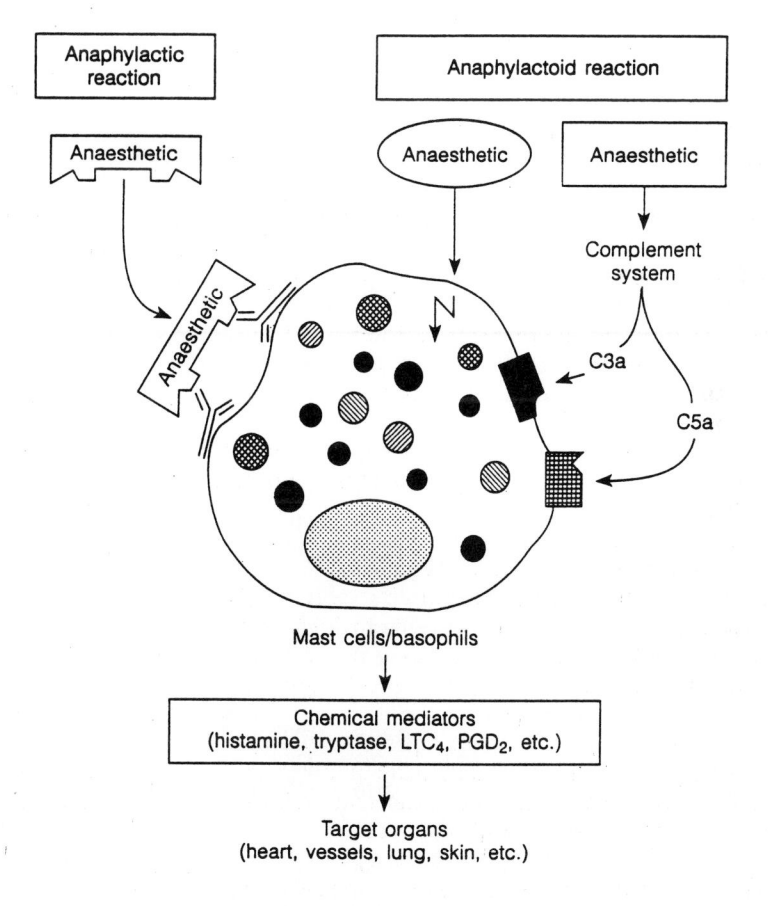

Fig. 13.2 Immune and non-immune mechanisms in anaesthesia giving rise to mast cell degranulation. (From Marone and Stellato.)

Chemical mechanisms

The mechanism of histamine release is usually considered to involve the direct binding of the anaesthetic drug(s), or other administered substances, to receptors upon the surface of the mast cell or basophil, stimulating the membrane and triggering degranulation. As might be expected the chemically active entity of the neuromuscular blocker shows particularly high avidity for such receptors, but a wide range of intravenously administered substances from pharmacologically active drugs[9, 17, 18, 20-25] to plasma substitutes and even saline[25] causes measurable histamine release without significant clinical manifestations (Table 13.4). The high frequency of such 'subclinical' (<3 ng ml^{-1}) histamine release adverse (cf. thiopentone 95% frequency[21] and saline[25] with an 18% frequency, Table 13.4) compared with an extremely low incidence of *life-threatening* reactions to such commonly used substances negates the view that *this* mode of histamine release predicts the *clinical* risk of individual drugs in patients. Of the few drugs which do not release histamine in clinical trials, most (excepting etomidate[26]) produce a similar incidence of serious anaphylactoid reactions to their histamine releasing peers.

Low level chemical plasma histamine release (approximately 1-3 ng ml^{-1}) may well be a self-limiting mechanism in the general population reaching significance perhaps with drugs such as tubocurarine and morphine formulations to which certain patients, such as those with mast cell syndromes, overreact due to abnormalities, both qualitative and quantitative of these

Table 13.4 A comparison of neuromuscular blockers with other drugs that cause elevated plasma histamine levels (incidence) as demonstrated in clinical trials

Drug	Incidence		Ref. Source
Neuromuscular blockers			
Atracurium	16/41	(39%)	9
Alcuronium	2/8	(25%)	19
Pancuronium	1/7	(14%)	19
Suxamethonium	3/8	(38%)	19
Tubocurarine	14/20	(70%)	17
Hypnotics			
Thiopentone	9/10	(90%)	21
Methohexitone	6/8	(75%)	22
Opioids			
Morphine	18/18	(100%)	23
Fentanyl	2/8	(25%)	24
Alfentanil	1/10	(10%)	24
Miscellaneous			
Atropine	6/36	(17%)	22
Saline	4/22	(18%)	25
Methylprednisolone	3/7	(43%)	25

cells. Nevertheless, the tendency to administer simultaneously several drugs rapidly into narrow veins increases the probability of additive waves of histamine, significantly increasing the chance of a hazardous clinical response.[9]

Complement mechanisms

Activation of complement components, notably complement C3 (approximately $1\,gl^{-1}$ in normal plasma), with the release of the biologically active anaphylatoxin fragments (C3a) may give rise to the clinical manifestations of shock through the subsequent reaction of the anaphylatoxin on mast cells and basophils. The clinical severity of the shock relates to the nature and to the extent of the activation, subclinical activation is highly beneficial in disease protection.

The adverse role of C3 activation was apparent in the late 1970s and early 1980s when two useful hypnotic drugs (propranidid, Althesin) formulated in the detergent Cremophor EL were in frequent use. The discovery that the high incidence of anaphylactoid shock was related to almost total C3 activation through interaction with Cremophor[27] led to the demise of these drugs. Such *in vivo* activation presenting at clinical level is now predominantly confined to gelatin plasma expanders and radiological contrast media.

Subclinical complement activation: In 1991 during inter-laboratory studies of complement activation on machine surfaces during cardiovascular surgery we observed that the *magnitude* of C3 activation by the induction drugs was frequently as high as that encountered on bypass (a membrane effect,[28]), although maintained for considerably less time (Fig. 13.3). Preliminary indications were that the C3 activation arose from the neuromuscular blocker

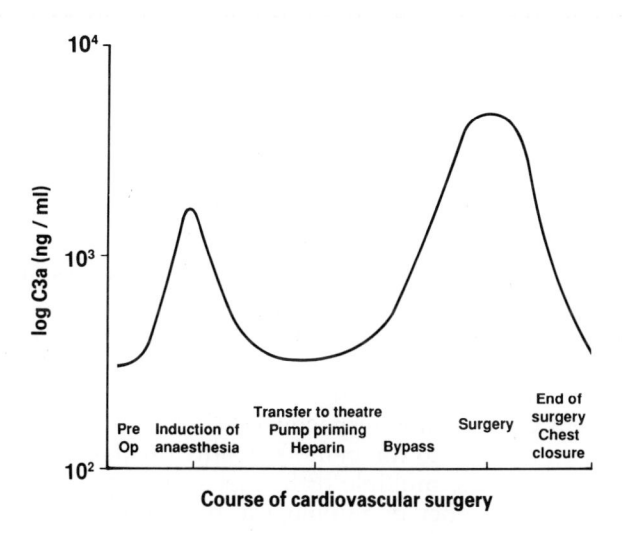

Fig. 13.3 Semi-diagrammatic representation of anaphylatoxin C3a release during the course of cardiovascular surgery. Mean of ten actual operations but horizontal axis deliberately distorted. Compare peak anaphylatoxin release at anaesthetic induction with the bypass procedure.

and even in the apparent absence of unforeseen clinical manifestations was reaching some 10% conversion. Apart from being a potent trigger for histamine release, the C3a anaphylatoxin levels suggested potential effects on haematological mechanisms, an important consideration in cardiovascular surgery. Choice of neuromuscular blocking drug may well be an important consideration in such surgery.

Aggregate anaphylaxis

In addition to increasing the risk of clinically significant histamine release, the small-vein, fast-administration technique is also largely responsible for the phenomenon of aggregate anaphylaxis. With a dead space of 0.1–0.2 ml, the usual 'butterfly' provides ample opportunity for physicochemical drug interactions, notably between thiopentone and all the newer neuromusuclar blockers. While the usual effect is a mild and temporary thrombophlebitis, manifest in reddening and urticaria around the injection site and along the injection vein, the inevitable production of colloid particles rather than visible precipitates can, on occasion, lead to life-threatening reactions and even death. High concentrations of colloid aggregates may be sequestered by the pulmonary microvasculature, triggering acute bronchospasm through local mediator release. In this syndrome acute bronchospasm precedes acute hypotention and cardiac arrest. Its presentation follows uneventful induction of anaesthesia and intubation. Increasing airways resistance gives rise to acute bronchospasm, followed by cardiac arrest and, not infrequently, death.[29,30] Reports almost inevitably state 'uneventful induction, increasing pulmonary resistance, followed by rock-hard lungs'. In the survivor of the severe reaction this mechanism may be further involved in the aetiology of disseminated intravascular coagulation (DIC) and shock lung syndromes (ARDS), the clinical severity of which depends upon both the initial trauma and whether or not this has been compounded by completed surgery. The prevention or reduction of the reaction is that described for histaminoid reactions, slow administration of drugs into large veins with adequate wash-out between drugs.

Immune mechanisms

These reactions predominantly involve IgE antibodies although there are some indications that IgG or IgM reactions implicating the classical complement cascade may also occur. There were indications for such specialized immune response to both thiopentone and to Althesin although the predominant mechanisms with these drugs are type I, IgE-mediated response, and complement C3 activation respectively.

IgE-mediated reactions to thiopentone follow (inadvertant) classic drug sensitization procedures: multiple drug exposure in patients with a generalized trait towards other immunological hypersensitization situations, e.g. pollens, penicillins. These patients usually have high levels of circulating IgE plasma immunoglobulin. A typical reactor might be a multiple (20 or more) check cystoscopy patient with carcinoma of the bladder, but life-threatening responses have been encountered in orthopaedic patients with only a few, closely spaced, general anaesthetic procedures. Rare reactions have been reported for methohexitone. The particular danger with the barbiturate reactions appears to lie in the persistence of the 'antigen' in the body, repeatedly subjecting the unfortunate reactor to further anaphylactic shock as the initially discharged mast cells are restored or replaced: continuous therapy with adrenaline under constant monitoring becomes necessary.

By far the most hazardous situation involves the neuromuscular drug suxamethonium. This does not always obey classic sensitization theory and an appreciable number of suxamethonium reactions occur in patients with no previous history of exposure to this drug.

Despite the evidence of skin testing in general terms the risks of crosssensitivity between different neuromuscular drugs and local anaesthetics seem small, despite the common molecular entity of these classes of substances. Alcuronium, like suxamethonium, is also allergenic but the evidence for antibodies to the other commonly used drugs is equivocal. This in no way of course reduces their dangers by other, non-immune mechanisms.

Management of anaphylactoid reactions; anaesthesia

Irrespective of their mechanism or causative agent, all anaphylactoid reactions, with the exception of cutaneous manifestations divorced from cardiovascular instability, demand immediate and aggressive action.

Appropriate clinical action

The Association of Anaesthetists of Great Britain and Ireland has published guidelines for the emergency management of acute major anaphylaxis under general anaesthesia (Appendix, p. 259). Initial management comprises the administration of 100% oxygen with mechanical ventilation if necessary. If it is impossible to palpate a femoral or carotid pulse, external cardiac massage should be instituted. Adrenaline should be given i.v. in an initial dose of 50–100 μg (0.5–1.0 ml 1:10 000) which may be repeated on up to five occasions. Occasionally, ventricular fibrillation may be precipitated, especially during halothane anaesthesia, and it may be necessary to defibrillate the heart. The plasma volume is severely depleted and i.v. fluid should be given rapidly. Colloid is preferable to crystalloid: 10 ml kg^{-1} should be given in the first instance and up to 2 l may be necessary. Bronchospasm is the most lifethreatening feature of the anaphylactoid response. There is considerable doubt concerning the usefulness of steroids and antihistamines in this situation. If bronchospasm is severe, salbutamol, terbutaline or aminophylline may be given cautiously. Great care must be taken to calculate doses accurately in the tense atmosphere of major anaphylaxis.

Once the patient is stabilized, thought should be given to whether or not the operation should be abandoned. On balance, for minor elective surgery, abandonment seems the sensible choice; otherwise, the further trauma of surgery may reinforce the haematological changes consequent to anaphylactoid shock giving rise to serious problems of intravascular coagulation or ARDS in the postoperative situation. Thought must also be given at this point to the thorough investigation of the possible causes of the reaction.

Laboratory investigation

Laboratory investigation forms an important part of patient management.[29, 31, 32] The majority of patients will require further surgical procedures throughout their lives, if not in the foreseeable future. A few patients, unfortunately, will be the subject of litigation. It is essential to contact a laboratory with specialized knowledge of drug reactions as soon as possible after a reaction and to have blood and urine sampling arrangements already in hand; unwanted samples can always be discarded.

Blood samples (5–10 ml) should be collected, preferably into tubes containing EDTA, as close to the reaction time as possible and then at convenient intervals over the next 24 hours: clotted blood is acceptable. Three or four samples are sufficient but the first sample should be taken within 1 hour of reaction or sooner if at all possible. The blood tubes should be centrifuged and the separated plasma stored at $-20°$ until required for analysis.

Blood samples always represent a significant souce of error since the peripheral circulation will suffer a variety of changes, both as a result of fluid movements (from the reaction and from exogenous fluid used in management) and as a result of activation of various plasma components such as complement, prostaglandins and IgE. These changes can be evaluated correctly only from a study of sequential samples taken over the 24 hours following a reaction. The evaluation requires considerable experience in relative 'movements' of different plasma components if artefacts and possibly hazardous conclusions are to be avoided.

Development of a decision making tree

We have recently developed a protocol[32] which does not require such expertise and deploys a simple decision tree based upon the demonstration of mast cell degranulation (Fig. 13.4). A positive finding indicates a drug-involved reaction and automatically excludes error such as oesophageal intubation, underlying pathology and vasovagal response. A negative finding excludes most drug-involved reactions with the exception of those to propofol, vecuronium and etomidate and to aggregate anaphylaxis, i.e. reactions where the lung is the sole or predominant shock organ.

Tryptase assays in decision making: Histamine release might appear to be the most suitable measure of mast cell degranulation. However, histamine is not specific for the mast cell and its short half-life of 2 minutes makes its assay in the emergency clinical situation almost impossible. A valuable alternative assay to histamine has proved to be that of the protease tryptase.[33] This is more than 99% confined to the mast cell, and parallels histamine

Management of anaphylactoid reactions; anaesthesia 247

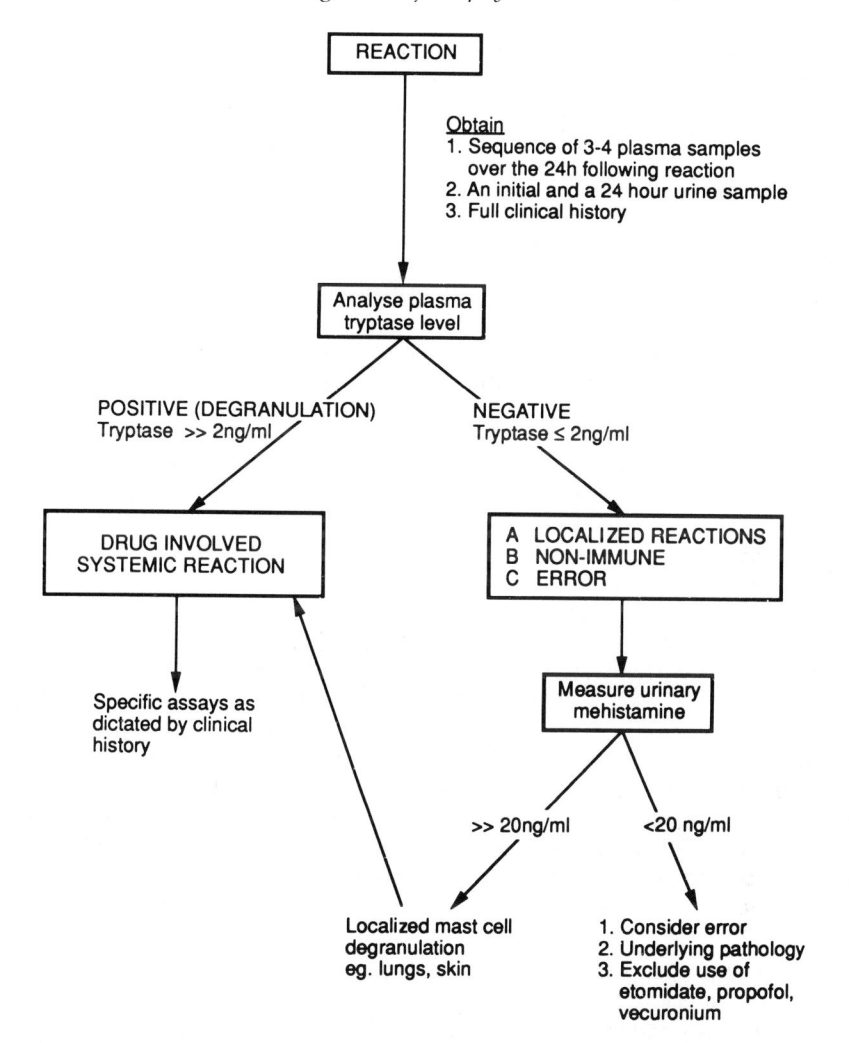

```
                    ┌─────────────┐
                    │  REACTION   │
                    └──────┬──────┘
                           │
                    Obtain
                    1. Sequence of 3-4 plasma samples
                       over the 24h following reaction
                    2. An initial and a 24 hour urine sample
                    3. Full clinical history
                           │
                           ▼
                    ┌─────────────┐
                    │Analyse plasma│
                    │tryptase level│
                    └──────┬──────┘
```

POSITIVE (DEGRANULATION) / NEGATIVE
Tryptase >> 2ng/ml   \ Tryptase ≤ 2ng/ml

```
┌──────────────────┐        ┌──────────────────────┐
│ DRUG INVOLVED    │        │ A  LOCALIZED REACTIONS│
│ SYSTEMIC REACTION│        │ B  NON-IMMUNE         │
└──────────────────┘        │ C  ERROR              │
                            └──────────┬───────────┘
                                       ▼
                            ┌──────────────────┐
                            │ Measure urinary  │
                            │   mehistamine    │
                            └──────────────────┘
```

Specific assays as
dictated by clinical
history

>> 20ng/ml   <20 ng/ml

Localized mast cell   1. Consider error
degranulation   2. Underlying pathology
eg. lungs, skin   3. Exclude use of
                                etomidate, propofol,
                                vecuronium

Fig. 13.4 Decision tree: systematic investigational procedure for clinically severe anaphylactoid reactions following induction of anaesthesia. (From Watkins and Wild.[32])

release. Unlike histamine it is highly stable in plasma once released (half-life *in vivo* >2 hours). The assay is now commercially available from Pharmacia (Uppsala, Sweden) as an elegant radio-immunoassay. In the absence of mast cell degranulation basal plasma tryptase levels are <2 ng ml^{-1} plasma. Following degranulation, tryptase levels increase rapidly. In anaphylactic shock levels >100 ng ml^{-1} are frequently encountered such that initial decision making (drug involved, yes or no?) is usually a straightforward matter (Fig. 13.4). Peak tryptase levels >25 ng ml^{-1} are associated with unmeasurably low systolic pressure[34] (Fig. 13.5).

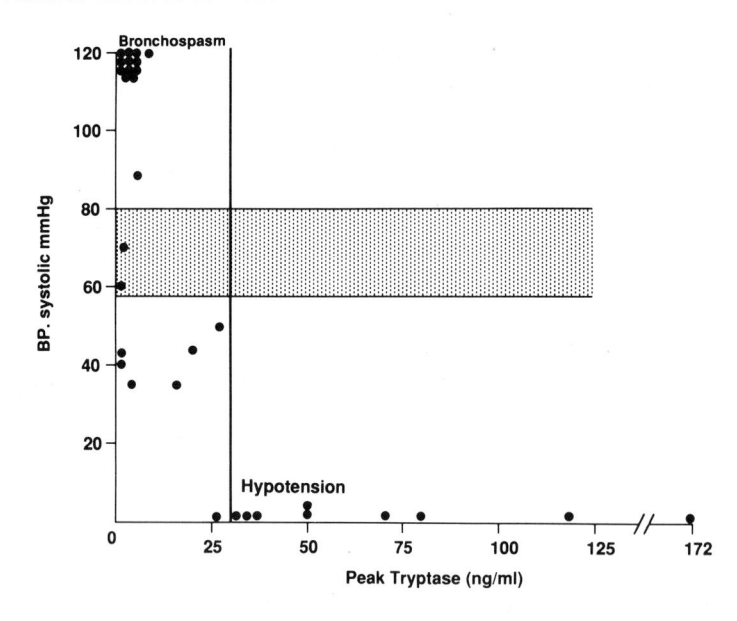

Fig. 13.5 Relationship between peak tryptase level and degree of hypotension expressed by lowest recorded systolic pressure (mmHg) for 30 life-threatening reactions to general anaesthetics (two deaths). The stippled area defines the levels between which most anaesthetists would intervene for the safety of their patients. Tryptase levels greater than 25 ng ml^{-1} are associated with unmeasurable cardiac output and usually with type I hypersensitivity response. Pure pulmonary reactions manifest in acute bronchospasm are not associated with plasma tryptase release, here curiously they refer to propofol implicated responses. Intermediary tryptase levels >2 ng ml^{-1} <25 ng ml^{-1} may infer chemical action on mast cells or indirect action via complement cascade activation. (Adapted from Watkins.[34])

Urinary methylhistamine assays in decision making: The availability of a urine sample (10 ml) taken near the time of the reaction, and a similar sample 24 hours later is necessary to ascertain the mechanism of reactions predominantly manifest in bronchospasm (Fig. 13.4). The urine samples are analysed for methylhistamine (Pharmacia, Uppsala, Sweden) and creatinine (to 'normalize' methylhistamine excretion). 'Normal' methylhistamine excretion lies between 10–20 ng ml^{-1} methylhistamine mmol creatine^{-1} l^{-1}. In patients exhibiting plasma tryptase release, excessive methylhistamine excretion is always present. However, in predominantly non-vascular reactions where the target organs are the lungs or skin (e.g. mastocytosis) histamine and tryptase become fixed in those organ tissues and the only metabolite of note to emerge is methylhistamine, concentrated in the urine.[32, 34] *Plasma* levels of tryptase (and histamine) are virtually normal. Reactions implicating propofol and vecuronium are predominantly bronchospasmic and these drugs may react only with mast cells specific to the lungs. They may also involve prostaglandin release, as probably does etomidate, in the gastrointestinal tract.

Specific tests: Once the involvement of the administered substance(s) has been established one can then call upon specific assays to identify the culprit. The most satisfactory way is to demonstrate both the presence of specific antibodies *and* their consumption in the sequence of blood samples taken immediately after anaphylactoid response. While such tests (RAST: Pharmacia, Uppsala, Sweden,) are now commercially available for a whole range of everyday allergens, e.g. dust mite, pollens, foodstuffs and insect venom, we are poorly served inthe anaesthetic field, although individual workers, notably Baldo and Fisher in Australia[35, 36] and Laxenaire, Moneret-Vautrin and Gueant in France[37, 38] have established 'in house' radio-allergosorbent assays to almost all of the commonly used drugs. Commercial assays are at present only available for thiopentone, suxamethonium and alcuronium (Pharmacia).

The alternative has always been skin testing but unfortunately with pharmacologically active drugs, notably those containing the quaternary ammonium groups, considerable experience is required if false positives are to be avoided.[39] The prick test is considerably more specific but less sensitive, than the intradermal test and the use of both simultaneously is recommended.[29]

The principal problem with both intradermal and prick tests, and predominantly with the former, is that high concentrations of avid drugs are presented to mast cells in the skin which are morphologically distinct from those in the vascular bed. The weal and flare response arises as well from direct chemical action on the skin mast cells as from genuine immune response – and appears identical.[40] Test conditions are arbitrarily chosen to reduce the risk of artefact.[39,40]

Unlike RAST assays all skin tests must be carried out days or weeks after the precipitating event. If 'positive' we must then accept the possibility that antibody may be present, thus raising the possibility that antibody may be present as a result of the reaction and not its cause. The combination of *specific* tests with the systematic investigation of anaphylactoid reactions scheme outlined here (Fig. 13.4), removes this objection as does the use of sequential samples to show not only the presence of antibodies but *also* their utilization in the reaction. There is of course no way 'utilization' can be demonstrated by skin testing.

Practical requirements for laboratory testing

Basic sample requirements

Blood samples (as EDTA plasma or serum)

Sample 1 as near to peak reaction as possible ≤ 1 h.
Sample 2 3 h (3–6 h).
Sample 3 24 h (18–30 h).

Urine

Sample 1 first or an early urine from catheter after reaction (10 ml).

Sample 2 24 h sample (10 ml). (It is important that urine Sample 2 is a discrete sample and *not* an aliquot of a 24 h collection.)

Case history

A concise case history, including previous anaesthetic history, the drugs given producing adverse reactions (ADR), the manifestations and management.

General observations

1. Blood samples should be separated as soon as possible and plasma or serum stored at −20° prior to analysis or despatch.
2. Blood samples collected 'overnight' can be safely stored until the morning at the bottom of a 4° refrigerator – do not freeze whole blood.
3. Urine samples can be stored at 4° without preservative if examination or despatch within 24 hours is anticipated, otherwise freeze and store at −20°.
4. Postal despatch of specimens requires *neither* dry ice nor 'Red Star'. Use first class post and secure packing to prevent leakage, and avoid despatch near weekends and bank holidays.

Skin testing requirements

It is emphasized that skin tests should best be carried out by experienced dermatologists. Full resuscitation facilities should always be available.

Intradermal

1. Test drug(s) diluted 1:1000 and 1:10 000 of 'ampoule' strength with 'saline'.
2. Inject 0.02 ml i.d. into volar aspect of forearm.
3. Inject 0.02 ml saline i.d. as control.
4. Inject a suitable 'histamine control' similarly as a measure of responsiveness of patient to histamine.
5. Inject similarly tubocurarine diluted 1:1000 as a *skin* histamine releasing drug control.
6. Observe and wait 30 minutes. A weal 1 cm or greater persisting 30 minutes or more is considered *positive*.

Prick

1. Test drug(s) diluted 1:10 and 1:100 of 'ampoule' strength with 'saline'.
2. Place one drop on opposite arm to that used for intradermal tests, prick through with lancet.
3. Controls as for intradermal testing.
4. Read any weal/flare reaction as positive.

General observations

1. For further details consult Watkins and Levy[29] or Wood *et al.*[40]
2. A positive prick test is almost inevitably associated with a positive intradermal test result: the converse is less frequently found.

Specific involvement of neuromuscular blockers in adverse reactions

General considerations of molecular structure and ADRs

Neuromuscular blocking drugs are widely considered to be more allergenic than the intravenous hypnotic and analgesic agents employed in anaesthetic induction. Indeed, the wide presence of substances containing the quaternary ammonium moiety (e.g. in foodstuffs and cosmetics) is often used to explain 'positive' skin test identifications in patients without history of previous anaesthetic exposure. However, the argument can be advanced that such natural exposure leads to immunological tolerance to molecules carrying this grouping. This observation is analogous to that encountered with some of the beta lactam antibiotics. The penicillins, cephalosporins and monobactams all possess the same active beta lactam nuclei. Nevertheless, monobactams can be safely administered to patients exhibiting genuine anaphylaxis to penicillins or cephalosporins, i.e. immunological reactivity is directed primarily against antigenic determinants other than the common 'active' molecular structure.

In vitro, there is no doubt that neuromuscular blockers, by virtue of their characteristic molecular entity, are capable of releasing mast cell histamine. The unique structure of suxamethonium makes it particularly avid for attachment to adjacent sites on the mast cell (a requirement for degranulation). Nevertheless, *in vivo* the highest histamine release is seen with tubocurarine, some four times greater than any other intravenous anaesthetic agent and dose related. Tubocurarine also releases considerable skin histamine, possibly by a neurological mechanism, cf. atracurium.

Analgesics and allergy

A useful warning of the problem in distinguishing pharmacological effects from genuine response is provided by Szczelik[41] who reviewed a second group of anaesthetic drugs, the analgesics, and their tendency to provoke symptoms of intolerance resembling those of allergy. Skin tests are generally positive in less than 25% of patients with a history of anaphylactoid reactions to analgesic drugs. Szczelik recognizes two groups of reactor patients: the first has the classic manifestations of type I hypersensitivity response and only an 8% frequency of asthmatic attack; the converse operates in the pseudo-allergic group, where the patients predominantly have a history of chronic rhinitis or asthma. In the former, histamine is the likely mediator of response, but the patients in the latter group implicate prostanoid release through their (tested) intolerance to oral aspirin.

Relative frequency of involvement

As has been emphasized in the preceding chapters, anaesthetic drugs are rarely injected in isolation. Adverse effects arising during general anaesthetic procedures will generally implicate the hypnotic as well as the neuromuscular blocking agent and possibly premedication and the effects of anaesthetic gases and vapours.[42] Nevertheless, it cannot be denied that many severe anaesthetic reactions follow the administration of the neuromuscular blocker, although this may simply be the last straw in an already pharmacologically overloaded system. However, while we are predominantly considering ADRs in response to bolus induction, the long-term use of infusions in patients with multisystem organ failure brings more frequent practical problems linked to residual curarization. The two non-depolarizing relaxants, atracurium and vecuronium, offer particular advantage in such situations.

The author's concept of the relative frequency of involvement of various neuromuscular blockers, by *all* mechanisms, is based upon a study first presented in 1987.[16] This comprised a survey of 250 patients who had presented with life-threatening anaphylactoid response (grades II–IV) and for whom a careful study of the clinical details, coupled with laboratory investigation and skin testing, subsequently implicated the drugs used for the induction of anaesthesia. The cases were drawn from those reported nationally from throughout the United Kingdom over a 4-year period (1983–1986). The principal hypnotics were thiopentone (>75% of cases) and Althesin, although the latter was withdrawn early in spring 1984. The frequency of appearance of the various neuromuscular blockers in these reactions is illustrated in Fig. 13.6 which effectively compares the relative safety of the various neuromuscular blockers. The data in Fig. 13.6 have been normalized in terms of individual drug usage. The actual incidence per annum throughout the UK will be considerably higher than the incidence shown here since this relates to reports to the Sheffield laboratory only. The data are nevertheless comparative for different drugs.

Alcuronium[16] and suxamethonium[16,43] emerge as the most hazardous drugs. Perhaps surprisingly, the high frequency of suxamethonium in Fig. 13.6 and in our later surveys[43] represented predominant elective use and rarely emergency. One must question the use of this drug in situations which do not require its unique properties.[43] When used with thiopentone, suxamethonium and alcuronium are likely to be associated with serious hazardous response once in 4000 inductions.[16]

Pancuronium, with its steroid structure, emerges as the safest neuromuscular blocker. Unfortunately, its physiological properties are not always satisfactory for anaesthetic purposes and indeed comprise the bulk of the adverse reactions encountered with this substance.

In terms of clinical adverse response, atracurium emerges with a safety record identical to that of vecuronium, and the properties of these two drugs make both of them admirable alternatives to pancuronium despite the slightly increased risk. At present there is no evidence to suggest that the new practical analogues of vecuronium and atracurium, rocuronium and mivacurium, are safer or less safe than their parents.

Despite its well known histamine release properties tubocurarine appears to pose few problems for the anaesthetist.[16,43] The answer is presumably that histamine release here is a *predictable* pharmacological property of the drug.

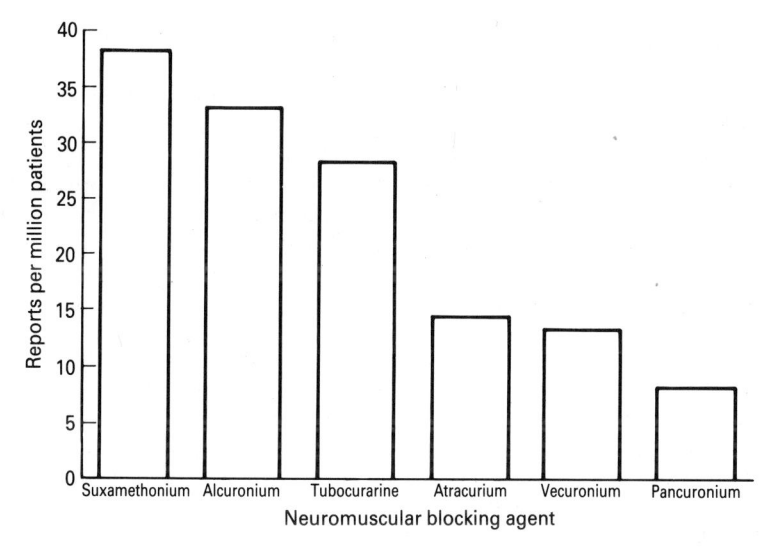

Fig. 13.6 The number of serious adverse reaction reports to the author's laboratory per million patients treated with various neuromuscular blockers (January 1984 to December 1986) and for which laboratory investigation indicated cause and effect. Normalization was achieved in terms of (a) estimates of drug ampoule sales from official sources and (b) estimated dose/ampoule use. The current trend for multidose vials for Intensive Care use has complicated such assessments in later years. Further, the use of alcuronium has declined rapidly in the UK since 1990 as a result of replacement by atracurium and vecuronium.[43] Nevertheless, the indications are that the *relative* safety profiles of the various neuromuscular blockers are unchanged.[43] (Taken from Watkins 1988.[16])

Comments and conclusions

The laboratory report

This should be a concise, written report, including both an interpretation of the measured data *and* suggestions for future care of the patient. By necessity it should be composed by a specialist immunologist with a working knowledge of anaesthetic practice or after consultation with a professional anaesthetist. The author is aware of several incidents in which the anaesthetist was assured by a laboratory that there was significant complement consumption in a reaction, implying drug involvement, when the 'consumption' related simply to haemodilution of sequential samples as a result of administered fluid to correct hypotension.

The original copy of the report should be sent to the *involved* anaesthetist for incorporation into the patient's 'notes', the front cover of which should then be clearly marked to indicate a patient with anaesthetic problems. In making the report the *immunologist* will certainly find it necessary to discuss aspects of the clinical reaction and its management (as an indication of reaction severity) with the anaesthetist and not infrequently with the surgeon (expected problems with the surgery?, abnormal anatomy? physiology?) and the patient too.

What the anaesthetist should do with the laboratory report

The future management of the patient always remains the fundamental responsibility of the anaesthetist. The recommendations of the laboratory report are suggestions only for him or her to either follow or to disregard. Dialogue between the anaesthetist and the immunologist should remain open and indeed further laboratory investigations, at even more specialized laboratories, may be agreed necessary. This would certainly be the case in reactions which although apparently anaphylactoid could well implicate biochemical abnormalities, e.g. metabolic disorders including malignant hyperpyrexia.[44, 45]

Who to notify

(a) The anaesthetist should notify the *patient* of the nature of his or her reaction and of its future implications. Immune reactions are the easiest to cope with since the patient can be issued with an explicit Medic-Alert bracelet or the like. In more complex situations a bracelet which indicates 'anaesthetic risk' may be self-defeating and a recommendation to 'consider fentanyl/etomidate/atracurium in emergency' would be more useful: the author has yet to encounter a severe reaction to this cocktail.

The patient can also be assured that there is no evidence of direct familial inheritance traits resulting in an adverse anaesthetic reaction. However, in the same way that breast cancer tends to run in families, sensitivity to suxamethonium may behave anomalously and prescreening should be afforded to immediate family members, if they wish. Inheritance of biochemical anomalies which cause problems during anaesthesia, including mastocytosis, malignant hyperpyrexia and pseudocholinesterase enzyme defects is well defined and should be excluded by general medical screening.

(b) The anaesthetist should notify the patient's *GP and surgeon* with at least copies of the laboratory report and to indicate to both that the immunologist is willing to discuss further if required.

(c) Central reporting using the existing Yellow Card system to the Committee on Safety of Medicines (CSM) remains important. Because of the complexities of general anaesthesia the author has fought over a number of years for close cooperation between specialized laboratories, such as that in Sheffield, and the CSM in the interpretation and reporting of *anaesthetic* reactions. This view has not, unfortunately, received the support of the professional bodies representing the anaesthetists. The author's personal view is that at the present time anaesthetic reactions reported to the CSM frequently lack validity since they are almost inevitably based on *clinical observation alone* which may easily implicate the wrong drug.

Drug-specific antibodies and prescreening

Skin testing and RAST assays appear to reveal the presence of circulating IgE antibodies directed against various anaesthetic drugs in those patients who have undergone serious *anaphylactic* response to these drugs. The majority of these antibodies are directed against neuromuscular blockers: a small proportion against thiopentone. The important question arises as to the origin of these antibodies. Did they arise by (a) classic sensitization following previously apparently uneventful (clinically speaking) surgery (as with thiopentone), or (b) were they caused as a consequence of the reaction itself, or (c) were they already preformed by some unforseen self-presensitization of the patient? Further, severity of reaction may depend not only on the presence of the antibody (i.e. its concentration in plasma and mast cell membranes) but its affinity (binding strength) to the antigen in question.

The classic situation (a) can be readily distinguished from (b) and (c) by a study of the patient's anaesthetic history and the laboratory demonstration of the consumption (and subsequent regeneration) of the antibodies following the onset of the reaction coupled with clear evidence of mast cell degranulation (tryptase release). The latter will not occur if the antibodies are purely an epiphenomenon, arising as a consequence of the reaction itself, (b). However, antibodies (IgE) to suxamethonium appear to be particularly important harbingers of serious adverse reactions[46] and frequently occur in the female patient (and in occasional male patients too) without previous anaesthetic history, (c).

Suxamethonium antibodies

The most commonly encountered IgE antibody in the population is that directed against suxamethonium. The second is against alcuronium. An apparent reversal of this was encountered by McSharry (personal communication) but in a population of patients with allergic skin disease. However, here it was possible that the patients were being sensitized by the topical creams used in their treatment.

It is conceivable that IgE anti-suxamethonium antibodies are actually autoantibodies against the neurotransmitter molecule acetycholine which cross-react with suxamethonium: this would parallel the presence of IgG autoantibodies against the neurotransmitter *receptor* in myasthenia gravis. As with myasthenia gravis, the presence of 'anti-suxamethonium antibodies' is heavily biased towards the female. Curiously, the most at risk patient to suxamethonium appears to be female between the active reproductive period of 15–35 years. This appears even when one considers groups (e.g. ENT surgery) which exclude the obvious nature of female surgery. The risk to this group is four to five times that of the male and parallels the situation with aspirin-sensitive asthmatics (Watkins, unpublished observations). There is considerable debate concerning the value of prescreening for suxamethonium antibodies.[47]

Finally, despite a vast growing literature to the topic, costly monitoring systems and postgraduate training, the frequency of both the anaphylactoid response and its clinical sequelae appear unchanged over the last 20 years. While anaesthetic mortality is low, the frequency of aborted and repeated procedures and of varying degrees of brain damage, not to mention 'shattered' anaesthetists, remains high. The more widespread use of systematic laboratory investigation viewed as an *essential* part of reaction management could substantially reduce the impact of anaesthetic morbidity in the future.

References

1. Hurwitz N. Predisposing factors in adverse reactions to drugs. *British Medical Journal* 1969; **1:** 536–539.
2. Watkins J. Intravenous therapy and 'immunological' disasters. *Theoretical Surgery* 1986; **1:** 103–112.
3. Ring J, Messmer K. Incidence and severity of anaphylactoid reactions to colloid volume substitutes. *Lancet* 1977; **i:** 466–469.
4. Watkins J. Heuristic decision-making in diagnosis and management of adverse drug reactions in anaesthesia and surgery: the case of muscle relaxants. *Theoretical Surgery* 1989; **4:** 212–222.
5. Watkins J. Immediate hypersensitivity-type reactions in anasthesia. Allergy or otherwise, a problem or an overstatement. *Theoretical Surgery* 1991; **6:** 229–233.
6. Lunn JN, Mushin WW eds. *Mortality Associated with Anaesthesia* 1982, London: Nuffield Provincial Hospitals Trust.
7. Laxenaire MC, Moneret-Vautrin DA. Immunology and intravenous anaesthetics. In Kaminski B, Rawicz M eds *Proceedings I, 8th European Congress of Anaesthesiology* 1990, Warsaw, pp. 190–197.
8. Pearce FL. Functional heterogeneity of mast cells from difference species and tissues. *Klinische Wochenschrift* 1982; **60:** 954–957.
9. Barnes PK, de Renzy Martin N, Thomas VJE, Watkins J. Plasma histamine levels following atracurium. *Anaesthesia* 1986; **41:** 821–824.
10. Watkins J. Histamine release and histamine-mediated adverse effects due to muscle relaxants. In Bowman WC ed. *Neuromuscular Blocking Agents: Past, Present and Future* 1990, Amsterdam: Excerpta Medica, pp. 87–99.
11. Mirakhur RK, Lyons SM, Carson IW, Clarke RSJ, Ferres CJ, Dundee JW. Cutaneous reactions after atracurium. *Anaesthesia* 1983; **38:** 818–819.
12. Lawrence ID, Warner JA, Cohan VL, Hubbard WC, Kagey-Sobotka A, Lichtenstein LM. Purification and characterisation of human skin mast cells: evidence for human mast cell heterogeneity. *Journal of Immunology* **1987; 139:** 3062–3069.
13. Mellinghoff H, Diefenbach C, Buzello W. The clinical pharmacology of doxacurium. In Bowman WC ed. *Neuromuscular Blocking Agents: Past, Present and Future* 1990, Amsterdam: Excerpta Medica, pp. 79–83.
14. Savarese JJ. The clinical pharmacology of mivacurium. In Bowman WC ed. *Neuromuscular Blocking Agents: Past, Present and Future* 1990, Amsterdam: Excerpta Medica, pp. 69–78.
15. Maestrone E. Pipecuronium, pancuronium and alcuronium: dose–response relationships and clinical profiles. In Bowman WD ed. *Neuromuscular Blocking Agents: Past, Present and Future* 1990, Amsterdam: Excerpta Medica, pp. 148–160.
16. Watkins J. Anaphylactoid response to neuromuscular blockers. In Jones RM, Payne JP eds *Recent Developments in Muscle Relaxation: Atracurium in Prospective*. International Congress and Symposium Series, No 131 1988, London: Royal Society of Medicine Services, pp. 13–20.
17. Moss J, Philbin DM, Rosow CE, Basta SJ, Gelb C, Savarese JJ. Histamine release by neuromuscular blocking agents in man. *Klinische Wochenschrift* 1982; **60:** 891–895.
18. Moss J, Rosow CE, Savarese JJ, Philbin DM, Kniffen KJ. Role of histamine in the hypotensive action of d-tubocurarine in humans. *Anaesthesiology* 1981; **55:** 19–25.
19. Marone G, Stellato C. Activation of human mast cells and basophils by general anaesthetic drugs. In Assem E-SK ed. *Allergic Reactions to Anaesthetics. Clinical and Basic Aspects. Monographs in Allergy* 1992; **30:** 54–73.
20. Lorenz W, Doenicke A, Schöning B, Neugebauer E. The role of histamine in adverse reactions to intravenous agents. In Thornton JA ed. *Adverse Reactions to Anaesthetic Drugs* 1981, Amsterdam: Elsevier/North Holland, pp. 169–238.

21. Lorenz W, Doenicke A, Meyer R, Reimann JH, Kusche J, Barth H, et al. Histamine release in man by propanidid and thiopentone: pharmacological effects and clinical consequences. *British Journal of Anaesthesia* 1972; **44:** 355–369.
22. Lorenz W, Seidel W, Doenicke A, Trauber R, Reimann HJ, Uhlig R, Mann G, et al. Elevated plasma histamine concentrations in surgery: causes and clinical significance. *Klinische Wochenschrift* 1974; **52:** 419–425.
23. Philbin DM, Moss J, Rosow CE, Akins CW, Rosenberger JL. Histamine release with intravenous narcotics: protective effects of H1 and H2 receptor antagonists. *Klinische Wochenschrift* 1982; **60:** 1056–1059.
24. Doenicke A, Ennis M, Lorenz W. Histamine release in anaesthesia and surgery: a systematic approach to risk in the perioperative period. In Sage DJ ed. *Anaphylactoid Reactions in Anaesthesia. International Anesthesiology Clinics* 1985; **23:** 41–66.
25. Lorenz W, Doenicke A, Schöning B, Mamorski I, Weber D, Hinterlang E, et al. H1 + H2 receptor antagonists for premedication in anaesthesia and surgery: a critical view based on randomized clinical trials with histamine and various anti-allergic drugs. *Agents and Actions* 1980; **10:** 114–124.
26. Watkins J. Etomidate: an 'immunologically safe' anaesthetic agent. *Anaesthesia* 1983; **38** (Suppl): 34–38.
27. Watkins J. Anaphylactoid reactions to i.v. substances. *British Journal of Anaesthesia* 1979; **51:** 51–60.
28. Anon. Editorial: complement activation in plasma exchange. *Lancet* 1988; **2:** 1464–1465.
29. Watkins J, Levy CJ eds. *Guide to Immediate Anaesthetic Reactions* 1988, London: Butterworths.
30. Wright PJ, Shortland JR, Stevens JD, Parsons MA, Watkins J. Fatal haemopathological consequences of general anaesthesia. *British Journal of Anaesthesia* 1989; **62:** 104–107.
31. Watkins M. Markers and mechanisms of anaphylactoid reactions. In Assem E-SK ed. *Allergic Reactions to Anaesthetics. Clinical and Basic Aspects. Monographs in Allergy* 1993; **30:** 108–129.
32. Watkins J, Wild G. Improved diagnosis of anaphylactoid reactions by measurement of serum tryptase and urinary methylhistamine. *Annales Française d'Anesthésie et de Réanimation* 1993; **12:** 169–172.
33. Schwartz LB, Metcalf DD, Miller J, Earle H, Sullivan T. Tryptase levels as an indicator of mast cell activation in systemic anaphylaxis and mastocytosis. *New England Journal of Medicine* 1987; **316:** 1622–1626.
34. Watkins J. Tryptase release and clinical severity of anaesthetic reactions. *Agents and Actions* 1992; (Special Conference Issue): C203–5.
35. Baldo BA, Harle DG, Fisher MM. *In vitro* diagnosis and studies of the mechanism(s) of anaphylactoid reactions to muscle relaxant drugs. *Annales Française d'Anesthésie et de Réanimation* 1985; **4:** 139–145.
36. Baldo BA, Fisher MM. Mechanisms in IgE-dependent anaphylaxis to anaesthetic drugs. *Annales Française d'Anesthésie et de Réanimation* 1993; **12:** 131–140.
37. Laxenaire MC, Moneret-Vautrin DA, Vervloet D. The French experience of anaphylactoid reactions. In Sage DJ ed. *Anaphylactoid Reactions in Anesthesia. International Anesthesiology Clinics* 1985; **23:** 145–160.
38. Guéant JL, Mata A, Masson C, Moneret-Vautrin DA, Laxanaire MC. Non-specific interactions in anti-agent IgE-RIA to anaesthetic agents. *Annals Française d'Anesthésie et de Réanimation* 1993; **12:** 141–146.
39. Fisher MM. Intradermal testing in the diagnosis of acute anaphylaxis during anaesthesia – results of five years' experience. *Anaesthesia and Intensive Care* 1979; **7:** 58–61.

40. Wood M, Watkins J, Wild G, Levy CJ, Harrington C. Skin testing in the investigation of reactions to intravenous anaesthetic drugs. A prospective trial of atracurium and tubocurarine. *Annales Française d'Anesthésie et de Réanimation* 1985; **4**: 176–179.
41. Szczelik A. Analgesics, allergy and asthma. *Drugs* 1986; **32** (Suppl): 148–163.
42. Gibb D. Drug interactions in anaesthesia. In Fisher MM ed. *Adverse Reactions. Clinica in Anaesthesiology* 1984; **2**: 485–512.
43. Clarke RSJ, Watkins J. Drugs responsible for anaphylactoid reactions in anaesthesia in the United Kingdom. *Annales Française d'Anesthésie et de Réanimation* 1993; **12**: 105–108.
44. Simpson KH, Ellis FR. Anaesthetic problems related to defects of enzyme function. In Watkins J, Levy CJ eds *Guide to Immediate Anaesthetic Reactions* 1988, London: Butterworths, pp. 68–84.
45. Denborough MA. Malignant hyperpyrexia. In Fisher MM ed. *Adverse Reactions. Clinics in Anaesthesiology* 1984; **2**: 669–675.
46. Watkins J. Anaesthetic drug allergies. *Anaesthesia* 1993; **48**: 639–640.
47. Roberts JA. Allergic reactions to anaesthetics: economic aspects of pre-operative screening. In Assem E-SK ed. *Allergic Reactions to Anasesthetics. Clinical and Basic Aspects. Monographs in Allergy* 1992; **30**: 207–221.

Appendix

Emergency management of acute major anaphylaxis under general anaesthesia

All doses for 70 kg patient – adjust as necessary.
Basic monitoring assumed (pulse oximeter, blood pressure, ECG).
Exercise caution if the diagnosis is not certain.

Immediate therapy

1. Discontinue administration of the suspect drug.
2. Summon help.
3. Discontinue surgery and anaesthesia if feasible.
4. Maintain airway with 100% oxygen (consider tracheal intubation and IPPV).
5. Give i.v. *adrenaline* especially if bronchospasm present, 50–100 µg (0.5–1.0 ml 1:10 000). Further 1 ml aliquots as necessary for hypotension and bronchospasm. Prolonged therapy may be necessary occasionally (5–8 µg kg^{-1} is the usual dose range).
6. Start intravascular volume expansion, preferably *colloid*, 10 ml kg^{-1} rapidly.
7. Consider external chest compression.

Secondary management

1. Adrenaline-resistant bronchospasm:
   Consider:
   Salbutamol 250 µg i.v. loading dose. 5–20 µg min^{-1} i.v. maintenance.
   or
   Terbutaline 250–500 µg i.v. loading dose. 1.5 µg min^{-1} maintenance.
   or
   Aminophylline 6 mg kg^{-1} i.v. over 20 minutes.
2. Bronchospasm and/or cardiovascular collapse:
   Steroids:
   Hydrocortisone 500 mg i.v.
   or
   Methylprednisolone 2 g i.v.
3. Antihistamines:
   Chlorpheniramine 20 mg i.v. diluted given slowly.
4. *Sodium bicarbonate* if acidosis severe after 20 minutes' treatment.
5. Catecholamine infusions:
   Adrenaline 5 mg in 500 ml (10 µg nl^{-1}). Start at 10 ml h^{-1}, up to 85 ml h^{-1}.
   Noradrenaline 4 mg in 500 ml (8 µg ml^{-1}). Start at 25 ml h^{-1}, up to 100 ml h^{-1}.
6. Consider the possibility of coagulopathy: clotting screen.
7. Measure arterial blood gas tensions for oxygenation and acid-base status.

Index

Abdominal muscles, monitoring 117–18
Accelerometry 115
Acetylcholine
　postjunctional receptors 16–19
　　channel protein macromolecule 16
　　occlusion 18
　　open/closed channel block 18–19
　prejunctional receptors 7–9, 19–20
　　endplate 7
　　release 6
　　and drug interactions 177
　　fate 9
　　role in reversal of neuromuscular
　　　blockade 135–6
　synthesis and storage 5–6
Acetylcholinesterase
　anionic site 136
　esteratic site 136
　in injured muscle 168
　role in neuromuscular
　　transmission 135–6
Acid–base balance
　changes, effect on muscle relaxants 180
　disorders of 168–9
　drug interactions 189–91
　and electrolytes 150
Acquired neuromuscular disorders 169–71
Actin 10
Action potential 4, 10
　compound muscle (CMAPs) 112–14
　evoked compound (ECAPs) 112–15
Adductor pollicis
　atracurium resistance 122
　blockade, loading dose 130
　and intubation 57–8
Adrenaline, for anaphylactic reactions under
　anaesthesia 258–9
　emergency administration 245, 258–9
Adverse reactions 234–58
　aggregate anaphylaxis 244
　decision-making tree 246–7
　drugs, relative safety 252–3
　immune and non-immune
　　mechanisms 241, 244–5
　neuromuscular blockers 251–8
　see also Anaesthetic reactions;
　　Anaphylactic/oid reactions
Age effects
　neuromuscular blocking drugs 38–9, 156–60

reversal of neuromuscular blockade 149
　see also Paediatric anaesthesia; specific
　　substances
Aggregate anaphylaxis 244
Agonists, ACh receptors see
　Decamethonium; Suxamethonium
Alcuronium 77, 78–9
　clinical use
　　effect of pregnancy 42
　　in intensive care practice 228
　　relative safety 252–3
　　side-effects 79
　pharmacodynamics 77, 78–9
　　age of patient 157, 160
　　interactions with other non-
　　　depolarizing relaxants 191–2
　　structure 78
　pharmacokinetics 78
　　hepatic failure 166
　　rate of reversal of neuromuscular blockade
　　　after neostigmine dosages 141
　　renal failure 163
Alimentary side-effects, anticholinesterase
　drugs 138–9
Allergy, and analgesics 251
Alpha 1 acid glycoprotein, in renal failure 160
Alpha adrenoreceptors, acetylcholine
　release 9
Althesin, drug interactions 182, 237, 238, 252
Aminoglycoside antibiotics, drug
　interactions 183–4
Aminophylline, adrenaline-resistant
　bronchospasm 259
Aminopyridines, contraindications 145
Aminosteroid muscle relaxants see
　Pipecuronium; Rocuronium
Anaesthetic agents
　choice 148
　comparisons, rate of reversal of
　　neuromuscular blockade after
　　neostigmine dosages 141
　drug interactions 180–1
　intravenous use 181–2
Anaesthetic Hazard Warning card 72, 254
　plasma cholinesterase: decreased activity 62
Anaesthetic reactions
　causes 236–7
　emergency action 245, 258–9
　incidence of common hypnotic drugs 238

Index

Anaesthetic reactions—*contd*
  relative safety of relaxant drugs 252–3
  reporting trends 237
  see also Anaphylactic/oid reactions
Analgesics
  and allergy 251
  drug interactions
    local 182
    opioids 182
Anaphylactic/oid reactions 234–5
  classification 235
  definition 235–6
  immune and non-immune
    mechanisms 241, 244–5
  laboratory testing, requirements 249–51
  management 245–9
    decision-making tree 246–7
    emergency management 258–9
    extent of problem 236
  mechanisms 239–45
    aggregate anaphylaxis 244
    drug-specific antibodies and pre-
      screening 254–5
  mortality 236
  relative safety of relaxant drugs 252–3
  specific involvement 251–3
  see also Anaesthetic reactions
Antagonist drugs *see* Anticholinesterase drugs
Antiarrhythmic agents, drug
  interactions 186
Antibiotics, drug interactions 183–4
Antibodies, drug-specific 254–6
Anticholinesterase drugs 136–50
  action at neuromuscular junction 136–7
  action at other sites 138–9
  reversal of neuromuscular blockade
    choice 140–6
    choice of anaesthetic agent 148
    choice of neuromuscular blocking
      drug 140
    clinical guidelines 150
    drug interactions 149–50
    effect of age 149
    effect of electrolyte and acid–base
      imbalance 150
    effect of renal failure 148–9
    measurement of return of
      neuromuscular function 139–40
    plasma concentration of relaxant 146–8
    priming 145
    recovery index 146–50
    relaxant drugs compared 141
  side-effects
    alimentary 138–9
    cardiac 138
    pulmonary 139
  see also Edrophonium; Neostigmine;
    Pyridostigmine
Anticonvulsants, drug interactions 184–5
Aprotonin, drug interactions 188

Arrhythmias
  neostigmine and atropine 138
  suxamethonium-induced 66
Aspirin, suxamethonium-induced myalgia 65
Asthmatic patients 139
Asystole, suxamethonium-induced 66
Atracurium 77, 79–81
  clinical use
    indications in MH 71
    in intensive care practice 228–9
    maintenance dose 132
    paediatric anaesthesia 203, 204, 210–11
    relative safety 252–3
    side-effects 80–1, 240
  pharmacodynamics 80
    hydrolysis 36, 37
    interactions with hypnotics 238
    interactions with other relaxants 191–2
    protein binding 33
    renal excretion 34
  pharmacokinetics 80
    Hofmann degradation 38
    of metabolites 37–8
    percentage recovered in urine 34
    rate of reversal of neuromuscular blockade
      after neostigmine dosages 141
  pharmacokinetics, effects of
    age of patient 39, 157, 158
    body size 40–1
    burns 46, 167–8
    hepatic cirrhosis 46, 47, 165
    hepatic failure 44
    infancy 40
    pregnancy 42
    renal failure 45, 161
    structure 82
    design 23
    stereoisomers 79
Atropine, with anticholinesterase drugs
  cardiac effects 138–9
  histamine release 242
  side-effects 163–4
AUC, AUMC, defined 28
Azathioprine, drug interactions 187

Barbiturates, drug interactions 181–2
Benzodiazepines, drug interactions 181
Benzylisoquinolium drugs
  pharmacokinetics 46–8
  see also Atracurium; Doxacurium;
    Mivacurium; Tubocurarine
Beta adrenergic blocking agents, drug
  interactions 185
Beta adrenoreceptors, acetylcholine release 9
Blood sample, tests 249
Body size
  infants and children 200–1
  pharmacokinetics, neuromuscular
    blocking drugs 40–1, 128
Bradycardia, suxamethonium-induced 65–6

Bradyphylaxis, defined 57
Bretylium, drug interactions 186–7
Bronchospasm
  adrenaline resistance 258–9
  aggregate anaphylaxis 244
  emergency action 245, 258–9
  hypotension, and tryptase assays 246–8
Burns 46, 167–8

Caesarean section, failed-intubation procedure 57
Caffeine, muscle biopsy: *in vitro* contracture test 71–2
Calcium antagonists, drug interactions 185–6
Calcium ions
  drug interactions 190
  malignant hyperthermia 70
Carbamazepine, drug interactions 185
Carbon dioxide, end-tidal concentration 69
Cardiac side-effects
  anticholinesterase drugs 138
  arrhythmias, suxamethonium-induced 66
Cardiovascular surgery, anaphylatoxin C3a release during 243–4
Catecholamines
  catecholamine receptors 20–1
  infusions, management of acute reactions 259
  release
    elevation, causes 70
    suxamethonium-induced 65
Children *see* Paediatric anaesthesia
Chlorpheniramine, adrenaline-resistant bronchospasm 259
Chlorpromazine, suxamethonium-induced myalgia 65
Cholinesterase, plasma
  action 35–6
    hydrolysis of suxamethonium 59–60
  clinical use
    administration 62
    contraindications to use 145–6
    effects of hepatic and renal failure 166
    in hepatic failure 160
    in renal failure 160
  decreased activity
    acquired deficiency 60
    causes 36, 60
    inherited deficiency 36, 61–2
    investigation and counselling 62–3
    in pregnancy 60
  structure 61
    synthetic 62
    variants, parameters 61
Cholinoceptors
  proliferation in hyperkalaemia 67
  see also Acetylcholine; Acetylcholinesterase
Chondodendron tomentosum, source of tubocurarine 90
Clearance, defined 27

Colistin, drug interactions 184
Compartment, defined 28
Compartmental analysis 29–30
Complement mechanisms
  C3 activation 243
  histamine release 243–4
  subclinical activation 243–4
Creatine kinase, release from injured muscle 65, 68
Cremophor EL 243
Cyclo-oxygenase inhibitors, drug interactions 188
Cyclosporin
  drug interactions 187
  effect of renal failure 43

Dantrolene
  dose make-up 71
  drug interactions 188
  in management of malignant hyperthermia 70–1
Decamethonium, pharmacology 22
Decision-making tree, anaphylactic/oid reactions 246–7
Denervation conditions 167, 168
Depolarization and desensitization 56
Dexamethasone, tubocurarine interactions 187
Diaphragm, monitoring 119
Dibucaine number, plasma cholinesterase 61, 62
Diclofenac, suxamethonium-induced myalgia 65
Dimethyltubocurarine *see* Metocurine
Direct muscle stimulation 110–11
Disopyramide, drug interactions 186
Disseminated intravascular coagulation (DIC) 244
Disuse atrophy 225
Diuretics, drug interactions 183
Double burst stimulation 62, 103–4, 107, 110, 140
Doxacurium 77, 81–2
  clinical use
    in intensive care practice 230
    maintenance dose 132
    paediatric anaesthesia 203, 204, 210
    side-effects 82
  pharmacodynamics 81
    interactions, other non-depolarizing relaxants 191–2
    protein binding 33
    structure 82
  pharmacokinetics 81
    elimination 48
    hydrolysis 36
    rate of reversal 141
  pharmacokinetics, effects of
    age of patient 39, 157, 159
    hepatic (cirrhotic) disease 166

Doxacurium—*contd*
  pharmacokinetics, effects of—*contd*
    hepatic failure 44
    renal failure 45, 163
Doxapram
  drug interactions 189
  interactions, and recovery index 150
Drug interactions 177–97
  list 178–80
  sites 177
  see also specific substances
Drug-specific antibodies 254–6
Duchenne muscular dystrophy 171
  and malignant hyperthermia 70
Dystrophia myotonica, anaesthetic hazards 172

Eaton–Lambert myasthenic syndrome 171
Ecothiopate, drug interactions 188–9
ECT, suxamethonium 59
Edrophonium
  clinical study 143
  pharmacokinetic variables 142
    plasma elimination half-time 148–9
  properties 143–4
Electrodes, types 108
Electrolytes
  and acid–base imbalance 150
  effect on muscle relaxants 180
Electromyography 10, 11, 112–14
  accelerometry 115
  attenuation during monitoring 114
  compound muscle action potential 112–14
  differences between electromyographic and mechanometric measurements 114–15
  ECAP 112–15
Elimination half-time, defined 28
Emergencies, management 258–9
Endotracheal intubation 222–3
  awake, contraindications 63
  rapid, indications for suxamethonium 57
Enflurane, drug interactions 180–1
Enflurane anaesthesia, reversal of mivacurium 147
Epinephrine *see* Adrenaline
Esmolol, interactions with suxamethonium 185
Esteratic site, acetylcholinesterase 136
Esters, hydrolysis 35–6
Etomidate, drug interactions 181, 238
Evoked compound action potential (ECAP) 112–15
Excitable membranes 3–5
Excitation–contraction coupling 10–11
  twitch and tetanic contractions 11
Eye, rise in intraocular pressure, suxamethonium-induced 68
Facial muscles, monitoring 120–1
Facilitation 98–9

Fade 98
Fasciculations
  pre-curarization 193–4
  suxamethonium-induced, paediatrics 63
Fazadinium
  pharmacodynamics, renal failure 163
  pharmacokinetics, hepatic cirrhosis 46, 47
Fentanyl, indications in MH 71
First order, defined 28
Fluoride number, plasma cholinesterase 61
Force transducer, measurement of muscle contraction 111–12

Gallamine 77, 83–4
  clinical use
    dose requirements 33
    intensive care practice 228
    pre-curarization 163
    prevention of fasciculations 63
    prevention of suxamethonium-induced myalgia 64
    side-effects 83–4
  pharmacodynamics 83
    hepatic uptake 35
    interactions with other non-depolarizing relaxants 191–2
    structure 83
  pharmacokinetics 83
    excretion 27, 34
    percentage recovered in urine 34
  pharmacokinetics, effects of
    extrahepatic biliary obstruction 46
    renal excretion 34
    renal failure 45, 163
Ganglionic blocking agents, drug interactions 186
Gas exchange, in intensive care practice 223
Gastrocnemius muscle, tubocurarine and pancuronium resistance 122
Gastrointestinal tract, rise in intragastric pressure, suxamethonium-induced 68
Glyceryl trinitrate, drug interactions 187
Glycogen storage diseases, anaesthetic hazards 172
Glycopyrrolate
  with anticholinesterase drugs, cardiac effects 138–9
  antisialogoguic activity 139

Halothane
  drug interactions 180–1
  enhancement of neostigmine blockade 143
  muscle biopsy: *in vitro* contracture test 71–2
  rate of reversal of neuromuscular blockade after neostigmine dosages 141
Hand muscles, indications for monitoring 117
Hazard cards and bracelets
  Anaesthetic Hazard Warning card 72
  plasma cholinesterase, decreased activity 62
  positive vs negative recommendation 254

Index

Hemiplegia 167
Hepatic disease
  cholestatic disease 164
  cirrhosis 44–5, 164–5
  effects of neuromuscular blocking drugs 44–6
  hepatic failure 160, 164–7
  ICU 226
  renal impairment in 161
Hepatic uptake of drugs 34
Hexamethonium, drug interactions 186
Histamine release
  chemical mechanisms 242–3
  complement mechanisms 243–4
  hypnotics, comparisons 242
  plasma levels 239–40
  and prostaglandin release 240
  reactions 226, 239–43
  situations involving 239
  specific drugs 81, 86, 91, 93, 240–1
    comparisons 242
Hofmann elimination 38
  in acidosis 169
Hydrocortisone, adrenaline-resistant bronchospasm 259
3-Hydroxy-phenyl-trimethyl-ammonium, metabolite of physostigmine 142
Hyperkalaemia
  in MH, management 71
  suxamethonium-induced 67–8
    disorders list 67
Hyperthermia *see* Malignant hyperthermia
Hypnotic drugs, incidence in anaesthetic reactions 238, 242
Hypotensive agents, drug interactions 187
Hypothermia, pharmacokinetics, neuromuscular blocking drugs 43

IgE antibodies 244–5
Immediate sensitivity *see* Anaphylactic/oid reactions
Immobility 167
Immune and non-immune mechanisms, anaphylactic/oid reactions 241, 244–5
Immunosuppressants, drug interactions 187
In vitro contracture test 71–2
Infants and children *see* Paediatric anaesthesia
Inherited neuromuscular disorders 171–2
Intensive care unit 221–33
  neuromuscular blockade
    choice of relaxant 226–31
    contraindications, relative 224–5
    indications 222–4
    monitoring 231
Intracranial pressure, suxamethonium-induced elevation 68–9
Intravenous anaesthetic agents, drug interactions 181–2
Isoflurane
  drug interactions 180–1

rate of reversal of neuromuscular blockade after neostigmine dosages 141

Ketamine, drug interactions 181

Laboratory investigations 246
  anaphylactic/oid reactions 249–51
  report 253–4
Larynx, monitoring 119–20
Laudanosine
  metabolite of atracurium 37, 38
  properties 79
Leg muscles, monitoring 121
Lincosamines, drug interactions 184
Lithium ions, drug interactions 190
Local analgesics, drug interactions 182

Magnesium ions, drug interactions 190
Malignant hyperthermia 69–72
  clinical features 69–70
  laboratory investigations 71–2
  management 70–1
    anaesthesia 72
    investigation of MHS patients 71–2
    safe drugs 72
  pathogenesis 70
Malouetine 23
Margin of safety, receptors 8
Masseter spasm
  paediatric anaesthesia 206–7
  suxamethonium-induced 66, 69
Mast cells, degranulation, immune and non-immune mechanisms 241–2
Mechanomyography 110
Membrane structure
  excitable membrane at rest 3
  receptors, at motor end-plate 8
MEPP, miniature end-plate potential 7
Meptazinol, drug interactions 182
Metabolic acidosis 168–9
Metabolites of neuromuscular blocking drugs 37–8
Methohexitone 238
Methylhistamine assays, urinary 248
Methylprednisolone, adrenaline-resistant bronchospasm 259
Metocurine (dimethyltubocurarine) 77, 84–5
  clinical use, side-effects 85
  pharmacodynamics 84
    hepatic uptake 35
    interactions with other non-depolarizing relaxants 191–2
    percentage recovered in urine 34
    protein binding 33
    structure 84
  pharmacokinetics 84
  pharmacokinetics, effects of
    age of patient 39
    burns 46
    pregnancy 42
    renal failure 45, 163

Metronidazole, drug interactions 184
Midazolam, suxamethonium-induced myalgia 65
Mivacurium 77, 85–6
  clinical use 129–31
    histamine release 240
    infusions 129–31
    in intensive care practice 230–1
    maintenance dose 132
    paediatric anaesthesia 203, 204, 212–13
    recovery index 148
    side-effects 86
  pharmacodynamics 86
    hydrolysis 35–6
    interactions with other non-depolarizing relaxants 191–2
    metabolism 38
    reversal of neuromuscular blockade 141, 147
    structure 82
  pharmacokinetics 85
    renal excretion 34
  pharmacokinetics, effects of:
    age of patient 157, 158–9
    hepatic cirrhosis 165
    hepatic failure 44
    renal failure 43, 162
Motor nerve terminals
  depolarization 137
  motor end-plate 7
Motor unit 1–2
Muscarinic receptors 20
Muscarinic stimulation 65
Muscle
  and body compartment size, infants and children 201
  damage, suxamethonium-induced 205–6
  direct stimulation 110–11
  indications for monitoring specific muscles 115–17
  physiology 1–12
  skeletal muscle compartment, infants and children 200–1
Muscle biopsy
  in vitro contracture test 71–2
  MHS patients 71–2
Muscle fibres, biology of 1–12
  contraction
    electromyography 112–14
    excitation–contraction coupling 9–10
    measurement with force transducer 111–12
    twitch and tetanic contractions 11
    visual and tactile asessment 110–11
  fast:slow fibres, infants and children 200
  fibre types 122
  in vitro contracture test 71–2
  infants and children, skeletal muscle compartment 200–1
  neuromuscular junction 5–10

Muscle relaxants, depolarizing *see* Suxamethonium
Muscle relaxants, non-depolarizing
  choice 170–1
    choice of reversal agent 140–6
    list 77
    properties of ideal relaxant 225–6
  classification 127–8
  clinical use 129–30
    adverse reactions 234–58
    applications and requirements 127
    following suxamethonium 58–9
    infusions 131–2
    in intensive care unit 227–31
    loading dose 130
    maintenance dose 130–1
    non-relaxant actions 23–4
    recovery index 146
  drug-specific antibodies 254–6
  intensive care unit 221–33
  paediatric anaesthesia 207–13
    characteristics 207–8
    intermediate-duration drugs 210–12
    longer-duration drugs 208–10
    newer drugs 212–13
  pharmacodynamics
    action on ACh 9, 21–2
    additional effects 21
    additive effects 191–2
    agonists 22
    antagonists 22–3
    combinations, interactions 191–4
    drug interactions 149–50, 191–2
    facilitation and fade 98–9
    list of interactions 178–80
    metabolites 37–8
    pathways of drug elimination 34–7
    plasma concentration of relaxant 146, 146–8
    protein binding 32–3, 160
    rate of onset 58
    reversal of neuromuscular blockade 150, 163–4
    synergistic effects 191–2
  pharmacokinetics 26–54, 77–96
    in advanced age 39, 129
    analysis, compartmental/non-compartmental analysis 29–31
    body size 40–1, 128, 200–1
    definitions and approaches 27–9
    elimination 27
    elimination pathways 34–7
    general principles 26–32
    hepatic disease effects 44–6, 160–7, 226
    hypothermia 43
    infants and children 3, 40, 128–9
    pregnancy 41–2
    renal failure effects 43–5, 148–9, 160–9
    specific drugs 32–48
    thermal injury 46, 69–72, 167–8

pharmacology 13–25, 97–9
    postjunctional receptors 16–19
    prejunctional receptors 19–21
    structure and function of
        neuromuscular junction 13–15
    'train-of-four', basis 9
  structure, design 23
Muscular dystrophy, anaesthetic hazards 172
Myalgia, suxamethonium-induced 63–5
Myasthenia gravis 169–71
Myasthenic syndrome 171
Myelin 1
Myoglobin release, suxamethonium-
    induced 68
Myopathies
  anaesthetic hazards 172
  contraindications to suxamethonium 171
  creatine kinase release from injured
      muscle 65, 68
Myosin 10
Myotonias
  anaesthetic hazards 172
  contraindications to suxamethonium 171
  and malignant hyperthermia 70

Neostigmine
  action 136–7
    effect on cholinoceptors 137
    release of acetylcholine 137
    side-effects 138–9, 163–4
  pharmacokinetic variables 142
  plasma elimination half-time,
      increase 148–9
  properties 142
Nerve axons 2
Nerve stimulation 4–5, 100–1
  electrodes, types 108
  'maximal current' 107
  muscle contraction, visual and tactile
      assessment 110
  optimum 107
  single stimuli 99–100
  stimulators, desirable features 108–9
  stimulus current 107–8
  stimulus duration 108
  see also Tetanic stimulation
Nerve terminal, motor 2
Neurological assessment 224
Neuromuscular blockade
  choice of drug 140
  extension 193
  facilitation and fade 98–9
  monitoring during anaesthesia 99–126
    indications for monitoring specific
        muscles 115–21
    measuring response 110–15
    patterns of stimulation 99–107
    techniques of stimulation 107–9
  muscle fibre types 122
  percent block 98

priming with anticholinesterase drugs 145
  reversal 135–55
    see also Anticholinesterase drugs;
        Edrophonium; Neostigmine;
        Pyridostigmine
    acetylcholine role 135–6
    anticholinesterase drugs 136–50
    clinical guidelines 150
  see also Muscle relaxants
Neuromuscular disorders
  acquired 169–71
  inherited 171–2
Neuromuscular junction 5–9
  action of anticholinesterase drugs 136–7
  interactions 177
  maturation, and paediatric
      anaesthesia 199–200
  postjunctional receptors 16–19
  prejunctional receptors 19–21
  structure and function 13–15
Neuromuscular transmission,
    pharmacological antagonism 17
Nicotinic receptors 19–20
Nifedipine, drug interactions 185–6
Node of Ranvier 1
Non-compartmental analysis 30–1
NONMEM program, population analysis 31
Noradrenaline, anaphylactic reactions under
    anaesthesia 259
NSAIDs, suxamethonium-induced
    myalgia 65

Obesity see Body size
Obstetric anaesthesia and relaxant drugs see
    Pregnancy
Opioid analgesics
  drug interactions 182
  and histamine release, comparisons 242
Opioid receptors 21
Orbicularis oculi, atracurium resistance 122
Org 6368, hepatic uptake 35
Organ failure 160–7

Paediatric anaesthesia 39–40, 198–220
  age effects on drug pharmacokinetics 40,
      128–9
  anatagonism of neuromuscular
      blockade 213–14
  cholinesterase, plasma, decreased
      activity 60
  neonates 40, 201–2
  non-depolarizing relaxants 207–13
  pharmacokinetics, neuromuscular
      blocking drugs 40
  physiological factors 198–202
  suxamethonium 202–7
Pancuronium 77, 86–8
  clinical use
    in intensive care practice 227–8
    maintenance dose 132
    neonates, 'pancuronium belly' 225

Index   267

Pancuronium—*contd*
  clinical use—*contd*
    paediatric anaesthesia 203, 204, 209
    prevention of fasciculations 63
    prevention of suxamethonium-induced myalgia 64
    recovery index 146–7
    relative safety 252–3
    side-effects 88
  pharmacodynamics 88
    acetylation 37
    design 23
    hepatic uptake 35
    interactions with other non-depolarizing relaxants 191–2
    metabolites 37–8
    protein binding 33
    resistance, soleus muscle 122
    reversal of neuromuscular blockade 141, 147
    structure 87
  pharmacokinetics 86–7
    elimination 27, 34
  pharmacokinetics, effects of
    age of patient 157, 159
    body size 40–1
    burns 167–8
    extrahepatic biliary obstruction 46
    hepatic cirrhosis 46, 47, 165
    hypothermia 43
    pregnancy 41–2
    renal failure 45, 162
PCHE *see* Cholinesterase, plasma
Penicillamine, drug interactions 188
Pentolinium, drug interactions 186
'percent block' 98
Pharmacokinetics 27–54, 127–8
  see also Muscle relaxants, non-depolarizing, pharmacodynamics; pharmacokinetics
Phase I and II blocks 22
Phenothiazine, drug interactions 188
Phenytoin, drug interactions 184–5
Phosphodiesterase inhibitors, drug interactions 187–8
Phospholipase A2, inhibition 65
Physiology of muscle 1–12
Physostigmine
  metabolite 3-hydroxy-phenyl-trimethyl-ammonium 142
  properties 142
Pipecuronium 77, 88–9
  clinical use
    in intensive care practice 230
    paediatric anaesthesia 203, 204, 209–10
    side-effects 89
  pharmacodynamics 89
    hepatic uptake 35
    interactions with other non-depolarizing relaxants 191–2
    protein binding 33

reversal 141, 147
  structure 23, 87
pharmacokinetics 88–9
  elimination 48
pharmacokinetics, effect of
  age of patient 39, 157, 159
  hepatic cirrhosis 166
  renal failure 45, 163
Placental barrier 42
Plasma cholinesterase *see* Cholinesterase, plasma
Polymixins, drug interactions 184
Population analysis, NONMEM program 31
Post-tetanic count (PTC) 104–7
Post-tetanic facilitation 98–9
Potassium ions
  drug interactions 189–90
  serum vs plasma, and suxamethonium administration 67
Pre-curarization 64–5, 193–4
Pre-synaptic receptors, acetylcholine 9
Pregnancy
  cholinesterase, plasma, decreased activity 60
  pharmacokinetics of muscle relaxants 41–2
Prejunctional receptors
  catecholamine receptors 20–1
  muscarinic receptors 20
  neuromuscular junction 19–21
  nicotinic receptors 19–20
  opioid receptors 21
Prick test 250
Procainamide, drug interactions 186
Propanidid, drug interactions 182
Propofol
  anaesthetic reactions 238
  drug interactions 182
  indications in MH 71
  reversal of mivacurium 147
Propranolol, interactions with suxamethonium 185
Prostaglandin release, and histamine release 240
Protein binding 32–3
Pulmonary function tests 170
Pulmonary side-effects, anticholinesterase drugs 139
Pulmonary syndromes, shock lung (ARDS) 244
Pyridostigmine
  action 136–7
  pharmacokinetic variables 142
  release of acetylcholine 137
  side-effects 138–9

Quinidine, drug interactions 186

Radio-allergosorbent (RAST) assays 249
Rapid sequence induction 57
Rate constant, defined 28

Recovery
  rate, measurement 146
  spontaneous 146
Recovery index 146–7
  defined 146
  effects of doxapram 150
  short-acting drugs 148
Recovery room, residual blockade 97
Relaxograph, EMG 113
Renal excretion 33
Renal failure 160–4
  effect on reversal of neuromuscular blockade 148–9
  effects of various drugs 45
  intensive care practice 226
  muscle relaxants 43–4
  serum potassium in 67
Respiratory acidosis 168–9
Resting potential 3–4
Reversal of neuromuscular blockade 135–55
  choice of neuromuscular blocking drug 140
  choice of reversal agent 140–6
  clinical guidelines 150
  measurement of return of neuromuscular function 139–40
Rocuronium 77, 89–90
  clinical use
    and early endotracheal intubation 63
    in intensive care practice 230
    maintenance dose 132
    paediatric anaesthesia 213
    side-effects 90
  pharmacodynamics 90
    design 23
    interactions with other non-depolarizing relaxants 191–2
    protein binding 33
    rate of reversal 141, 147
    structure 87
  pharmacokinetics 89–90
    elimination 48
  pharmacokinetics, effects of
    age of patient 39, 157, 159
    hepatic cirrhosis 46, 47, 165
    renal failure 45, 162
Ryanodine receptor, malignant hyperthermia 70

Salbutamol, adrenaline-resistant bronchospasm 258–9
Schistosomiasis 33
Sedation level 225
Shivering, management 71
Shock lung syndromes (ARDS) 244
Skin tests 249, 250–1
Soleus muscle
  pancuronium resistance 122
  tubocurarine resistance 122
Starling's Law 111
Steroid anaesthetics see Althesin
Steroid muscle relaxants 187
Steroid muscle relaxants see Pancuronium; Pipecuronium; Rocuronium; Vecuronium
Strabismus, and MMR 206
Stroke patients 167
Succinylmonocholine see suxamethonium
Suxamethonium
  antibodies 255–6
  clinical use 57–73, 129–30, 193
    action and onset 59–60, 129
    before non-depolarizing relaxant 193
    contraindications to use 73
    electroconvulsive therapy 59
    following non-depolarizing relaxant 194
    indications 57–9
    infusions 59, 130–2
    maintenance dose 132
    maintenance of neuromuscular blockade 59
    myotonias, contraindications 171
    non-depolarizing muscle relaxants following 58–9
    paediatric anaesthesia 201–7
      neuromuscular effects 202–5
      side-effects 205–7
    pre-curarization 193–4
    rapid sequence induction 57–8
    relative safety 252–3
  complications of use
    'apnoea' 62
    bradycardia 65–6
    fasciculations 63
    hyperkalaemia 67–8, 227
    intracranial pressure 68–9
    intragastric pressure 68
    intraocular pressure 68
    malignant hyperthermia 69–72
    masseter spasm 66
    myalgia 63–5
    myoglobin release 68
    notification of adverse reactions 254
    resistance in myasthenia gravis 170
    tachyphylaxis and bradyphylaxis 57
  pharmacodynamics 22, 55–7
    depolarizing blockade 55–6
    drug interactions
      hypnotics 238
      propranolol 185
    fade 98
    hydrolysis by plasma cholinesterase 35–6, 59–60
    interactions 193
    phase II blockade 56–7
    train-of-four fade 56
  pharmacokinetics, renal excretion 34
  pharmacokinetics, effect of
    age of patient 157–8
    hepatic failure 164
    renal failure 161

Synapsin I 7
Synaptophysin 7

Tachyphylaxis, defined 57
Taurocholate, hepatic uptake of
  vecuronium 35
Terbutaline, adrenaline-resistant
  bronchospasm 259
Tetanic stimulation
  disadvantages 100–1
  double burst stimulation 62, 103–4, 107,
    110, 140
  fade 100–1
  neostigmine blockade 143
  post-tetanic count 104–7
  post-tetanic facilitation 98–9, 100
  tetanic count 103
  Train-of-Four ratio 101–2
  Train-of-Four stimulation 100–1
  see also Nerve stimulation
Tetanus 224
Tetracyclines, drug interactions 184
Tetrahydropapaverine 38
Thermal injury 46, 69–72, 167–8
Thiopentone 237, 238, 252
Thumb accelerometry 115
Time constant, defined 28
Toxiferine, and alcuronium 23
Train-of-Four count 102–3
Train-of-Four ratio 101–2, 107, 139–40
Train-of-Four stimulation 100–1, 110–11
Trauma, multiple 223–4
Trimetaphan, drug interactions 186
Tryptase assays 246–8
Tubocurarine 77, 90–1
  clinical use
    histamine release 241
    in intensive care practice 228
    maintenance dose 132
    paediatric anaesthesia 203, 204, 209
    prevention of suxamethonium-induced
      myalgia 64
    relative safety 252–3
    resistance, soleus muscle 122
    side-effects 91
    for suxamethonium-induced myalgia 65
  dimethyltubocurarine see Metocurine
  pharmacodynamics 91
    hepatic uptake 35
    pre-curarization 64–5
    protein binding 33
    reversal 147
    source 90
    structural determination 23, 84
  pharmacokinetics 91

elimination 27, 34
pharmacokinetics, effects of
  age of patient 39, 157, 160
  burns 46
  hepatic failure 166–7
  hypothermia 43
  renal failure 163
Urinary methylhistamine assays 248
Urine sample, tests 249–50

Vecuronium 77, 92–3
  clinical use
    histamine release 240–1
    indications in MH 71
    in intensive care practice 229–30
    maintenance dose 132
    myotonias, hypnotics 238
    paediatric anaesthesia 203, 204,
      211–12
    prevention of suxamethonium-induced
      myalgia 64
    prolonged infusion 146
    relative safety 252–3
    side-effects 92–3
  pharmacodynamics 92
    acetylation 37
    blockade, measurement with force
      transducer 111
    design 23
    hepatic uptake 35
    interactions with other non-
      depolarizing relaxants 191–2
    metabolites 37–8
    protein binding 33
    reversal 146, 147
    structure 87
  pharmacokinetics 92
    excretion 34
  pharmacokinetics, effects of
    age of patient 39, 157, 158
    body size 41
    hepatic cirrhosis 47, 165
    infancy 40
    pregnancy 41–2
    renal failure 45, 161
Ventilation
  in intensive care practice 222–4
  disconnection 224–5
Verapamil, drug interactions 185–6
Volatile agents
  drug interactions 180–1
  see also Anaesthetic agents
Volume of distribution, defined 28

Zero order kinetics, defined 28